9th Topical Conference on Gas Utilization 2009

Topical Conference at the 2009 AIChE Spring National Meeting

Tampa, Florida
26 – 30 April 2009

ISBN: 978-1-61567-258-5

Printed from e-media with permission by:

Curran Associates, Inc.
57 Morehouse Lane
Red Hook, NY 12571

Some format issues inherent in the e-media version may also appear in this print version.

Printed by Curran Associates, Inc. (2009)

For permission requests, please contact AIChE
at the address below.

AIChE
3 Park Avenue
New York, NY 10016-5991

Phone: (203) 702-7660
Fax: (203) 775-5177

www.aiche.org

Additional copies of this publication are available from:

Curran Associates, Inc.
57 Morehouse Lane
Red Hook, NY 12571 USA
Phone: 845-758-0400
Fax: 845-758-2634
Email: curran@proceedings.com
Web: www.proceedings.com

9th Topical Conference on Gas Utilization 2009

Topical Conference at the 2009 AIChE Spring National Meeting

Tampa, Florida
26 - 30 April 2009

ISBN: 978-1-61567-258-5

AIChE

TABLE OF CONTENTS

9TH TOPICAL CONFERENCE ON NATURAL GAS UTILIZATION

AIChE Spring Meeting, April 26-30, 2009, Tampa, Florida

Utilization of natural gas continues to gain more attention because of higher oil price. The use of LNG is gaining popularity for power generation, industrial basic materials and household usage. Governments and companies are looking for more opportunities to convert the alternative carbonaceous feed-stocks into valuable products such as fuels and chemicals. This year's program consists of six sessions on LNG and one session on coal, biomass and natural gas to liquids and expected to cover the following sessions and one Keynote Speaker:

LNG -- This covers advances in one session each for LNG liquefaction technology or LNG liquefaction technology and LNG regasification, and two sessions each for LNG equipment and LNG safety. The sessions on liquefaction technology and regasification will focus on both the fundamentals and operational aspects of the baseload LNG and LNG receiving terminal plants. The sessions on equipment and safety will cover recent advances in equipment and technical interpretations and new results for safety issues.

Coal, Biomass and Natural Gas to Liquids -- This covers the fundamentals and processes for the conversion of coal, natural gas and other carbon-bearing materials to fuels and chemicals, and product upgrading.

Keynote Speaker

This year's Keynote Speaker is Dr. M. Sam Mannan, a Regent Professor and Director of Mary Kay O'Conner Process Safety Center at Texas A&M University. The topic of his speech is "Recent Advances in LNG Field Experiments and Modeling of LNG Release and Mitigation Scenarios".

Speakers and Session Chairs

We would like to thank the speakers for their efforts and contributions to the meetings and the Session Chairs, Chen-Hwa Chiu, Yoshitsugi Kikkawa, J. C. Kuo, Joseph Cho, Joel Madison, and Eginhard Berger for their hard work in chairing the six LNG sessions, and Richard Mallinson and Belma Dimirel for their hard in chairing coal, biomass and Natural Gas to Liquids session.

Topical Conference Chairs are Chen-Hwa Chiu of Chevron Energy Technology Company, Richard G. Mallinson of University of Oklahoma and Belma Demirel of Rentech, Inc.

Chen-Hwa Chiu

On the Hydrodynamics of Bubble Columns with Internals for Liquid Fuels Syntheses

Ahmed A. Youssef[1], Milorad P. Dudukovic[1] and Muthanna H. Al-Dahhan[2]

1. Chemical Reaction Engineering Laboratory (CREL),
Energy, Environmental and Chemical Engineering Department (EECE),
Washington University, St. Louis, MO, 63130-4899 USA
2. Currently at Chemical & Biological Engineering Department
Missouri University of Science and Technology, Rolla, MO, 65409 USA

Abstract

Most industrial applications involving bubble columns require the addition of internals for heat exchanging. However, most academic research on bubble columns has been performed in hollow (empty) ones. Hence, the impact of internals on the hydrodynamics of a bubble column has been insufficiently studied in the open literature. Gas holdup and bubble dynamics including local bubble holdup, gas-liquid interfacial area, bubble length, bubble velocity, and others are important parameters on which the performance of bubble columns depends. Accordingly, this study focuses on investigating the effect of internals that mimic the ones used in methanol synthesis (5% covered cross-sectional area) and Fischer-Tropsch (FT) process (22% covered cross-sectional area) on the local gas holdup, gas-liquid interfacial area, the bubble chord length and bubble velocity distributions. Measurements were made via a four point optical probe in a 8" diameter column using an air-water system. Gas superficial velocities applied were 3, 8, and 20 cm/s. An increase in gas holdup and interfacial area was obtained upon insertion of the internals. It is to be noted that, the impact of the less dense internals arrangement (5% covered area) was not significant. However, the bubble chord length and the bubble velocity decreased in the case of dense internals arrangement.

Keywords: Bubble columns; internals; gas holdup; bubble dynamics; Fischer-Tropsch; optical probe

1. Introduction

Bubble columns are considered the reactors of choice for many processes, such as oxidation, fermentation, wastewater treatment, methanol synthesis, and Fischer-Tropsch synthesis. Many of these applications, like methanol and Fischer-Tropsch syntheses, are exothermic and hence heat exchanging internals are needed to maintain the desired process temperature (Chen et al., 1999 and Hawthorne et al., 2006).

The performance of bubble columns depends on several parameters, including the gas holdup and the bubble properties (local bubble holdup, gas-liquid interfacial area, bubble frequency, bubble length, bubble velocity), which are considered among the key factors in determining the performance of these reactors. However, most of the reported studies in the literature do not investigate the effect of internals on the hydrodynamics and transport phenomena in bubble columns. Therefore, an understanding of the impact of heat exchanging internals on these parameters and the performance of bubble columns is, in general, lacking in the open literature.

Among the few reported studies, Chen et al. (1999) investigated the effect of internals on gas holdup, liquid recirculation, and turbulent parameters using gamma ray Computed Tomography (CT) and Computer Automated Radioactive Particle Tracking (CARPT) techniques available in the Chemical Reaction Engineering Laboratory (CREL) at Washington University in St. Louis. They simulated the internals of the 18" diameter column of the methanol synthesis pilot plant unit in La Porte, Texas. Air-water and air-drakeoil (Drakeoil® 10) systems were used in a Plexiglas 18" diameter bubble column. Their internals geometry comprised two concentric circular bundles of 1" OD aluminum tubes, each containing 8 tubes equally spaced, with a covered cross sectional area of 5% of the column's total cross sectional area. Chen et al. (1999) concluded that such internals, covering ~5% of the total column cross-sectional area, have no significant effect on liquid recirculation velocity, while gas holdup increases slightly (~10% at the column's center and less towards the wall region). The turbulent stresses and eddy diffusivities were lower with internals present.

Forret et al. (2003) used a basic tracer technique in a 1 m diameter column with internals. They observed a decrease in the liquid fluctuating velocity and an enhancement of large scale liquid recirculation with internals. They first used a 1D Axial Dispersion Model (ADM) to fit the dispersion coefficient in the column without internals. However, large scale recirculation (which increased with internals) contributes to the liquid backmixing, and hence, the 1D ADM was not able to fit the tracer response with the presence of internals. They developed a 2D model to account for the effect of internals on the liquid mixing, including by an axial dispersion coefficient, $D_{ax,2D}$, a radially dependent axial velocity profile, and a radial dispersion coefficient, $D_{rad,2D}$.

No study in the open literature has investigated the effect of internals on bubble properties. Therefore, this study focuses for the first time on the experimental investigation of the effect of internals on the gas holdup radial profile, gas-liquid interfacial area, bubble chord length, and bubble velocity of bubble columns, using a four-point optical probe, and presents such new data.

2. Experimental setup

The experiments were carried out in a Plexiglas column of 8 inch (19 cm) diameter with a height of 2 m. Figure 1a is a schematic diagram of the column used in this work.

Compressed air as the gas phase was introduced from the bottom of the column, while tap water was used as the liquid phase.

In all experiments, the dynamic height of the bed was maintained constant at a level of 160 cm above the gas distributor. The gas distributor had 225 holes of 1.32 mm diameter, arranged in a triangular pitch, with a total free area of 1.09%

A four-point optical probe advanced by Xue (2004) was used to measure the bubble properties within the fully developed flow region at a height of 0.75m above the gas distributor. The optical probe was first developed by Frijlink (1987) at Delft University and then validated and further developed in CREL at Washington University by Xue et. al. (2003), Xue (2004), and Xue et. al. (2008 (a) and (b)). This probe, with its developed algorithm, gives direct values for the local gas holdup, bubble chord length, specific interfacial area, bubble velocity, and other properties.

The experiments were carried out over a range of superficial gas velocities from 3 to 20 cm/s and at three percentages of covered cross-sectional area: 0 (no internals), 5% (typical to the percentage used in the liquid phase methanol synthesis), and 22% (typical to the percentage used in the FT synthesis).

Local probe measurements were taken at seven dimensionless radial positions: ±0.9, ±0.65, ±0.3, and 0 (center of the column). The ±0.65 locations were included since for the bubble column without internals they are within the inversion point between the time averaged liquid upward flow in the central region of the column and the time averaged liquid downward flow in the wall region of the column. Rados (2003) reported this value to be ±0.65-0.7, while Sannaes (1997) found it to be ±0.60-0.65. In this study, the probe tips were pointing downwards, yielding bubble property measurements for bubbles moving upwards. In addition, the probe was rotated 180° to be facing upward to account for the bubbles moving downwards (referred to as negative velocity). Hence, the upward and downward velocities are both reported.

The internals were vertical Plexiglas rods of 1.27 cm (0.5") outer diameter. Two different configurations of the internals geometry were investigated: 1) Two circles of six rods each covered 5% of the total cross section of the column. The two concentric circles were of 8 and 14 cm diameter, as shown in Figure 1b, and 2) 48 rods placed in a triangular (equilateral) pitch of 2.4 cm that covered 22% of the total cross-sectional area, as shown in Figure 2c.

3. Results and discussion

Local gas holdup

Figure 2 shows the effect of internals on the gas holdup radial profile for the studied air-water system at a superficial gas velocity of 20 cm/s, calculated based on the empty column's cross-sectional area. A slight increase occurred when 5% of the column's cross sectional area was covered, as has also been shown in the work by Chen et. al. (1999). However, when the internals covered 22% of the cross-sectional area, a significant increase in the gas holdup was observed. For example, at r/R=0, the gas holdup increased from 0.34 to 0.43, an increase of ~26%. When the internals occupied 22% of the column's cross-sectional area, the interfacial area between the bubbles and the liquid phase increased, the bubble chord length decreased, and the bubble rise velocity decreased at the column's center, as will be illustrated below. In other words, smaller bubbles were moving upwards with lower velocities in the central region compared to the empty column. This led to a longer bubble residence time, causing a higher gas holdup.

Specific interfacial area

Figure 3 shows how the radial profile of the interfacial area was affected by the presence of the internals at superficial gas velocity of 20 cm/s. Trends similar to those demonstrated in this figure were found at superficial gas velocities lower than 20 cm/s. One can see that the more internals covering the column's cross sectional area, the more the interfacial area provided. Furthermore, the increases in bubble frequency and interfacial area are due to the rapid bubble breakup and coalescence, as will be explained later.

It was found that a larger interfacial area existed at the column's center than in the region near the wall which is similar to the findings of Xue (2004) and Xue et. al. (2008). This difference is due to enhanced rates of breakup and coalescence among bubbles in the central region of the column in the churn turbulent flow regime, which was confirmed by the bubble frequency measured by the probe. An increase in bubble frequency leads to an increase in specific interfacial area. The same trend was observed and explained by Wu (2007) for the case of hollow bubble columns. At low velocity, small bubbles uniformly distribute across the column and move at low frequency, which causes low local gas holdup both in the center and wall region of the column. With an increase in superficial gas velocity, bubbles, including the newly formed large bubbles from coalescence that move towards the center of the column, move at high frequency, which means more bubbles appear in the column, causing an increase in gas holdup. Most small bubbles still stay in the wall region and move at relatively low frequency. Hence, the gas holdup, the specific interfacial area, and the bubble frequency in the center become larger than those in the wall region.

Bubble chord length

The bubble chord length distribution follows a lognormal distribution as shown in Figure 4. This trend is similar to what has been reported in literature (Akita and Yoshida, 1974; Yasunishi et. al., 1986; Luewisutthichat et. al. 1997; and Pohorecki et. al., 2001). The results clearly show a decrease in the bubble chord length as the internals density increases, which confirms the findings of increased specific interfacial area and local gas holdup with internals.

The bubble chord length distributions have been analyzed statistically by providing the mean and the variance, as shown in Table 1.

The chord length distribution with 22% internals exhibits a mean value of 0.68 cm, compared to 0.88 cm and 0.91 cm for the cases of 5% internals and an empty column, respectively. This finding suggests that the bubble size gets smaller when high density internals are used. Moreover, the mean chord lengths for the 5% internals and empty column cases are close to each other, implying that the effect of low density internals is nearly negligible. It is noteworthy that the variance was lower for the 22% internals case than for the empty column and 5% internals cases, revealing that a narrower range of bubble chord lengths exists with the dense internals configuration. Also, Figure 4(c) shows larger chord length bubbles in the cases of the empty column and 5% internals.

The observed fact that internals density directly impacts the hydrodynamics is supported by the results presented earlier, showing that smaller bubbles are generated in the 22% internals system due to enhanced bubble breakup. In summary, it has been experimentally as well as statistically proven that high density internals substantially affect the bubble characteristics.

Bubble velocity

In the literature, the bubble velocity in bubble columns has been erroneously confused with the bubble rise velocity, but bubbles in bubble columns move downwards as well (Xue, 2004). Hence, in this work we distinguish between upward bubble velocity and downward bubble velocity. To account for the downward bubble velocity, a separate measurement was conducted where the probe was oriented with the tips pointing upwards at the same point where it had previously been oriented downwards. The downward and upward bubble velocity distributions with and without 22% internals at the column's center and the wall region for 20 cm/s gas velocity are shown in Figure 5. In the cases of with and without internals, at the column center it is clear that the small bubbles dragged downward with the liquid phase recirculation resulted in a downward velocity distribution (represented in terms of negative velocity). The latter observation becomes a greater significant with the presence of internals.

Furthermore, the figure qualitatively indicates a greater fraction of bubbles moving downwards without internals, as compared to 22% internals; this indicates that less recirculation exists with this dense internals structure. However, at r/R=0.9 (the wall region), a greater fraction of bubbles moving downwards ("negative" velocity) was observed. This was due to the overall flow pattern of the bubble column, where the liquid moves downward in the wall region, dragging along the small bubbles. Figure 7(a) and (b) show that the downward bubble velocity without internals has a wider range of negative velocities than that with 22% internals. In addition, a greater fraction of smaller bubbles with downward bubble velocity was obtained with 22% internals. This could be attributed to the reduction in small scale-liquid recirculation intensity with 22% internals in which small bubbles are dragged downwards at a lower velocity compared to those without internals. This is also confirmed by the wide range of upward bubble velocity compared to that with 22% internals. Such findings necessitate the need for detailed studies in bubble columns with high intensity internals.

4. Remarks

The bubble properties (local gas holdup, interfacial area, bubble chord length, and bubble upward and downward velocity) were investigated in an 8 inch diameter bubble column for an air-water system with and without internals. For the first time, the effects of vertical internals on these bubble properties have been reported. An increase in local gas holdup and gas-liquid interfacial area has been observed as the area covered by internals is increased. Statistical analyses for the bubble chord length and bubble velocity distributions indicated that when the internals covered 22% of the total cross sectional area, the bubble chord length and the bubble upward velocity both decreased due to enhanced bubble breakup. The investigation of the bubble downward velocity has indicated that less recirculation would exist with the dense internals arrangement compared to the empty column.

The results of this work necessitate the need to investigate and properly understand the impact of internals on liquid recirculation, backmixing, flow field, and turbulent parameters.

5. Acknowledgments

The authors are grateful for the financial support provided by the consortium of Clean Alternative Energy using Slurry Bubble Column Reactors (CAE-SBCR) [ConocoPhillips (USA), Eni (Italy), Johnson Matthey (UK), Sasol (South Africa), Statoil (Norway)] which made this research possible. Dr. Fadha Ahmed and Sean Mueller provided invaluable help with the optical probe.

Nomenclature

A_{tot} Column's cross sectional area

a Specific gas-liquid interfacial area

r/R Dimensionless radius

U_g Superficial gas velocity (based on the empty column's cross sectional area)

Greek letters

ε_g Local gas holdup

References

- Akita, K., and Yoshida, F., Gas Holdup and Volumetric Mass Transfer Coefficient in Bubble Columns, Ind. Eng. Chem. Process Des. Dev., 12, 76-80 (1973).
- Akita, K., and Yoshida, F., Bubble size, interfacial area, and liquid-phase mass transfer coefficient in bubble columns. Ind. Eng. Chem. Process Des. Develop., 13(1), 84-91 (1974).
- Chen, J., Li, F., Degaleesan, S., Gupta, P., Al-Dahhan, M. H., Dudukovic, M. P. and Toseland, B. A., Fluid Dynamic Parameters in Bubble Columns with Internals, Chem. Eng. Sci., 54 (13-14), 2187-2197 (1999).
- De, S. K., Ghosh, S., Parichha, R., K. and De, P., Gas hold-up in two phase system with internals, Indian Chem. Engr. Section A, 41(2), T54-T58 (1999)
- Degaleesan, S., Turbulence and Liquid Mixing in Bubble Columns, D.Sc. Thesis, Washington University, Saint Louis, Missouri, USA (1997).
- Forret, A., Schweitzer, J.M., Gauthier, T., Krishna, R., Schweich, D., Liquid dispersion in large diameter bubble columns, with and without internals, The Canadian Journal of Chemical Engineering 81, 360–366 (2003)
- Frijlink, J. J., Physical Aspects of Gassed Suspension Reactors, Ph.D. Thesis, Delft University of Technology, the Netherlands (1987).
- Hawthorne W. H., Ibsen M. D., Pedersen P. S. and Bohn M. S., US Patent 7,108,835 B2 (Sep. 19, 2006)
- Luewisutthchat, W., Tsutsumi, A., and Yoshida, K., Bubble characteristics in multiphase flow systems: bubble sizes and size distributions. Journal of Chemical Engineering of Japan, 30(3), 461-466 (1997).
- Pohorecki, R., Moniuk, W., Zdrojkowski, A., Bielski, P., Hydrodynamics of a pilot plant column under elevated temperature and pressure, Chem. Eng. Sci., 56, 1167-1174 (2001).
- Rados, N., Slurry bubble column hydrodynamics, D.Sc. Thesis, Washington University, Saint Louis, Missouri, USA (2003)
- Sannaes B. H., Solids movement and concentration in column slurry reactors, Dr. Ing. Thesis, Norwegian University of Science and Technology, Trondheim, Norway (1997).
- Shaikh, A., Mixing and Flow Behavior in Slurry Bubble Column Reactors, D.Sc. Thesis, Washington University, Saint Louis, Missouri, USA (2007).
- Wu, C., Heat Transfer and Bubble Dynamics in Slurry Bubble Columns for Fischer-Tropsch Clean Alternative Energy, Ph.D. Thesis, Washington University, Saint Louis, Missouri, USA (2007).
- Xue, J., Al-Dahhan, M., Dudukovic, M.P., and Mudde, R.F., Bubble dynamics measurements using four-point optical probe, The Canadian Journal of Chemical Engineering, 81, 1-7 (2003)
- Xue, J., Bubble velocity, size and interfacial area measurements in bubble columns, Ph.D. Thesis, Washington University, Saint Louis, Missouri, USA (2004).

- Xue, J., Al-Dahhan, M., Dudukovic, M., P., and Mudde, R., F., Bubble velocity, size, and interfacial area measurements in a bubble column by four-point optical probe, AIChE J., 54(2), 350-363 (2008).
- Yasunishi, A., Fukuma, M., and Muroyama, K., Measurement of behavior of gas bubbles and gas holdup in a slurry bubble column by a dual electroresistivity probe method, Journal of Chemical Engineering of Japan, 19(5), 444-449 (1986)

Table 1 Statistical quantities of the chord length distribution for different internals arrangements

Internals	Mean	Variance
No	0.911	0.939
5%	0.874	0.960
22%	0.676	0.669

List of Figures:

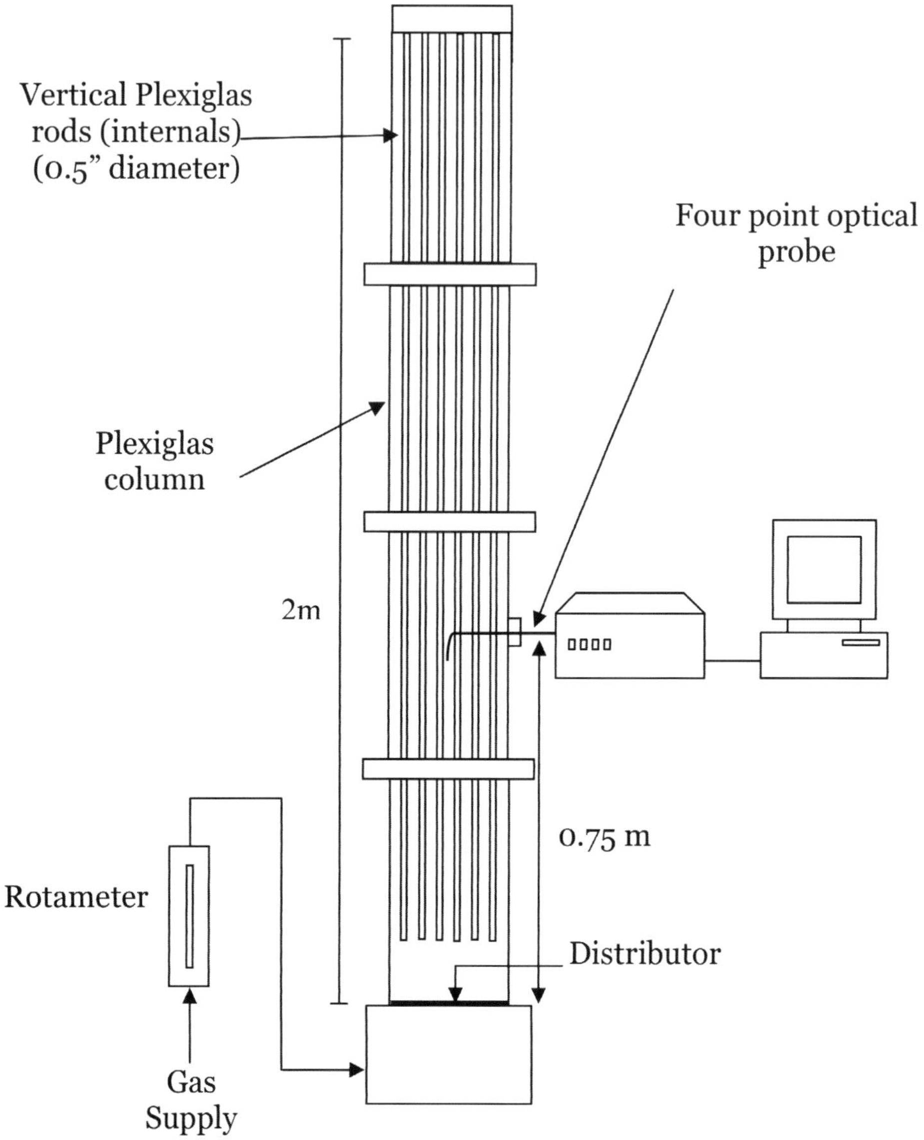
Vertical Plexiglas
rods (internals)
(0.5" diameter)
Four point optical
probe
Plexiglas
column
2m
0.75 m
Rotameter
Distributor
Gas
Supply

(a)

(A) (B)

Figure 1 – (a) Schematic diagram of the bubble column setup (b) Internals configuration. (A) 5% covered cross-sectional area and (B) 22% covered cross-sectional area.

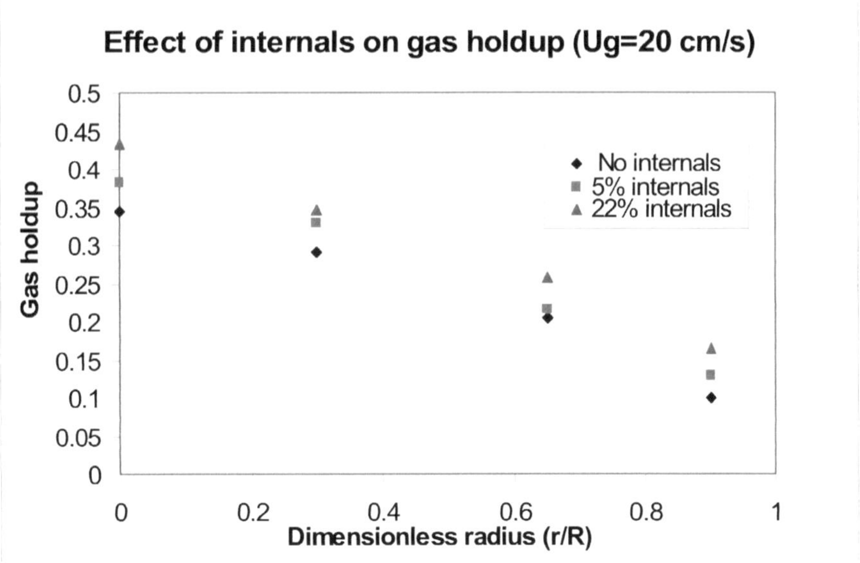

Figure 2 - Effect of vertical internals on the local gas holdup

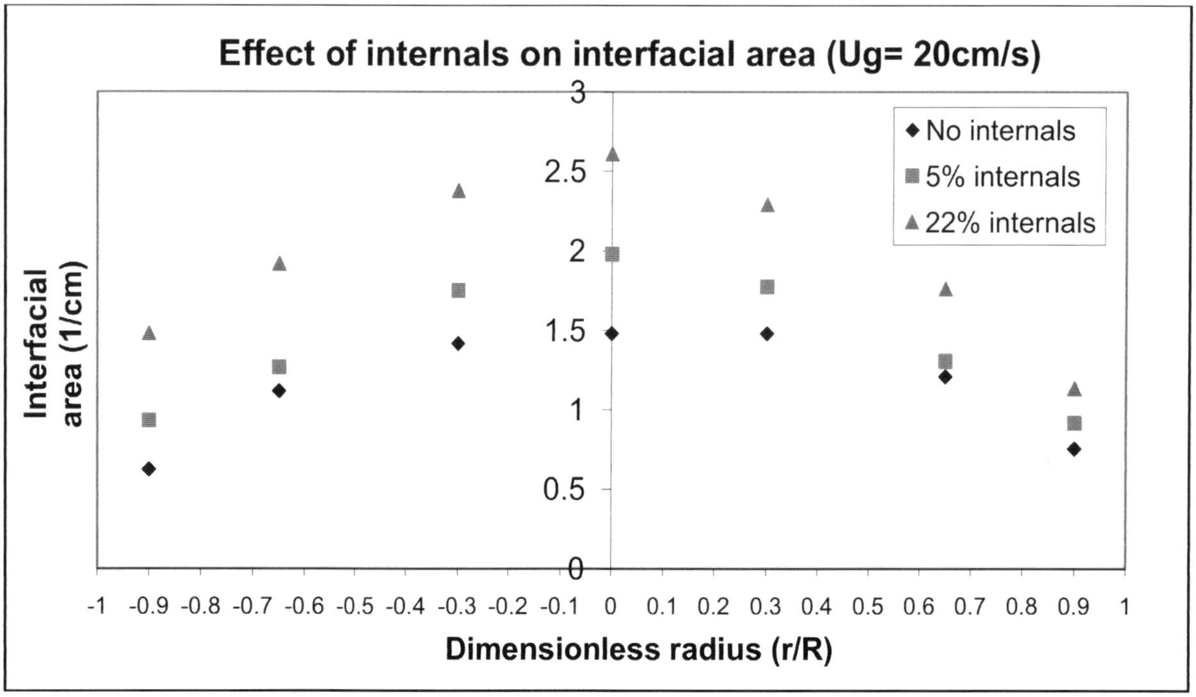

Figure 3 - Effect of vertical internals on the interfacial area radial profile\

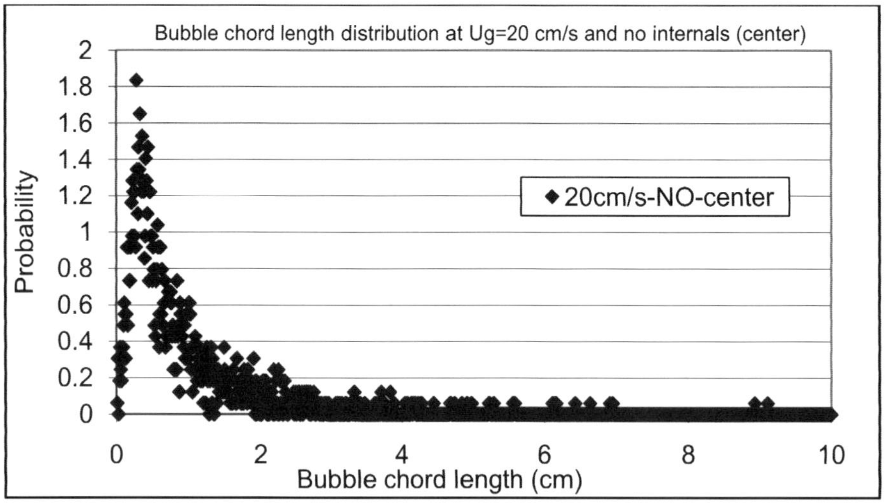

(a)

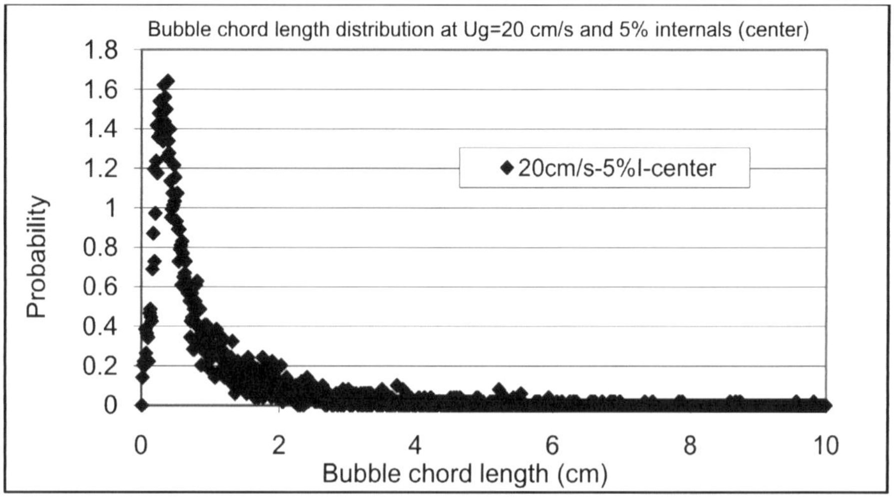

(b)

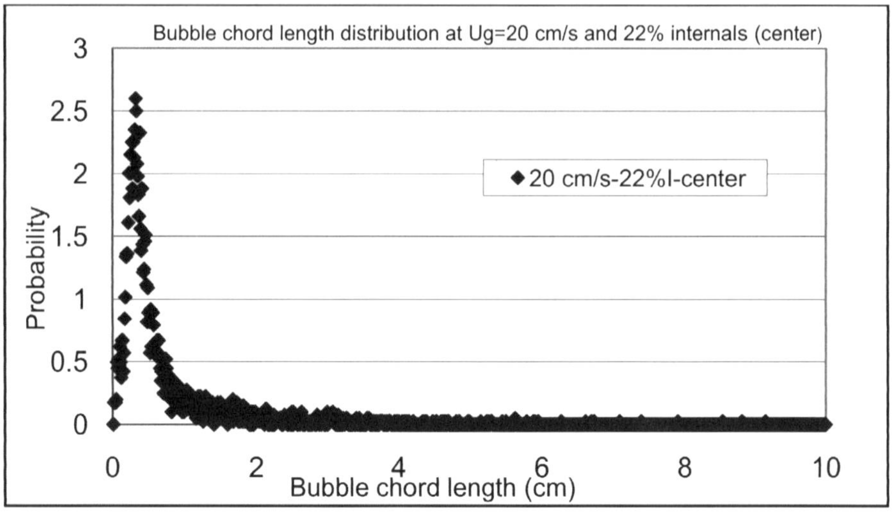

(c)

Figure 4 - Bubble chord length distribution at the column's center at Ug=20 cm/s and different percentage of internals; (a) no internals, (b) 5% internals, and (c) 22% internals

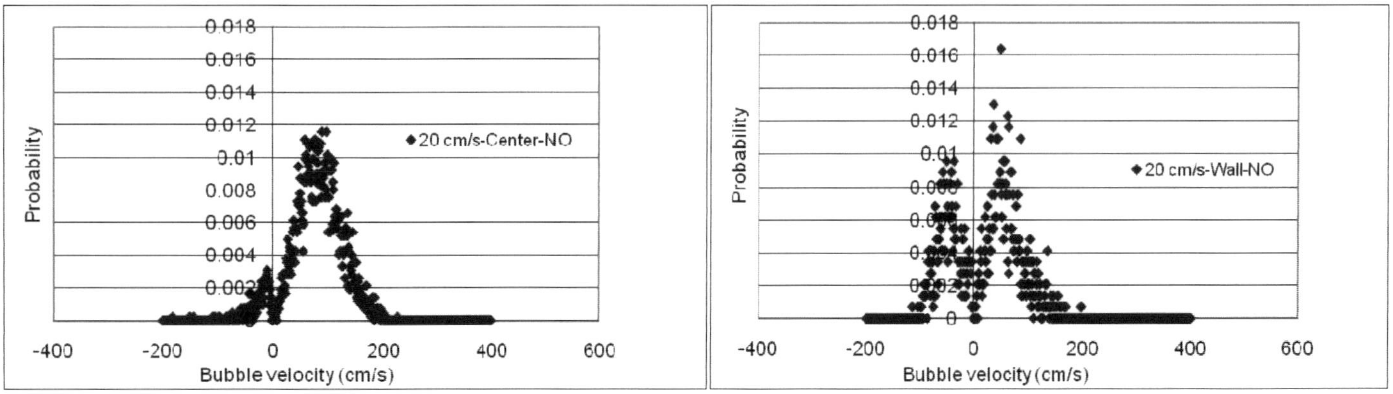

(a)

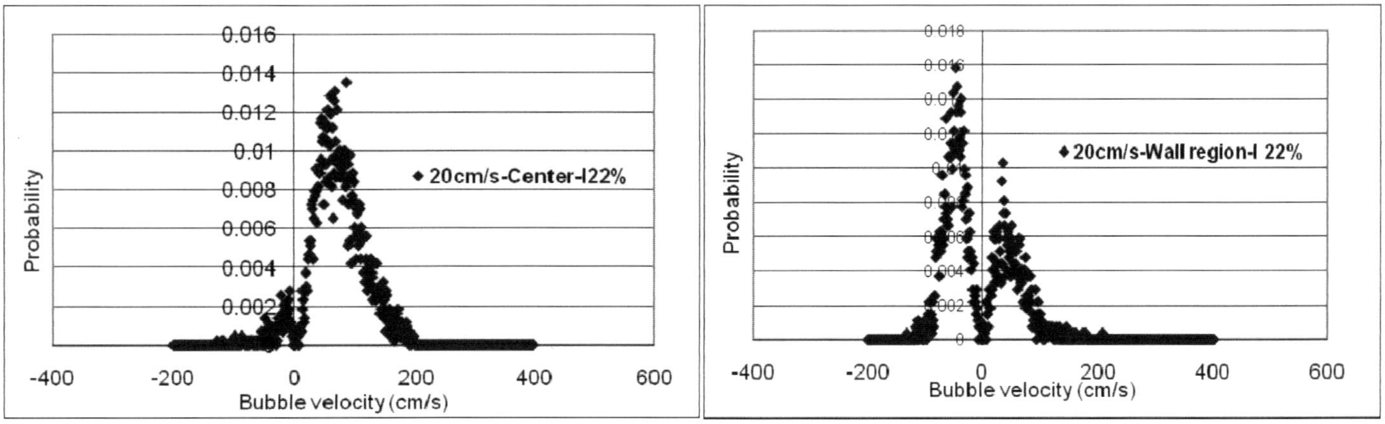

(b)

Figure 5 – Bubble velocity distribution at 20 cm/s in the wall region and the column's center with different percentage of internals; (a) no internals (empty column), (b) 22% internals

3b Application of Pressure Swing Adsorption in the Production of Ammonia

Felix Jegede[1], James Ritter[2], Armin D. Ebner[2], Shubhra Bhadra[3], and Charles E. Holland[3]. (1) SmartKoncept Technology, 11999 Katy Frwy,, Houston,, TX 77079, (2) Department of Chemical Engineering, University of South Carolina, Columbia, SC 29208, (3) Chemical Engineering Department, University of South Carolina, Columbia, SC 29208

Pressure swing adsorption (PSA) technology enables the energy-efficient recovery of specific compounds from a gas under pressure. In this project, an ammonia production process that incorporates PSA for the recovery of ammonia from the product gas and purification of the process reactants, nitrogen (from air) and hydrogen (from steam methane reforming, SMR) is being developed. The process converts methane, air, and water to ammonia, with carbon dioxide and a small amount of argon as byproducts. Current ammonia processes react air and syngas (gas mixture of carbon dioxide and hydrogen) to form ammonia, which is recovered via energy-intensive refrigeration and condensation. The new approach will increase ammonia yield and reduce natural gas feedstock and fuel consumption. The improved ammonia process will find application in the U.S. chemical industry, which produced 23.7 billion pounds of ammonia in 2004. Ammonia is also used at power plants in selective catalytic reduction (SCR) and selective non-catalytic reduction (SNCR) systems, to reduce NOx emissions. The new process is scaleable and enables on-demand ammonia production for such applications. This presentation will provide an overview of the proposed ammonia production flow sheet and give an update on the status of the project with respect to PSA process design for ammonia purification.

Rotating Machinery for Liquefied Ammonia

Steve Rush
Vice President
Sales & Marketing
Ebara International Corporation
Sparks, NV, USA
Phone: (775) 356-2796
E-mail: srush@ebaraintl.com

INTRODUCTION

Ammonia is one of the most widely produced chemicals in the world and is mainly made by synthesizing natural gas. It is utilized in multiple applications, for fertilizers, refrigerants, explosives and waste water treatments. It is also employed in the industrial production of rubber, leather and paper. Ammonia, like hydrogen, is also a carbon free liquid fuel.

Liquefied Ammonia, or LNH_3, is (like Liquefied Natural Gas) a cryogenic fluid and production and transportation require technologies that are analogous to those used in LNG. Rotating machinery is an important part of this cryogenic fluid technology, in particular if the fluids are also flammable and corrosive like ammonia.

Ammonia is a particularly difficult liquid to handle, as it is very hazardous to personnel if the gas is present and can cause severe asphyxiation, even at very low concentrations. It is also very flammable, is highly corrosive to copper and copper alloys and has a huge affinity for water, which can make it highly conductive unless it is in its purest form.

When ammonia is stored in very large volumes, it is normally stored at atmospheric pressure with all inlets and outlets preferably made via the roof of the tank in accordance with BS EN 14620 Part 1 or API 620.

Traditionally, ammonia has been pumped using “canned” type motors where the motor is encased in a sealed membrane to prevent contact of the ammonia with the motor windings. Since ammonia is both corrosive and conductive, a fully submersible type motor cannot be used. In addition, shaft seal type pumps are also used, but any leakage through the seal will obviously create a hazardous situation.

Many ships and land based storage tanks use a “line shaft” type pump where the motor is located outside the liquid at the top with a long flexible shaft going down into the tank at the bottom to the pump. This type of pump also has seals that can leak and if there are any problems, they are not easy to remove and perform maintenance.

Some of the earlier submersible, in-tank designs used an electric motor in which both the rotor (rotating portion) and the stator (static, winding portion) were encased in a stainless steel membrane that was filled with an epoxy compound. Those designs were somewhat successful but the manufacturing process was difficult and any air entrained in the rotor portion of the motor would expand from heat generated by the rotor causing premature failure.

DESIGN SOLUTIONS

To meet the challenge of pumping LNH_3, Ebara International Corporation (EIC) has developed a submersible, magnetic coupling design which is in use today in several locations (Figure 1). This design uses a sealed stainless steel casing to house the motor, which has inert nitrogen gas fed into the casing from the outside of the pump vessel or tank to keep the motor section dry and free of ammonia (Figure 2). The nitrogen pressure can be maintained to prevent ammonia from entering the casing and also to provide the proper differential pressure across the magnetic coupling membrane.

Figure 1: Installation of EIC In-Tank Ammonia Pump

The key feature of the design is safety. The pump has no penetrations through the vessel or outer casing with a shaft seal, so there is no possibility of leakage into the surrounding atmosphere. In addition, since the motor is located within the tank or vessel along with the pump hydraulic section, there is no need for an explosion-proof housing around the motor since the interior of the tanks under operating conditions have no oxygen present.

These pumps can be installed in a suction vessel to act as an in-line design that can be used in almost any location where it is necessary to transfer liquid or boost pressures, including from tanks where a few still specify bottom or side connections. The liquefied gas industry almost universally recognises the enhanced safety of having no penetrations below the liquid level, where pumps are installed in storage tanks in a pump column similar to a submersible deep well design to transfer liquid from the storage tank to another location such as a ship, or other process area.

For larger storage tanks, the retractable, or in-tank, type of magnetic coupling pump is an excellent choice for the application as it is installed through a column from the top of the tank. In many ammonia tank applications, where the liquid is stored at near atmospheric conditions, an external shaft seal type pump is used, which requires a side or bottom penetration from the storage tank. The penetrations, with valves in the piping, are potential leakage areas and the pump type used would have the same issues as noted above with the external motor and shaft seals.

For the electrical portion of the Ebara design, the power cables from the motor are enclosed in a flexible conduit/hose arrangement that is seal welded on both ends. This conduit, which is connected from the headplate at the top of the tank to the motor casing at the bottom, allows nitrogen gas to be applied into the hose and casing to keep those spaces inert and free from moisture. This space can also be monitored to determine if any leakage is present, and if such an event happens, the pump electrical system can be shut down immediately for safety purposes.

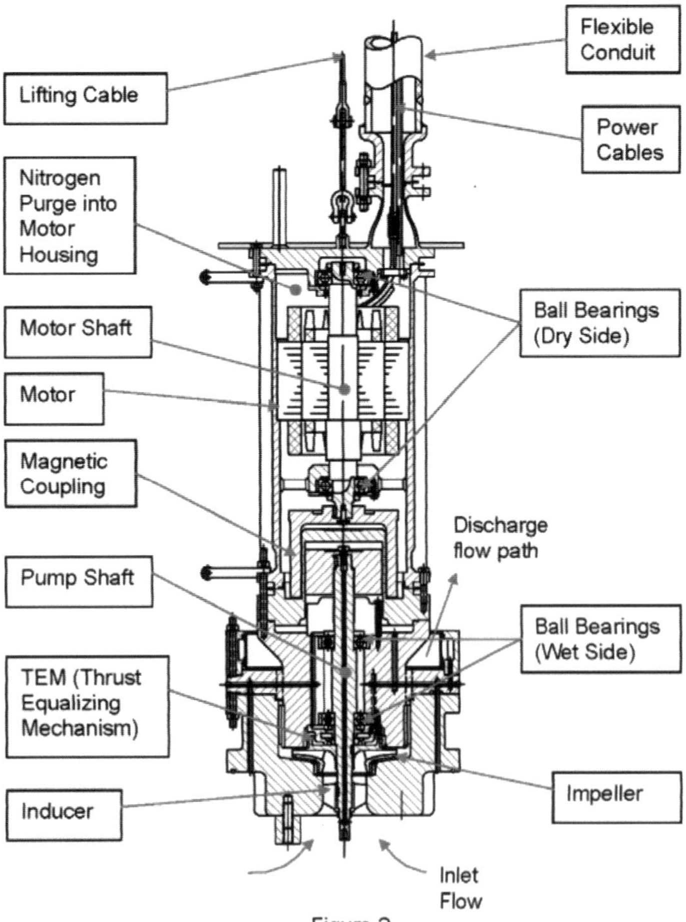

Figure 2

At the top of the electrical system, to enable the power cables to exit the tank, a dual seal penetration is used, which is the same design used for pumps and liquid expanders used in LPG, LNG, ethylene and other liquefied gas applications. The design meets all of the more common codes such as NFPA 59 and 59A, and meets the requirements of NFPA 70, National Electrical Code, for equipment used in Class 1, Division 2, Group C/D atmospheres. The equipment also meets applicable IEC codes for Class I, Zone 1 or 2 hazardous areas. This system has been in regular use in thousands of applications and has proven to be extremely safe and reliable.

For the pump design itself, many features from EIC's standard submerged motor pumps are used. The hydraulic end of the pump carries the impeller(s) and inducer as well as the wet end coupling, and the bearings are product lubricated (Fig. 2). This is a stiff shaft design which uses low mass rotating components made from aluminum to ensure minimal radial loads and low vibration, thereby extending bearing life.

The motor section, or dry end, uses typical shielded grease filled bearings with the magnetic coupling attached at the lower end.

Another key feature in the design is the use of EIC's Thrust Equalizing Mechanism (TEM), which is a built-in axial balancing system that eliminates axial thrust loads on the bearings at all flow rates using the pumped fluid. When using the pumped product, which in this case is a liquefied gas, for lubrication, it is imperative that the axial thrust loads are totally balanced to prevent vaporization of the fluid in the bearings to ensure reliability. This system has been used in many thousands of liquefied gas pumps all over the world with very high reliability.

EIC's pump design also uses an inducer, or axial flow impeller, in the inlet to provide very good NPSHR characteristics (Fig. 2). The inducer is designed to enable operation of the pumps in very low liquid level situations and is perfectly suited for storage tanks on land and in marine applications.

The current design was developed in the mid 1990's and eleven units are now installed and operating in Taiwan and mainland China with the most recent units just installed in a plant near Shanghai. The overall design, with the electric motor, magnetic coupling and TEM is also patented (US Patent number 6,213,736 B1).

NEW DEVELOPMENT

As the magnetic coupling pump for ammonia service has now been in operation for a number of years, it has become obvious that the design can be adapted for other rotating machine applications.

In the process and manufacture of LNH_3, the processes used are very similar to those used to liquefy natural gas or other liquefied gases which are at cryogenic temperatures at atmospheric pressure.

In the cryogenic liquefaction process, the process gas must be cooled and compressed in order to create the liquid form for compact transportation and storage. As part of the process, the liquid pressure must be dropped to allow the liquid to go into another stage of compression, or for transfer into low pressure storage. Typically, the pressure drop is accomplished using a Joule-Thomson (JT) valve which is basically an orifice type device. The drawback to this type of pressure reduction is that the enthalpy of the liquid does not change and depending on the inlet and outlet pressures, the temperature of the fluid may increase or decrease.

In order to provide a much more efficient drop in pressure, as an isentropic expansion, a rotating machine such as a turbine expander can be used in place of a JT valve. In LNG liquefaction plants, EIC developed an expansion turbine several years ago, and that type of expander is now used in many liquefaction systems today. These machines are now well proven and are known to increase the overall process efficiency by as much as 4% to 7%.

The design of the machine uses many of the same features found in the submerged motor cryogenic pumps, but uses nozzle rings to direct the flow into the rotating turbine runners, which are connected to a generator on a common shaft. The torque created by the generator is then used to extract energy from the fluid stream, thereby creating the thermodynamic efficiency that essentially produces more liquid versus vapor in the expansion process. The EIC design uses fixed vane nozzle rings directing the flow into Francis type radial inflow runners which are designed to extract the maximum amount of torque from the fluid as possible. The machine is also often enhanced using variable speed technology where the generator speed is controlled using a Variable Speed Constant Frequency (VSCF) controller, thereby providing flexibility in the operating range to optimize the efficiency of the process system.

For ammonia liquefaction, EIC is developing a turbine expander that uses the same hydraulic and generator technology as the machines used for LNG liquefaction coupled with the technology of EIC ammonia pumps to produce a unique machine for this application using proven technology. By marrying these technologies together, the typical ammonia plant can now benefit in a much more efficient process which will lead to an increase in production while reducing required power for compression. In fact, by replacing the traditional Joule-Thomson type expansion system in an existing plant with the turbine expanders, the same volumetric flow can be achieved with a 5% reduction in compression and 5% reduction in overall liquefaction energy.

CONCLUSION

For normal ammonia pump applications, Ebara International Corporation has developed a unique, proven design that enhances safety in a very difficult environment. The design not only improves safety but in doing so, also increases reliability by removing troublesome components that are prone to leakage.

In addition, by marrying the ammonia pump technology with well proven turbine expander technology, a new turbine expander with a magnetic coupling can be applied for ammonia liquefaction to increase process efficiency while keeping the same inherent safety philosophy intact.

BIOGRAPHY

Steve Rush has worked for Ebara International Corporation, Cryodynamics Division, for over 27 years and has held many positions in pump design, project management, engineering management, field service management and sales. Steve has authored several papers and articles related to submerged cryogenic pumps and expanders and has extensive experience in trouble-shooting problems in the field. His design work and calculations have led to many cryogenic pump engineering standards used by the company today. Steve attended various colleges in California with a major in Mechanical Engineering. Steve has been a member of ASME since 1994.

HYDROGEN PRODUCTION FROM NON-THERMAL PLASMA ETHANOL AND GLYCEROL REFORMING

T. Hoang, X. Zhu, R. Mallinson, and L. Lobban

Center for Biomass Refining
School of Chemical Biological and Material Engineering
University of Oklahoma, 100 E. Boyd St., SEC-T335, Norman, OK

Introduction

Production of hydrogen for fuel cell vehicles, mobile power generators, and for hydrogen enhanced internal combustion engines from biomass sources is a potential path to a green hydrogen economy component. In recent years, ethanol blended with gasoline known as E85, is showing potential as a renewable fuel. However, USDA reports have shown a relatively low energy balance of corn-based U.S. ethanol (about 1.3). Most of the intensive energy use is found from distillation to produce pure ethanol. Steam reforming of ethanol can take advantage of this step to produce hydrogen from a lower bio-ethanol concentration feedstock. In addition, as a major byproduct of biodiesel production, the huge increase of glycerol in the market makes it a good platform molecule for valuable chemicals and fuels production.

Hydrogen is an excellent fuel with high efficiency and no CO_2 emission during combustion. However, hydrogen has its own limitations, low energy density: 0.011MJ/l compared to 34 MJ/l of gasoline. The poor storage property has been a significant drawback in development of technology for hydrogen fueled transportation. In-situ hydrogen production is one of the approaches to minimize this issue. Steam reforming of ethanol and glycerol provides the highest hydrogen yield (total 6 H_2 from ethanol or 7 H_2 from glycerol) compared to partial oxidation and auto-thermal reforming.

Additionally, it has been shown that addition of hydrogen to a spark ignition (gasoline type) engine, can improve its combustion efficiency. Under some circumstances, converting some of the liquid fuel to hydrogen can have a net improvement of the combustion efficiency over the penalty of producing the hydrogen. Ethanol is well suited to do this because of its high hydrogen yield and high efficiency for reforming [10].

Ethanol and glycerol steam reforming are endothermic processes, while water gas shift is mildly exothermic; carried out by following reactions:

$$C_2H_5OH + H_2O \rightarrow 2CO + 4H_2 \qquad \Delta H = 206 \text{ kJ/mol} \qquad (1)$$
$$C_3H_8O_3 + H_2O \rightarrow 3CO + 4H_2 \qquad \Delta H = 350 \text{ kJ/mol} \qquad (2)$$
$$CO + H_2O \rightarrow CO_2 + H_2 \qquad \Delta H = -40 \text{ kJ/mol} \qquad (3)$$

Several works have been focused in this area [1-3]. However, conventional hydrogen production with catalytic steam reforming technology is subject to several problems for mobile applications such as heavy weight, high temperature, and deactivation of catalyst due to coke or impurities in crude feedstocks. Hydrogen production from non-thermal plasmas is an alternative approach that takes advantage of low temperature, quick start-up, and less catalyst deactivation. Thus, utilization of non-equilibrium plasmas in fuels conversions has received increased attention in recent years [4-15].

Experiments

A pulse corona discharge is generated by high voltage amplifier (TREK – 20/20C) and pulse generator (HP – 8011A). The power consumption is analyzed by an oscilloscope (Tektronix – TDS750C). The liquid aqueous ethanol (or glycerol) solution is fed directly into the plasma zone. Thus, no external heater is used and the feed solution is vaporized *in situ*. Active species such as O, OH, H generated by electron initiated reactions in the plasma have been found to enhance energy efficiency [6-8]. The products are then passed through the cold-trap for collecting un-reacted water and feed. The dry gases are analyzed by GC (Carle – Series 400). In the integrated system with WGS catalyst, the catalyst bed is placed after the plasma zone. There is no external heater used for the WGS reaction. The catalysts for WGS are Pt/TiO_2 and $Pt\text{-}Re/TiO_2$ that were synthesized with incipient wetness impregnation method. The TiO_2 (Degussa P25) powder was impregnated for 1 h with NH_4ReO_4 (Aldrich) solution, then impregnated with H_2PtCl_6 (Aldrich) solution for another 12 h. The resulted sample was dried at 110 °C for 12 h and then was calcined in air flow (100 ml/min) at 400 °C for 4 h. It was denoted as Pt-Re. The system schematic diagram is shown in Figure 1.

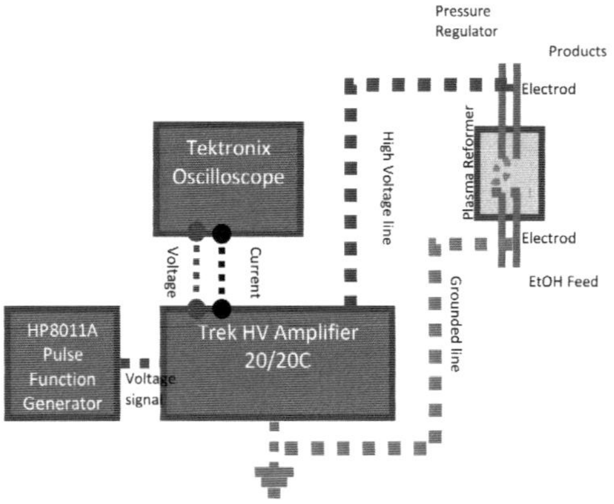

Figure 1. Pulse corona discharge system for ethanol reforming

Results and discussion

Ethanol conversion in pulse corona discharge

The effect of gas gap, water-to-ethanol ratio, and pressure were studied. The summary results are shown in Table 1. It is observed that larger gas gap increased both hydrogen and CO selectivity while reduce by-product hydrocarbons C_2s and methane. Higher water ratio significantly reduced formation of hydrocarbon product and boosted hydrogen selectivity. Higher pressures also yielded higher hydrogen and CO selectivity and significantly reduced hydrocarbon products. The plasma has performed the ethanol reforming reaction without addition of catalyst and at relatively low temperature, ca. 200 ^oC. Byproducts such as C_2s and CH_4 act as diluents but there is no poisoning effect for later processes. However, it has been found that there was greater tendency for coke formation with higher levels of hydrocarbon in the product gas phase.

Table 1 – Summary results of different gas gaps, water-to-ethanol ratios, and pressures.

Discharge gas gap (mm)	Feed flow rate (mmol-min^{-1})	H_2O/Ethanol molar ratio	H_2+CO production rate (sccm)	Pressure (psig)	Ethanol Conversion (%)	Gas phase selectivity* (%)			
						H_2	CO	CO_2	C_2s+CH_4
3	0.86	3.2	70.6	0	87	88	63	2	35
5	0.86		71.3		76	102	69	10	21
10	2.16		223.0		100	96	67	6	26
15	1.30		153.3		100	110	77	10	13
	1.34		160.2		100	111	78	7	15
15	1.20	1.4	119.0	0	83	110	79	6	15
	0.96	3.2	115.0		100	114	75	12	13
	0.58	7.6	70.0		97	127	66	26	7
	0.37	29.2	48.0		100	126	76	18	5
10	1.08	3.2	119.3	5	100	101	74	5	20
	1.00		120.5	10	100	112	78	8	13
	1.14		146.0	15	100	117	86	5	9
15	1.15		139.2	5	100	111	82	7	11
	1.00		131.4	10	100	120	87	7	6

Coupling pulse corona discharge and WGS catalyst

Figure 2 illustrates the gas phase compositions during plasma ethanol reforming with integrated WGS. The main products are H_2 (73.3%) and CO_2 (23.6%), respectively. Only trace byproducts are observed: C_2H_6 (1.4%), CH_4 (0.9%), and CO (0.8%). The H_2 concentration is close to thermodynamic equilibrium concentration (75%, if C_2H_5OH is completely converted to H_2 and CO_2), which is favorable for fuel cell applications. Meanwhile, CO concentration is as low as 1%. This means that the gas can directly proceed to CO preferential oxidation. In addition, as mentioned previously, the trace C_2H_6 and CH_4 are non-toxic for electrodes of fuel cells. Ethanol conversion attains 96% in this experiment. During a one week stability test with daily start-stop, no significant changes in ethanol conversion and gas phase composition are observed. It should be noted that the catalyst is left intact in open air in the reactor after daily operation. This shows the capability of the plasma to regenerate the catalyst *in situ* during operation.

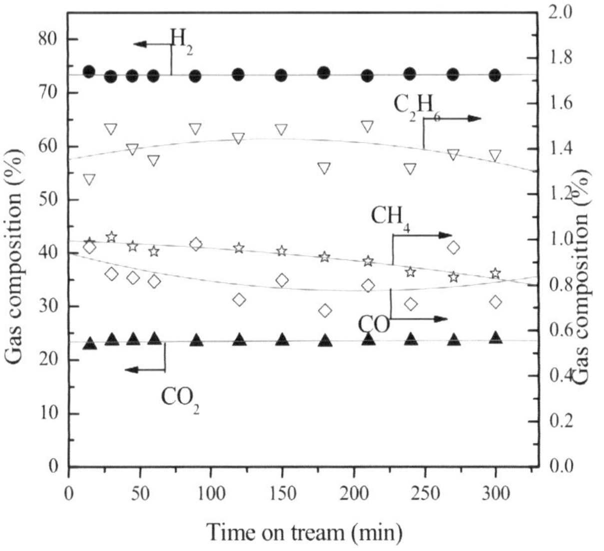

Figure 2. Gas phase compositions during ethanol plasma reforming with WGS catalyst

Glycerol conversion in pulse corona discharge

The results of the effect of feed flow rate on glycerol conversion and selectivities‡ to gas phase products with a 92 wt.% glycerol (H_2O:C_3H_8O=0.44), (which is a similar water content to that found in crude glycerol from biodiesel production) are shown in Fig. 3a. The glycerol conversion to gas phase products decreases from 79.6% to 53.9% with the feed flow rate increasing from 0.068 to 0.135 mL/min. However, the feed flow rate has little effect on selectivities. H_2 and CO have a similar high selectivity (~85%), that declines slightly with increasing feed flow rate, indicating that the major reaction route is glycerol decomposition. The effect of feed flow rate on glycerol conversion and selectivities using "pure" glycerol is shown in Fig. 3b. Due to the high viscosity of pure glycerol, the feed flow rate was tested only up to 0.078 mL/min. The trends for glycerol conversion and selectivities for gas phase products are similar to those observed with 92% glycerol feed. Compared to 92% glycerol, the selectivities to H_2 and CO are lower by ~4% for pure glycerol. The selectivity to C_2 increases by ~5% at the same conditions. These results indicate that a small amount of water helps inhibit C_2 formation and improves the glycerol decomposition to H_2 and CO. Note that the H_2 and CO selectivity is still very high even for pure glycerol.

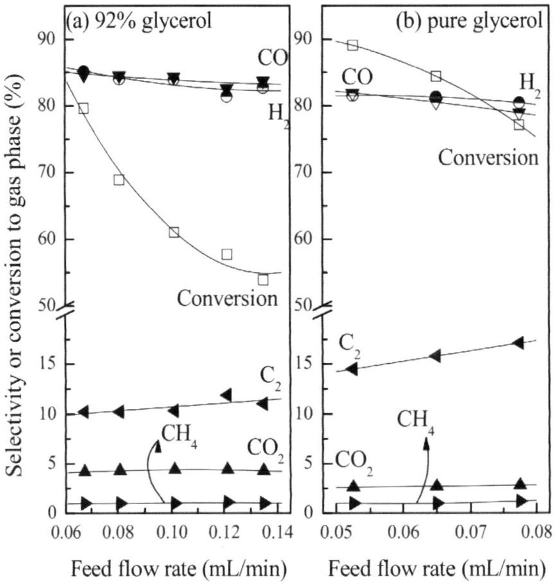

Figure 3. Effect of feed flow rate on the glycerol conversion and selectivities to gas phase products, (a) 92 wt% glycerol; (b) pure glycerol.

Conclusions

The results have demonstrated that non-thermal plasma applications are relevant for mobile application of hydrogen production from biomass derived resources: high H_2 yield, high energy efficiency, system compactness, short-transition time during cold-start, and no reforming catalyst use. Although water gas shift has not yet been optimized and no preferred oxidation has been used, the demonstrated integration of our developed WGS catalyst with the plasma has shown a potential process to further produce H_2 at low CO level (0.8%). Beside ethanol and glycerol, other alcohols or polyols such as sorbitol can possibly be used as feedstock. These feedstocks are renewable resources and largely available from fermentation processes resulting in a step closer to achieving the ultimate goal of sustainable energy.

Acknowledgement

This work has been supported by SEMGREEN, LLC., which the authors gratefully acknowledge

References

(1) Rostrup-Nielsen, J.R. Phys. Chem. Chem. Phys., 2001, 3, 283.

(2) Deluga, G.A.; Salge, J.R.; Schmidt, L.D.; Verykios, X.E. Science, 2004, 303, 993.

(3) Mavarro, R.M.; Pena, M.A.; Fierro, J.L.G. Chem. Rev., 2007, 107, 3952.

(4) Petitpas, G.; Rollier, J.D.; Darnon, A.; Conzalez-Aguilar, J.; Metkemeijer, R.; Fulcheri, L. Int. J. Hydrogen Energy, 2007, 32, 2848.

(5) Gordon, C. L.; Lobban, L. L.; Mallinson, R. G. Catalysis Today 2003, 84, 51-57.

(6) Hoang, T.; Pham, T.; Lobban, L. L.; Mallinson, R. G. Abstracts of Papers of the American Chemical Society 2006, 231.

(7) Larkin, D. W.; Caldwell, T. A.; Lobban, L. L.; Mallinson, R. G. Energy & Fuels 1998, 12, 740-744.

(8) Larkin, D. W.; Lobban, L. L.; Mallinson, R. G. Catalysis Today 2001, 71, 199-210.

(9) Le, H.; Lobban, L. L.; Mallinson, R. G. Catalysis Today 2004, 89, 15-20.

(10) Bromberg, L.; Cohn, D.R.; Rabinovich, A; Alexeev, N., Samokhin, N.; Hadidi, K. PSFC JA-06-03, 2006.

(11) Kalra, C.S.; Gutsol, A.F.; Fridman, A.A.; IEEE Trans. Plasma Sci., 2005, 33, 32.

(12) Sekine, Y.; Urasaki, K.; Asai, S.; Matsukata, M.; Kikuchi, E.; Kado, S. Chem. Comm., 2005, 78.

(13) Sekine, Y.; Asai, S.; Urasaki, K.; Matsukata, M.; Kikuchi, E.; Kado, S.; Haga, F. Chem. Lett., 2005, 34, 658.

(14) Paulmier T, Fulcheri L. Chem Eng J, 2006, 116, 13-18

(15) Rollier JD. Dissertation, Center for Energy and Process, Paris, 2006

IN SITU CATALYTIC FLUID BED PYROLYSIS of POPLAR WOOD

Sedat H. Beis and Foster A. Agblevor.

Virginia Polytechnic Institute and State University, Biological Systems Engineering
Blacksburg, VA, USA

1. Introduction

Rising global economic, strategic and environmental concerns putting pressure on the use of domestic and renewable energy options. On the other hand, the utilization nonfood lignocellulosic biomass is becoming more and more important as the negative effects of those already in the human and animal food chain as far as availability and costs are concerned.

Pyrolysis is one of the thermochemical conversion method for producing solid, liquid and gas products in the utilization of biomass. Fast pyrolysis is particularly important for bio oil production that can be further processed into fuels and chemicals. However, the direct substitution of bio oils with petroleum derived fuels and chemicals derived via pyrolysis is not likely due to its complex, highly oxygenated nature and higher viscosity without applying any upgrading. The early bio oil upgrading research focused on the catalytic cracking of the biooils over an acid type fixed bed of catalyst resulting in quick catalyst deactivation due to coke formation.

2. Experimental

In this research the fractional catalytic pyrolysis of poplar wood chips (<1.0 mm) into bio oil was achieved by an efficient and one step in-situ catalytic process on bench scale. An acid type catalyst supplied from Sud-Chemie, KY was used after size reduction and activation. A 100 g catalyst (-20+35 mesh) was used as for 150 g of biomass. The bed was fluidized with nitrogen preheated to 450 ^{o}C. A metal plate having 100µm porosity was used as gas distributing medium. The reactor was externally heated with a three-zone electric furnace. Biomass samples were fed into a feed hopper and conveyed by a twin-screw feeder into an entrainment zone where high-velocity nitrogen gas entrained the sample into the fluidized bed. Nitrogen flow rate was set to 15 L/min at room temperature. The pyrolysis temperature was maintained 450±5°C. The biomass feed rate was 120-150 g/h. The feed rate, nitrogen flow rate, and reactor temperature were kept constant during each run. The nitrogen and pyrolysis gases and vapors exiting the reactor passed through an externally heated hot gas/vapor filter to separate char/ash and any entrained sand. The transfer lines were maintained at 400 ^{o}C to avoid condensation of the pyrolysis vapors before they reached the first condenser. The pyrolysis gases and vapors were then passed through a cooling system consisting of two double pipe condensers, connected in series and maintained at -15 ^{o}C by circulating chilled water. An electrostatic precipitator and an coalescent filter were used to trap any aerosols leaving the second condenser. The amount of the liquid and solid products were determined by measuring the weight gain in the corresponding unit after the system cooled down to room temperature. Gas product yields were calculated by difference.

^{13}C NMR spectra of bio-oils were recorded in DMSO-d_6 solutions at 100.6 MHz and 25°C using a VARIAN 400 MHz spectrometer. About 32768 scans were accumulated for each sample. The ^{13}C spectra were divided into side chain aliphatics, methoxy/hydroxy-bound carbon, carbohydrate derivatives and lignin derivatives shift ranges for integration.

The molecular weight distribution of the bio oils were determined by gel permeation chromatography (GPC). Bio oil samples (~80 mg) were dissolved in BHT inhibited THF (20.0 mL). After 24 hr dissolution, the solutions were filtered through a 0.45 mm Teflon syringe filter. Eight polystyrene standards with peak molecular weights of 22000, 7000, 5050, 3570, 1700, 1050, 510, 162 (supplied by American Polymer Standards) were used to construct a universal calibration curve. Approximately 100 µL aliquots of each standard and the samples were individually injected through a 20 µL Rheodyne 7125 injection port for analysis. Separation was performed using Waters HR1 and HR3 columns connected in series. The THF mobile phase containing 0.025 wt% BHT inhibitor was pumped at 1.0 mL/min from a Waters 515 HPLC pump connected to a Li-ChromeIII flat coil pulse dampener. Inline solvent degassing was performed with a Phenomenex DG4000 vacuum degasser. A viscotek TDA 302 Triple array detector consisting of right angle and low angle light scattering detectors, an inline viscometer and a differential refractometer was used for signal detection. Viscotek Omnisec 3.1 Software was used for data analysis.

Karl-Fisher titration was used for determination of the water content of the bio oils.

A Brookfield Cap-2000 cone viscometer was used for the measurement of the viscosity of the bio oils. Measurements were taken by using spindle #3 at 25 °C at 1000-12000 shear rates.

The gaseous products were analyzed by gas chromatography using an on-line SRI Multigas#2 GC instrument (SRI Instruments-USA). High purity helium was used as carrier gas 20 mL/min in both of the auto sampling loops which were connected to Hayeseep-D 6' column, MXT-1 60 m capillary column and molecular sieve-13X 3' column. Thermal conductivity detector (TCD) and methanizer coupled flame ionization detector (FID) were used for detection. Gas stream were analyzed online for determination of mainly CO_2, H_2, CO, CH_4, C_2H_2, C_2H_4, C_2H_6, C_3H_6, C_3H_8.The gas components were identified by comparing their retention times to those of a mixture of standard compounds supplied from Scott Specialty Gases-USA.

3. Results and Conclusion

The average product distribution was 27.3 ±1.2 for liquid, 20.6 ±1.4 for the char and 52.2 ± 0.2 for the gas based on dry and ash free basis of the feed. On the other hand, the liquid, char and gaseous product yields obtained from conventional fast pyrolysis of poplar chips by using sand as the fluidization medium were 43.6±2.0, 14.9±2.5 and 41.5±1.2 respectively.

The composition of the fractional catalytic pyrolysis (FCP) bio oil differed greatly than the non-catalytic fast pyrolysis (NCP) bio oil. The lignin derivatives, carbohydrate derivatives and methoxy/hydroxy contents were in the 70%, 11% and 5% range for the FCP bio oil, while

49%, 20% and 15% for the NCP biooil. Aliphatic side chains were in the 11% and 16% range for the FPC and NPC bio oils respectively.

Molecular weight distribution of poplar wood FCP bio oil was determined as the number average molecular weight (Mn):120, the weight average molecular weight (Mw):257 and Mz:555, Mp:134. Dispersity was (Mw/Mn):2.13.

The pH of the NCP bio oil was 2.1 whereas the FCP biooil was 3.5. Viscosities of the bio oils produced by NCP and FCP was 800 cP and 150 cP at 25°C.

The ^{13}C nmr data of the produced bio oil revealed that the in-situ catalytic process was capable of producing a narrower range of liquid chemicals by converting the carbohydrate fraction of the poplar wood to gases while simultaneously cracking the lignin fraction into a mixture of low molecular weight phenols such as simple phenol, alkylated phenol and methoxylated ones. Moreover, FCP bio oil exhibited better physical properties when compared to NCP bio oil.

Gasification Technology as Alternative Hydrogen Production System: A Comparative Study of Fuels Available in Mexico

Emilio Manzanares-Papayanopoulos*, Manuel Fernández-Montiel, and Agustín Moisés Alcaraz-Calderón

*Gerencia de Procesos Térmicos, Instituto de Investigaciones Eléctricas, Reforma No. 113, Col. Palmira, CP62490, Cuernavaca, Morelos, México. * e-mail: emp@iie.org.mx*

ABSTRACT

The technology of gasification transforms any fossil fuels such as natural gas, coal, liquid or solid refinery residues or heavy oils into a synthesis gas, also called syngas, for subsequent utilisation as combustible gas. Also, biomass and wastes can be used as feed fuel.

In the past years, the Instituto de Investigaciones Electricas (IIE) in Mexico, technological research arm of the Energy sector in Mexico, has been working on economical and technical evaluation studies of the gasification technology as an alternative for power generation. However, in response of the national needs, a study of the gasification technology for hydrogen production by the shift-gas reaction process in a traditional refinery scheme has been performed.

The use the gasification technology as hydrogen production system is quite attractive in many ways since it offers several advantages, and allows synergies to other industries. One is a clean path of hydrogen source, due to sulphur and heavy metals can be removed from the syngas. Secondly, the CO_2 obtained can be sequestered and stored in special places, avoiding the traditional contamination to the atmosphere. In addition, there is flexibility on the feedstock, giving as a result a wide fuel diversification to be used in the hydrogen production.

The aim of this study is to present an analysis of the hydrogen production by the gasification technology as a function of fuel compositions. The fuels used are bituminous coal and refinery residues such as the liquid vacuum distillation residues and the petroleum coke. It is important to notice that all the feedstock are Mexican fuels, in which the refinery residues have relative high sulphur and metals concentration and the bituminous coal is from MICARE mine, at the north of the country, with relative high ash concentration. Also, potential imported fuels are considered in the study as the Australian coal, and Colombian coal, however, the results of the imported fuels are not presented.

The results shown that the application of the gasification technology for hydrogen production is very attractive and offers several advantages over other hydrogen production systems.

Keywords: Electric power generation, IGCC, Gasification, Fossil fuels, Hydrogen production

INTRODUCTION

The Instituto de Investigaciones Electricas (IIE), technological research arm of the Energy sector in Mexico, has followed on past years the status of the gasification technology as Integrated Gasification Combined Cycle (IGCC) for power generation (Fernández-Montiel *et.al.*, 2003; Fernández-Montiel and Alanis, 1993 and 1994; Fernández-Montiel, 1994), because it offers flexibility of the fuel used and with a required pollution reduction systems allows clean power generation (Higman and Der Burgt, 2003; Rezaiyan and Cheremisinoff, 2005).

In a previous paper (Manzanares-Papayanopoulos *et. al.*, 2006), the study of the syngas production analysis as a function of fuel compositions, temperatures and pressures by simulation carried out by commercial computational packages, and the comparison of the energy produced as function of the fuel composition at the same operational parameters was presented. Additionally, a design of an atmospheric pressure gasification system at work-bench scale to gather information of the kinetics gasification process has been also presented (Altamirano-Bedolla *et. al.*, 2007). Consequently, an analysis of the kinetic processes inside of a gasification reactor was published (Villatoro-Camposeco and Manzanares-Papayanopoulos, 2008). Furthermore, a study of the gasification technology versus traditional combustion technologies for power generation was performed in order to present the cost-benefit of them, also, to verify the applicability of the gasification technology in the power generation in Mexico (Manzanares-Papayanopoulos *et. al.*, 2008).

It is well known that the electricity demand have been growing around the world. In Mexico, that situation is not an exception, and as a result, there is a permanent scheme for an expansion in the current power plant fleet (SENER, 2008).

Although, in a similar situation, the growing demand for cheap-clean fuels, based in fossil fuels, is permanent. Actually, the main oil company, Petroleos Mexicanos (PEMEX) own and operate six refineries around the country. These refineries processed on 2007 about 1,270 thousand barrels per day of crude oil and produced about 1,312 thousand barrels per day of refined products, including dry gas by-products of the refining process (PEMEX Web Page, 2009). Fuel oil for power generation, automotive unleaded gasoline and diesel represent the bulk of PEMEX refined production. However, the internal demand of fuels is higher than its capacity of production of those fuels. As a consequence, there is a need to import fuels to supply the internal demand of the country.

Actually, as respond of the internal fuel needs, PEMEX has planned to build a new refinery plant. In the initial studies, PEMEX is interested in the potential use of the gasification technology, since it allows synergies to other oil refining processes. Other advantages of the gasification technology is the diversification in the use of different fuels, ultra low contaminant emissions, the used of the capture and sequestration of CO_2 technology, power generation for self-consumption, and as main attraction to obtain hydrogen as feedstock for its refinery processes.

The aim of this study is to present an analysis of the hydrogen production by the gasification technology as a function of fuel compositions, in order to supply the hydrogen demand in a typical Mexican refinery. Additionally, the study shows that other feedstock as medium vapour pressure and power energy can be supply to the refinery.

CASE DESCRIPTION

The analysis of the hydrogen production as a function of fuel compositions were carried out using commercial packages. The design of the IGCC simulation models were in base of the type of fuel feeding (dry or slurry), and the hydrogen separation technology.

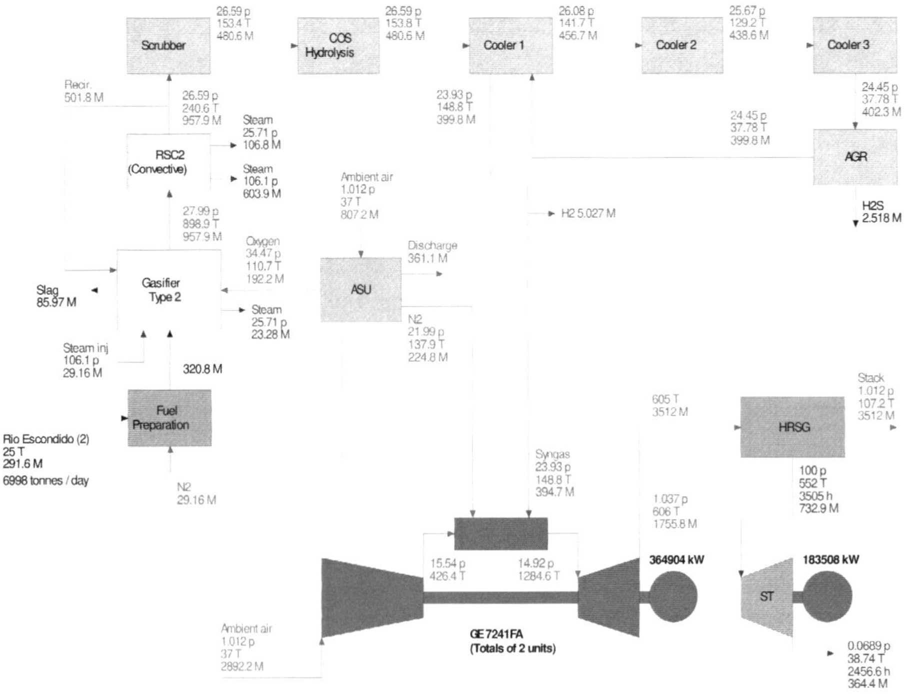

Figure 1. IGCC Model for hydrogen production, dry feeding and H_2 membranes separation technology.

The IGCC simulation models are presented in **Figures 1-4**. The fuels used in the analysis were liquid residuals and petcoke from Mexican refineries, and Mexican coal.

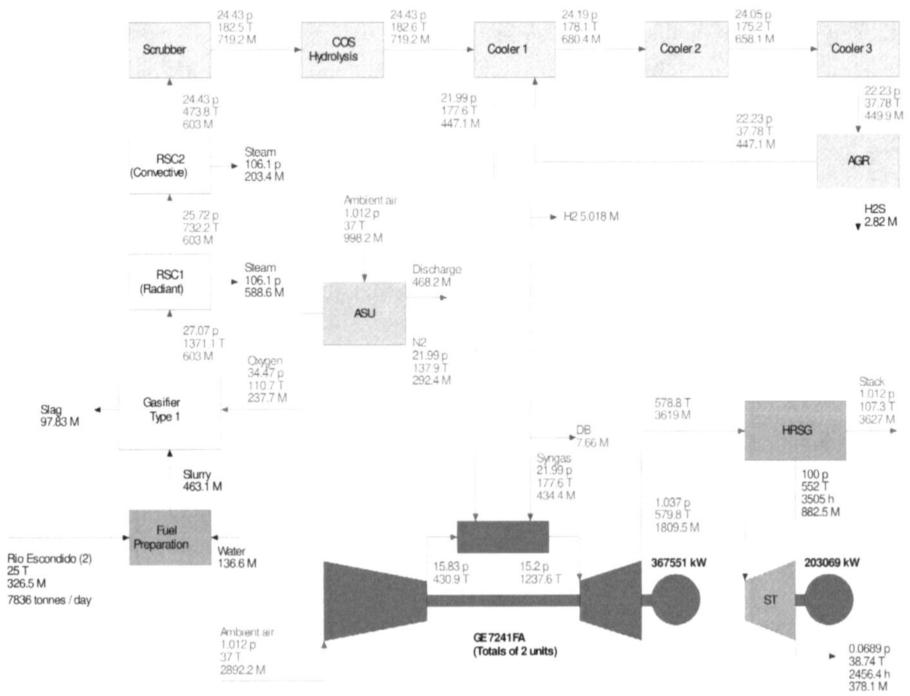

Figure 2. IGCC Model for hydrogen production, slurry feeding and H_2 membranes separation technology.

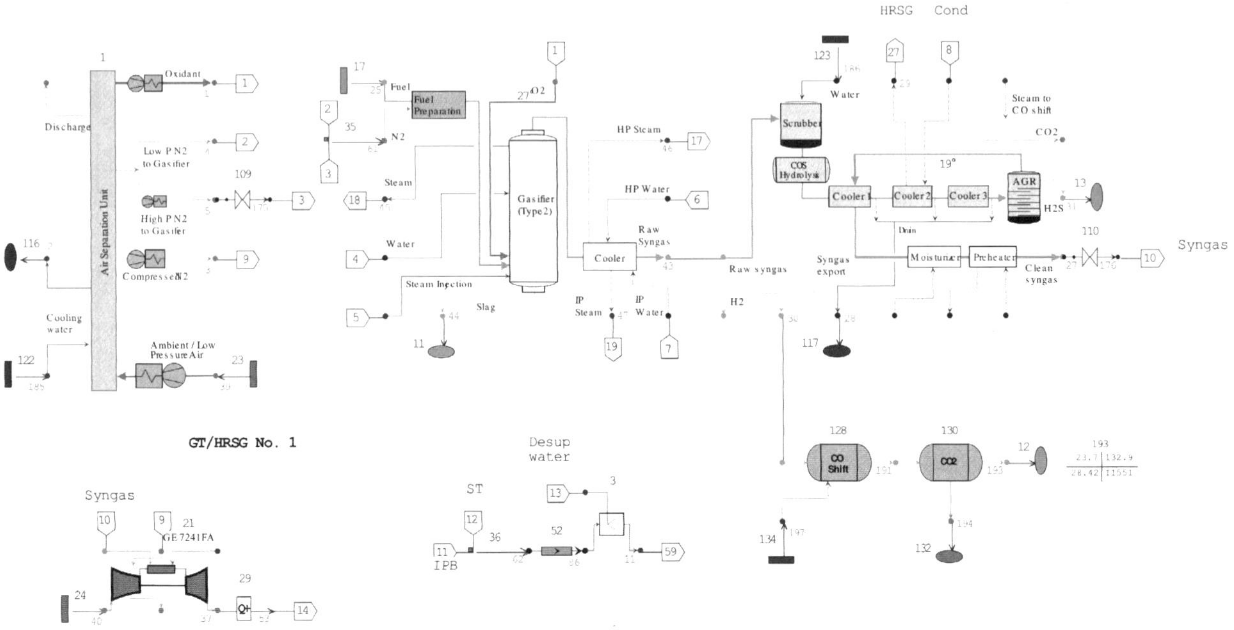

Figure 3. IGCC Model for hydrogen production, dry feeding and CO Shift separation technology.

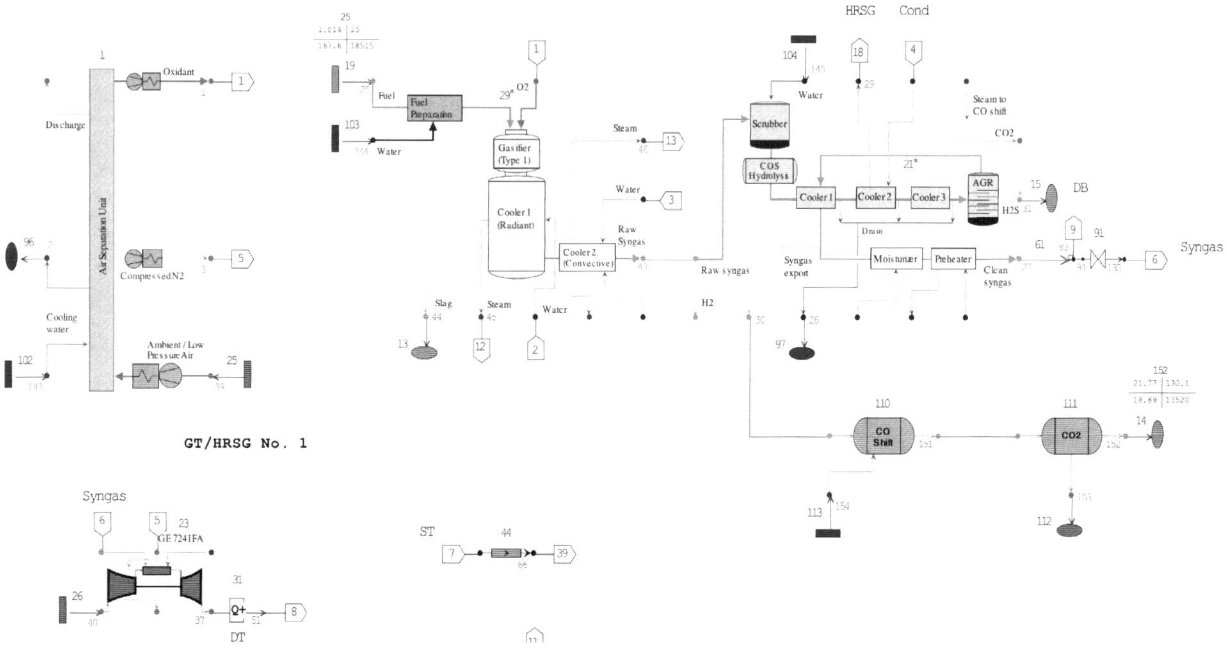

Figure 4. IGCC Model for hydrogen production, slurry feeding and CO Shift separation technology.

The basis premises for the IGCC simulation were to produce 50 MMPCSD of hydrogen, 500 T/h of medium vapour i.e. vapour at 20 kg/cm$_2$ y 350 ^oC, and 100 MW, to be used by the refinery. The site conditions were considered as ISO.

The hydrogen separation technologies considered in the models were two. One of them was by the use of membranes, and the second was by the use of the CO shift reaction technology.

SIMULATION RESULTS AND DISCUSSION

The results of the simulations for the study of the hydrogen production by an IGCC as function of the type of fuel were quite interesting. As it is shown in the following figures, the use of liquid residual as fuel was only considered in the slurry feeding model. The required fuel to achieve the premises described in the previous section is shown in **Figure 5**.

As shown in **Figure 5**, the used of Mexican coal increased the use of fuel since it has about 40% of ash. This fuel demand should be considered in an economical analysis to establish the feasibility of the Mexican coal as feeding fuel in an IGCC for hydrogen production. Also, the membrane technology for H_2 and CO_2 separation required less fuel than the CO shift reaction technology, especially in the case of liquid residual feeding fuel.

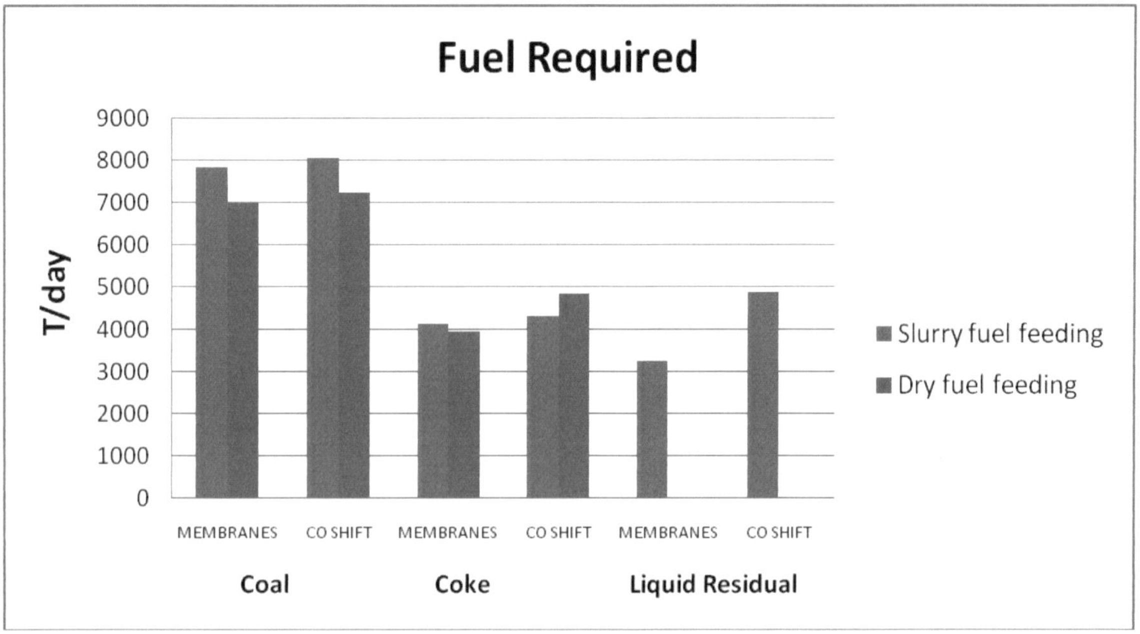

Figure 5. Fuel required by the IGCC to achieve the hydrogen production needed for a typical Mexican refinery.

The hydrogen generation achieved for the IGCC of function of fuel type and feeding technology is shown in **Figure 6**.

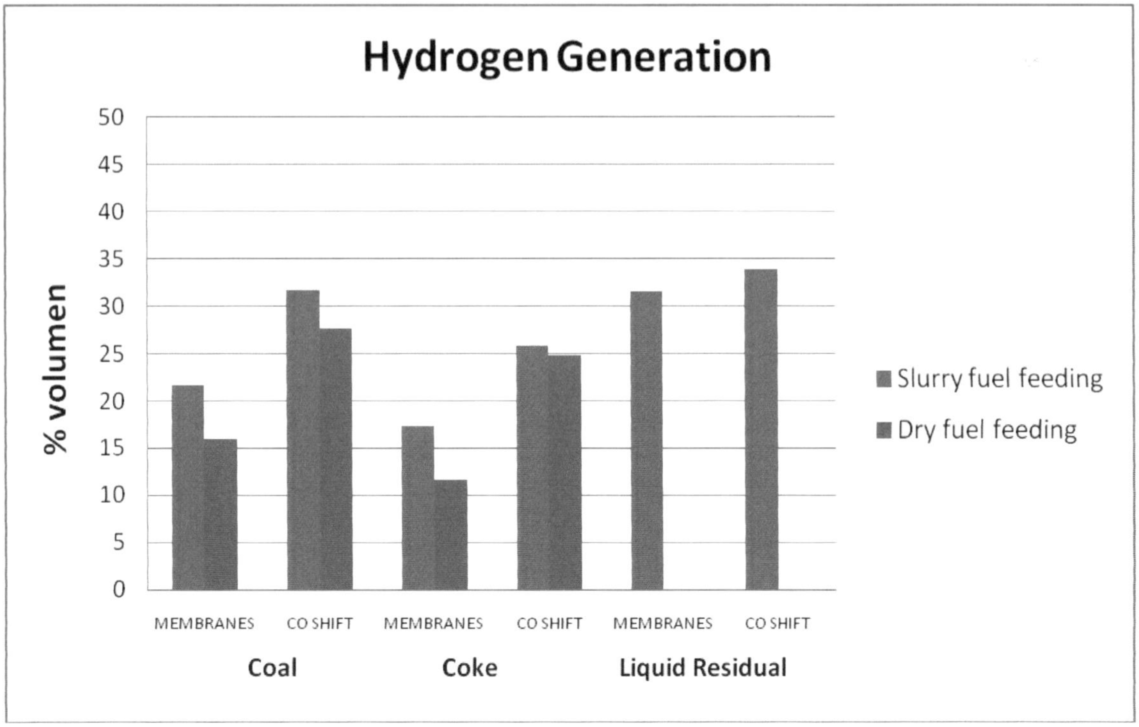

Figure 6. Hydrogen generation by the IGCC of function of fuel type and feeding technology.

Figure 6 shows that the slurry fuel feeding technology produces higher hydrogen volumen. However, in **Figure 7** is shown the amount of water that is required by the IGCC scheme with the slurry and dry fuel feeding systems to achieve the hydrogen required for the refinery. The water reported for the slurry fuel feeding technology are both, the one required for the IGCC scheme, and the one required for hydrogen production.

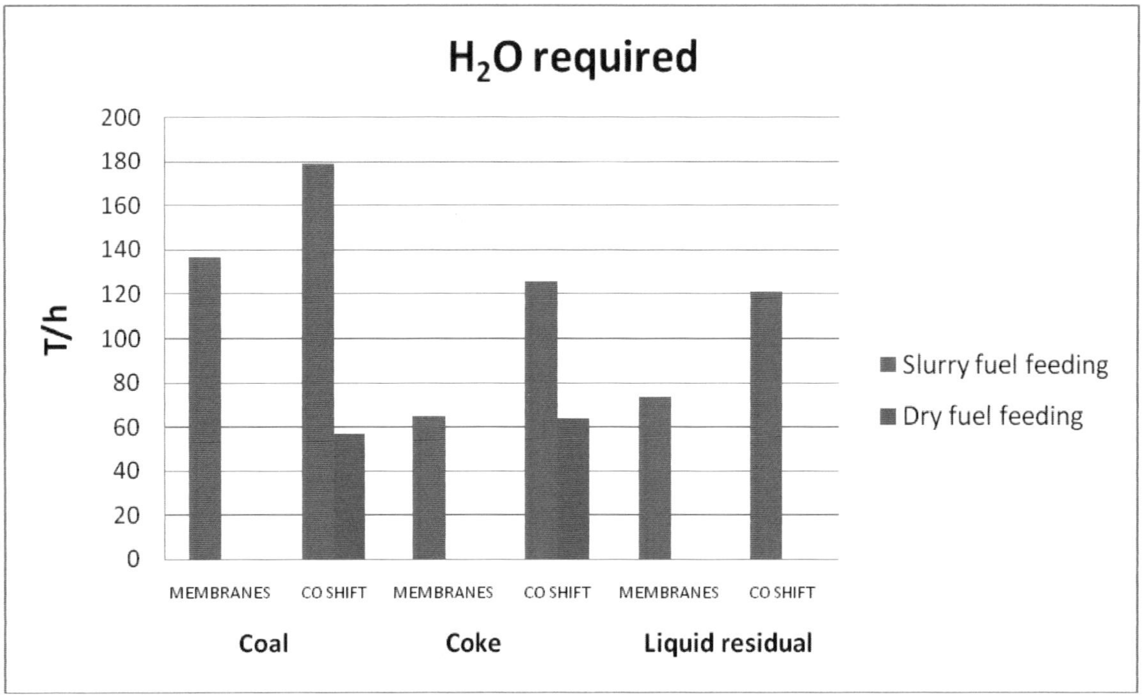

Figure 7. Water required for hydrogen generation by the IGCC of function of fuel type and feeding technology

In contrast, as shown in **Figure** 7, the dry fuel feeding technology used in an IGCC consumed just the water required in the CO shift reaction module to obtain the hydrogen requested for the refinery. It is important to mention that the situation of the use of water for industrial processes in Mexico is quite expensive in mainland, and also in many places there is not enough water to supply to the industry and the cities. Therefore, the location of the IGCC and the new refinery should be on-shore.

Figure 8 shows the CO generation in function of the feeding technology. The behavior presented is consistent as it has been reported previously (Manzanares-Papayanopoulos *et. al.* 2006). In **Figure 9**, the Syngas heating power of the resulting stream after the hydrogen extraction/production is presented. In both Figures, the dry-feeding technology has much better performance than the slurry feeding technology.

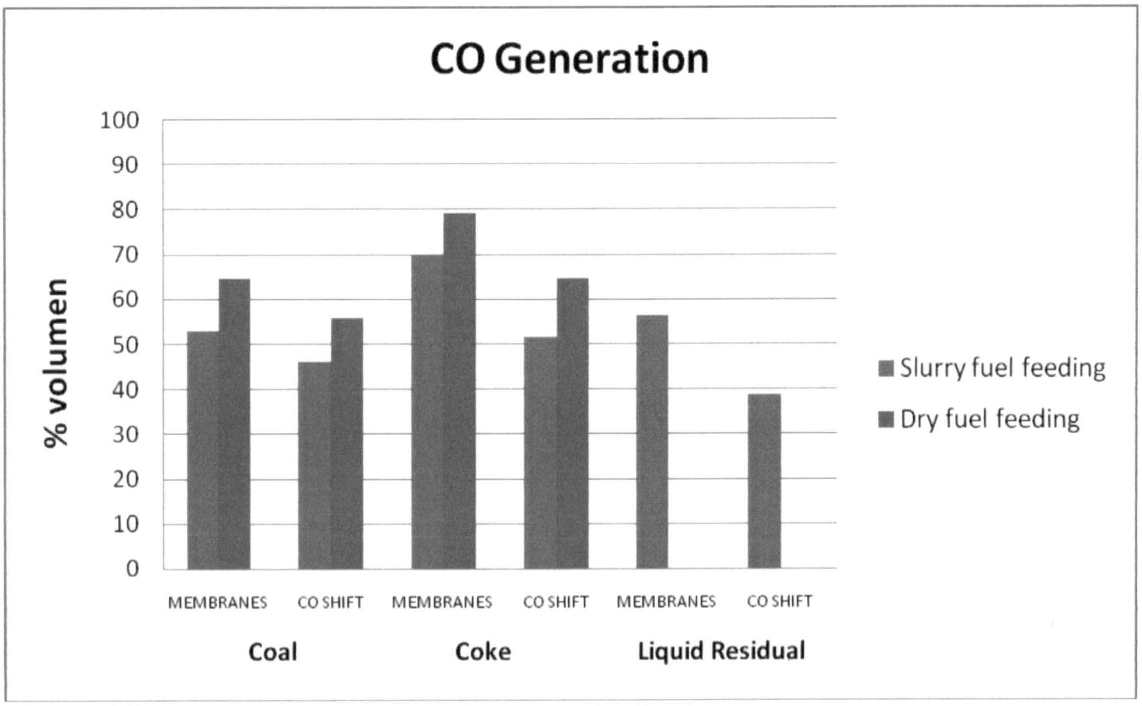

Figure 8. CO generation in function of the feeding technology.

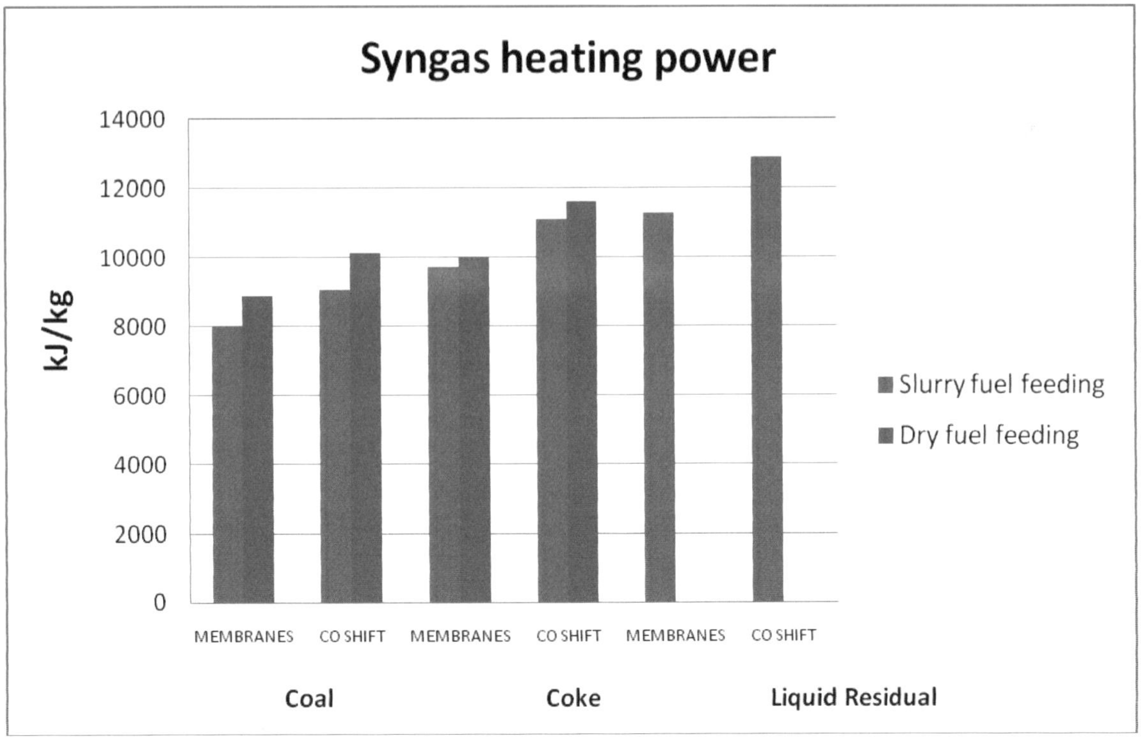

Figure 9. Syngas heating power of the stream to the gas turbine in function of the feeding technology.

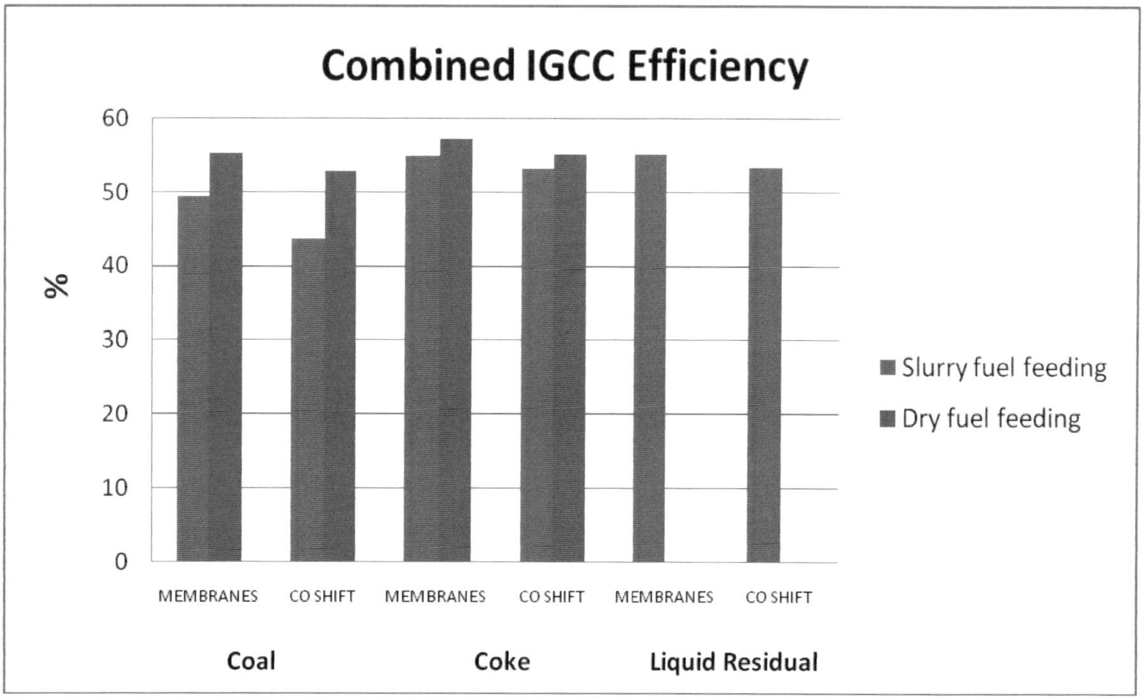

Figure 10. Combined IGCC efficiency without considering the hydrogen extraction in function of the feeding technology.

Figure 10 show the efficiency of the process without considering the hydrogen extraction. The IGCC dry-feeding technology shows better performance than the IGCC slurry-feeding technology as it has been also reported previously (Manzanares-Papayanopoulos *et. al.* 2006).

CONCLUSIONS

It is well know that the technology of gasification is still under study around the world. In this particular study, the understanding of the hydrogen production as a function of both the feedstock composition and fuel feed system was the main goal.

The simulation results pointed out in this work, open the discussion about the possibility of using the gasification technology as an attractive alternative as main supplier of hydrogen, power and steam to a typical Mexican refinery. More studies and research on the gasification technology and all its complementary systems, i.e. a separation air unit, feedstock systems, clean Syngas system, CO_2 sequestering and storage system, Syngas turbine systems, and particularly in separation technologies of hydrogen should be performed.

The feeding system is also important in the results, since the dry-feeding system has better performance than the slurry-feeding system. However, a second stage comprising an economical evaluation must be performed to confirm or reject these conclusions, and also to complete this study.

ACKNOWLEDGEMENT

The authors wish to thank to the Comision Federal de Electricidad (CFE) and to Petroleos Mexicanos (PEMEX) for the financial support of this project.

REFERENCES

Altamirano-Bedolla J.A., Mani-González A., **Manzanares-Papayanopoulos E.***, Fernández-Montiel M., and Romo-Millares C.A. (2007), "Preliminary Study of the Gasification Process of Mexican Fuels at Atmospheric Pressure", AIChE Spring Meeting 2007, Paper 9e, April, 2007, Houston, TX., USA.

Fernández-Montiel, M., Alcaraz-Calderón, A. M. y García-Chávez, L.; "Estudios y Desarrollo de Alternativas de Gasificacion para el Sector Eléctrico", Informe 12585, Instituto de Investigaciones Eléctricas, Diciembre 2003, México.

Fernández-Montiel, M. y Alanís, F.J. "Reporte del Estado Actual de las Tecnologias de Gasificacion Aplicadas a la Generacion de Energia Electrica", Informe 44207, Instituto de Investigaciones Eléctricas, Octubre 1993, México.

Fernández-Montiel, M. y Alanís, F.J.; "Estudios de Factibilidad para Plantas de Gasificacion de Carbon y de Combustoleo Integradas a Ciclos Combinados", Informe 44288, Instituto de Investigaciones Eléctricas, Marzo 1994, México.

Fernández-Montiel, M; "Reporte de actualizacion sobre el estado que guardan las tecnologias de gasificacion aplicadas a la generacion de energia electrica", Informe 44256, Instituto de Investigaciones Eléctricas, Octubre 1994, Mexico.

Higman C. and Der Burgt M. Van, "Gasification", Elsevier, 2003.

Manzanares-Papayanopoulos, E.*, Fernández-Montiel, M., Altamirano-Bedolla, J.A., Alcaraz-Calderón, A.M., and Romo-Millares, C.A., (2006), "Development and Evaluation of the Gasification Technology as Alternative Power Generation", AIChE Spring Meeting 2006, Paper 128e, April, 2006, Orlando, Fl., USA.

Manzanares-Papayanopoulos, E*, Fernández-Montiel, M., Altamirano-Bedolla, J. A. Mani-González, A. and Arriola-Medellín, A. (2008) "A Comparative Study of the Gasification Technology *versus* Traditional Combustion Technologies for Power

Generation in Mexico", AIChE Annual Meeting 2008, Paper 761b, November, 2008, Philadelphia, USA.

PEMEX Web Page, (2009) http://www.ref.pemex.com/

Rezaiyan J. and Cheremisinoff, N. P.; "Gasification Technologies: A Primer for Engineers and Scientists", Taylor & Francis, 2005.

SENER (Secretaría de Energía) (2008), "Prospectiva del Sector Eléctrico 2008-2017", Dirección General de Planeación Energética, Primera Edición.

Villatoro-Camposeco, N.N. and **Manzanares-Papayanopoulos, E***., (2008), "Análisis de los procesos químicos en un reactor de gasificación de una central de IGCC para generación de energía eléctrica", CIINDET 2008, 6º Congreso Internacional en Innovación y Desarrollo Tecnológico, Octubre 2008, Cuernavaca, Morelos., México.

3g Gasification Characteristics of CWM with Petroleum Coke and Glycerin in a 1T/D Entrained-Flow Gasifier

See Hoon Lee, Jae Go Lee, Young Chan Choi, Sang Jun Yoon, Ho Won Ra, and Jai Chang Hong.
Gasification research center, KIER, 71-2 Jang-dong Yusung-gu, Daejeon, South Korea

Gasification processes have been widely used in various industrial applications and the use of various feedstocks is being developed. In this study, the gasification characteristics of coal and petroleum coke water mixture and glycerin were investigated in 1 T/D entrained-flow gasifier (I.D. 0.2m x Height 1.7m). The gasification system consisted of oxygen input system, fuel mixing tank, dual mixing burner, entrained-flow gasifier, water scrubber and gas analyzer. At the gasification of coal and coke water mixture, the syngas flow and H2/CO ratio are higher than that at the gasification of petroleum coke. The calorific value of syngas from mixture shows about 1,600-1,800kcal/Nm3. At the gasification of glycerin, the average gasification temperature is about 1,000oC and the H2/CO ratio is much higher than that of mixture.

A SORPTION CATALYST COMPLEX WITH NANO $CaCO_3$ PRECURSOR FOR HYDROGEN PRODUCTION USED IN ReSER PROCESS

Su F. Wu, Lian B. Li, Qi Shi, Xie Q. Wang
Department of Chemical Engineering and Biochemical Engineering, Zhejiang University, Hangzhou, 310027, China

Abstract

As hydrogen becomes more and more important as clean fuel and basic raw material for petroleum refining and chemical processes, a new process is being under investigation for hydrogen production named ReSER (Reactive Sorption Enhanced Reforming) Process, which is based on Ni as catalytic material and CaO as adsorbent aimed at shifting the reaction equilibrium prefer to hydrogen yield. This paper describes the preparation and evaluation of a sorption complex catalyst containing Ni and CaO. A nano-$CaCO_3$ material used as precursor of CaO followed by CO_2 adsorption during the steam reforming reaction showed its advantage in a significant improvement of the sorption capacity and durability for cyclic runs in carbonations and calcinations. The physical properties tested by SEM and XRD indicated a nano- CaO can resist the sintering than the conventional CaO at high temperature carbonations and calcinations. The CO_2 sorption properties measured by a thermal gravimetric analyzer showed a sorption rate over 68% after 50 cyclic runs at 600°C carbonations and 750°C calcinations. The evaluation of the sorption complex catalyst for ReSER resulted in the high hydrogen yield over 95% (v/v) both in the fixed bed and fixed fluidized bed reactor in bench scale. The sorption catalyst has an attractive prospect in application on ReSER process for hydrogen production, especially in the fluidized mode where reactor and regenerator combined in a cyclic process.

Introduction

Steam methane reforming is a traditional and mature industrial process which has been used since 1826 to produce hydrogen from nature gas. An innovative technique based on the Le Chatelier principle has been developed within the decades. With the use of hydrotalcite as the adsorbent to capture CO_2 released from the reforming and water shift reaction, some new processes named sorption enhanced reforming process (SERP)[1-2], SE-SMR[3] or adsorption enhanced reforming(AER)[4-5] were under developing. Our new process is characterized by using nano CaO as adsobent for CO_2 capture combined with Ni as catalyst in one particle with in a reactor- regenerator cycle process and named reactive sorption enhanced reforming (ReSER), in which hydrogen can be highly yielded at lower temperature.

In the ReSER process, CO_2 sorpent properties have a significant impact on the reforming efficiency. CO_2 sorption rate, sorption velocity and sorption durability have been regarded as the key elements for evaluation a CO_2 sorbent in the reforming process. CO_2 captured by CaO forming $CaCO_3$ should have a fast reaction rate with large sorption capacity . Among some CaO precursors, $CaCO_3$ is a suitable candidate with its rich natural resources and lower cost. But the disadvantages of high calcination temperature of $CaCO_3$ to form CaO hindered the reuse due to the sintering and thus, resulted in the decrease of the sorption capacity in cyclic runs [6-7]. Many researchers tried to develop a modified $CaCO_3$ aimed at to increase of the durability using steam hydration of limestone[8-9], precipitation $CaCO_3$[10], adding $Ca_{12}Al_{14}O_{33}$ [11], Calcium acetate and $Ca(OH)_2$ as precursors[12,13] and so on. In recent years, the nano $CaCO_3$ production is mature in a large production scale with controlled size range[14]. The author attempts to use a nano size of $CaCO_3$ as the CaO precursor has an obvious increase in sorption capactity, sorption rate and durability[15-16].

In this paper, studies related to nano $CaCO_3$ powder as precursor of CaO for CO_2 adsorption, the spraying technique to preparation a micro particle size of sorption catalyst complex containing nano CaO and NiO have been carried out and the sorption catalyst complex has been characterized and evaluated for hydrogen production by ReSER process.

Experimental

Agents and instruments

Nano $CaCO_3$ powder with a particle size of 70 nm (Huzhou Linghua Co. Ltd., China) was used as CaO-based sorbent precursor. A thermo-gravimetric analyzer (TGA, Pyris1, PerkinElmer,USA) for sorption properties test. The crystalline phase of compositions of sorption catalyst were determined by X radial diffractor (XRD, D/MAX-RA, Rigaku, Japan). The morphology of sorption catalyst was investigated by scanning electron microscope (SEM, S-4800, HITACHI, Japan). The contents of metal in the surface of sorption catalyst were measured by energy-dispersive X-ray (EDX, EX350, HORIBA, Japan). The distribution of the size of sorption catalyst was analyzed by a particulate size description analyser (Mastersizer 2000, Malvern instruments, USA).

Preparation of sorption catalyst

The preparation process for the catalyst was illustrated as Figure1. CaO precursor, inorganic binder and Ni salt was mixed and stirred in a beaker according to the stechiometric ratio amount forming a slurry. The slurry was sprayed and dried to form the catalyst complex in a suitable size distribution and calcinated in a furnace and ready to use.

Test of the sorption properties by TGA

TGA evaluation system was shown in Figure 2. About 3mg catalyst complex was filled in a nacelle, CO_2 gas was switched on for adsorption and then switched off when $CaCO_3$ was formed and followed by calcination under the certain temperature. The data of weight change were collected in situ by a computer, sorption capacity and sorption ratio have been calculated with equation (1) and (2). Sorption ratio was defined as sorption CO_2 mole amount divided by the CaO mole amount of sorption catalyst complex, and sorption capacity was defined as sorption CO_2 mole amount divided by initial mass amount of sorption catalyst.

$$\text{sorption capacity} = \frac{\text{sorption } CO_2 \text{ mol}}{\text{sorption catalyst weight}} (\text{mol/kg}) \quad (1)$$

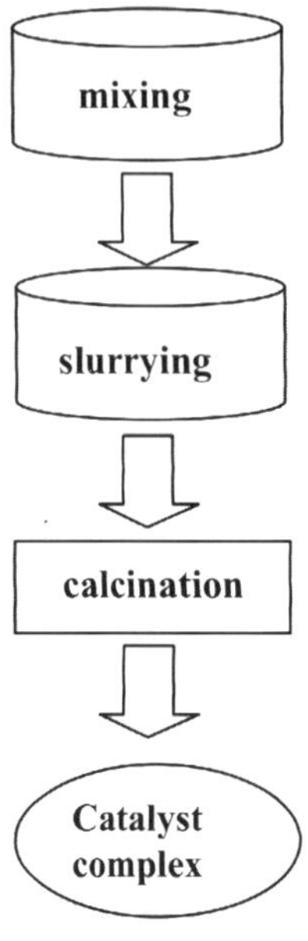

Fig. 1 The process of sorption catalyst preparation

$$\text{sorption ratio} = \frac{\text{sorption } CO_2 \text{ mol}}{CaO \text{ mol in catalyst}} \times 100\% \qquad (2)$$

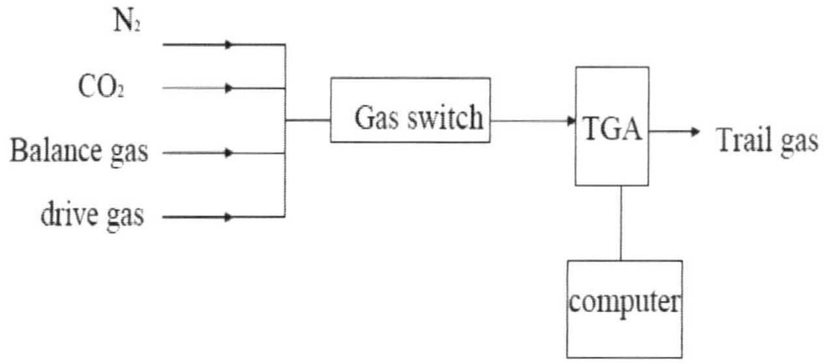

Fig.2. Scheme of the TGA analysis

Evaluation of ReSER process for hydrogen producing

The evaluation system of the sorption catalyst complex by ReSER process was shown in Figure 3. A fixed bed reactor with a size of Φ25mm×50mm stainless tube was used. Catalyst complex was filled in at the temperature constant area of the reactor. A PLC control system has been used to collect and summerized the data of reaction temperature, reaction pressure and the flow rate of feed and effluent gas. An online system of GC used to analysis the components of the production gas. Multiple runs of the ReSER process were carried out in a mode of reaction and regeneration consecutively. Methane conversion and hydrogen concentration was defined as normal definition.

Multiple runs the ReSER process and regeneration can test the cyclic ReSER hydrogen producing experiments. Methane conversion and hydrogen concentration was defined as normal definition.

A fixed fluid bed reactor with the size of Φ108mm×50mm made by stainless instead of fixed bed reactor to test the ReSER process for hydrogen production.

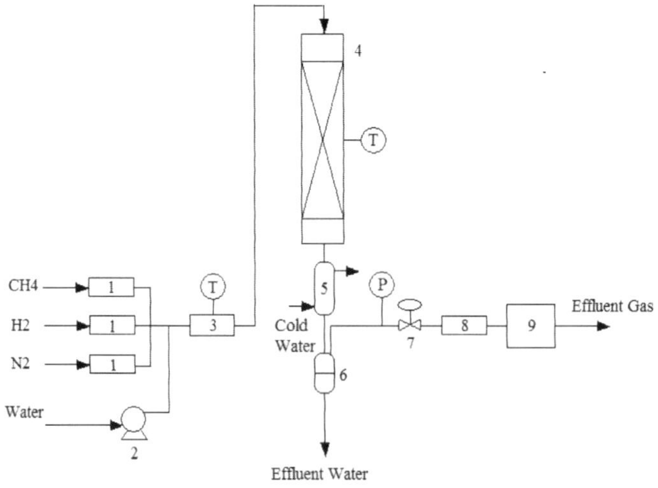

Fig 3. The diagram of ReSER process for hydrogen pruducing (1-FMC 2-pump 3-water evaporator 4-fixed bed 5-condensor 6-seperator of gas and liquid 7-valve 8-FM 9-GC)

Results and discussions

Characteristics of sorption catalyst complex

Nano $CaCO_3$ has a wide uses in the polymer industry and chemical engineering industry as a packing materials. Nano $CaCO_3$ has a lower decomposition temperature will be benefit to save energy. Figure 4 shows the SEM image of nano $CaCO_3$, the cubic crystalline type can be seen clearly. Figure 5 shows a SEM image of particle of sorption catalyst complex with the size around 60μm in average. Figure 6 shows a XRD test results for three sorption catalyst complex samples. Three samples all contain the $CaCO_3$, $Ca(OH)_2$ and NiO components.

Fig.4 SEM image of nano $CaCO_3$ particle

Fig. 5 SEM image of sorption catalyst

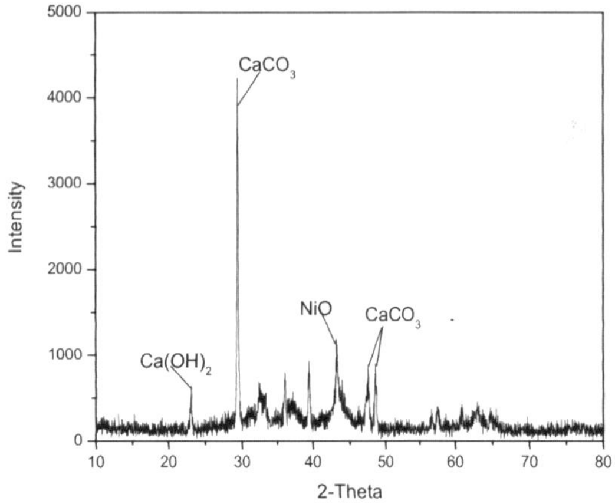

Fig.6 XRD test results of the sorption catalyst

The sorption properties of micro particle catalyst complex

The idea of combine the Ni base reforming catalyst and the CaO based CO_2 sorbent in one particle to form a sorption catalyst comes from the advantages of the following:

(1) Heat efficient: Inside of the sorption catalyst particle, it takes an endothermic reforming reaction and CaO reactive with CO_2 exothermic reaction at the same time as equations of (3) and (4).

$$CH_4+2H_2O \rightleftharpoons CO_2+4H_2,\quad \Delta H_{298}=165.1kJ/mol \tag{3}$$

$$CaO+CO_2 \rightleftharpoons CaCO_3,\quad \Delta H_{298}=-178.3kJ/mol \tag{4}$$

The heat can be used efficiency and decrease the lost of heat as much as possible.

(2) Easy operation: Since the density of reforming catalyst and the CaO based CO_2

sorbent have different particle densities, the mixing situation will be changed by feed flow during the operation. One particle can solve the problem and can be used in fixed or fluid bed reactor or the other moving bed reactors.

(3) Increase the reaction activity: Sorption catalyst can realize the CO_2 capture in site to enhance the reforming reaction as fast as possible.

The particle size distribution has been analyzed. The average of size is 98μm. The main composition of sorption catalyst is Al_2O_3 20%, NiO 15%, the remain is CaO. The density of the catalyst is 0.78g/cm^3.

Nano $CaCO_3$ as precursor of CaO sorption material provides a much better durability. TGA analytic results of sorption ratio and sorption capacity of catalyst complex were shown in Figure 7 and Figure 8 respectively. It can be seen from Figure 7 and Figure 8 that the sorption capacity remains almost unchanged after 40 cyclic runs. The stable sorption ratio is over 68%, and the sorption capacity is 6.0 mmolCO_2/g sorption catalyst complex. The figure showed a better result in a higher sorption capacity and the durability compared to the published data.

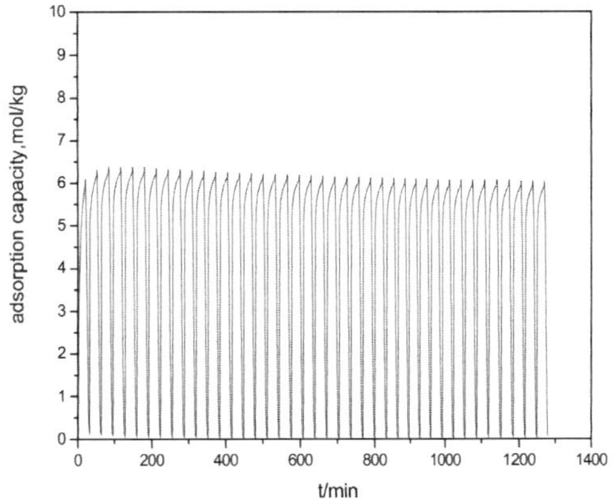

Fig.7 CO_2 sorption ratio of the sorption catalyst (600°C carbonations and 750°C calcinations)

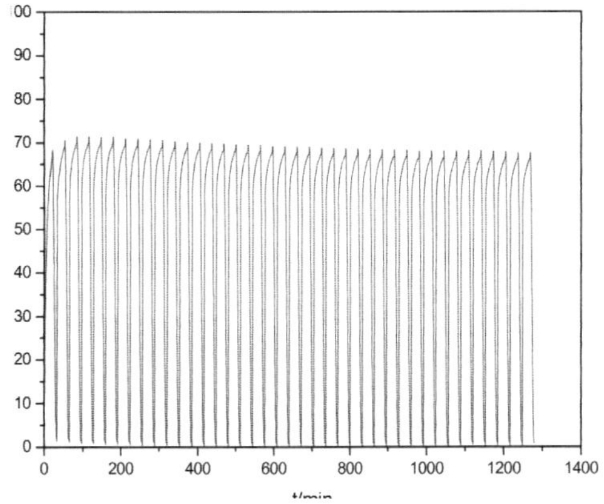

Fig. 8 CO_2 sorption capacity of the sorption catalyst(600°C carbonations and 750°C calcinations)

The evaluation of sorption catalyst complex in a fixed bed reactor

Using the average size of 98μm sorption catalyst complex to evaluation the properties in the ReSER process for hydrogen production. The results shown in Figure 9 indicate that the concentration of hydrogen is over 95% in the sorption enhanced process.

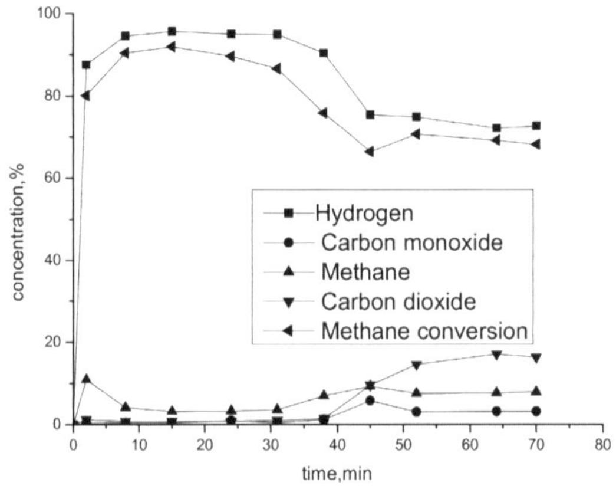

Fig.9 The ReSER test results of sorption catalys(temperature 600℃, reaction pressure 0.1MPa, steam methane ratio 4, W/F=0.05g min ml^{-1})

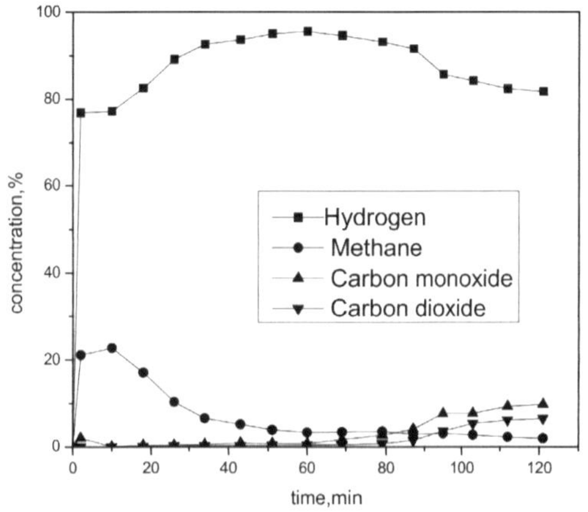

Fig 10. Sorption catalyst of ReSER process in a fluid bed (Reaction temperature T 600□, reacion pressure 0.1MPa, steam/carbon mol ratio 4, W/F=0.16g min ml^{-1})

The evaluation of sorption catalyst complex in a fluidized bed reactor

Using an average particle size of 80μm, the ReSER process for hydrogen producing results are shown in Figure 10. From Figure 10, the concentration of H_2 95.5% can be obtained. In the sorption enhanced process, the average concentration of CO_2 and CO is 0.22% and 0.71% respectively.

The experiments were carried out at following operation conditions: temperature of 600□, atmospheric pressure, steam/carbon mol ratio 4, and different Weight of catalyst/Feed (W/F) ratio. The concentration of H_2 can be seen in Table1.

Table 1 The H_2 concentration under different W/F ratio

W/F, g min ml^{-1}	0. 16	0.13	0.08
H_2 concentration,%	95.5	86.8	75.2

Table1 indicated that the higher the W/F, the higher the hydrogen yield. However, an optimal ratio should be found both for high hydrogen yield and energy saving.

Conclusions

(1) Nano $CaCO_3$ is a best choice for making CaO which can avoid the sintering of CaO at high temperature carbonation and calcinations procedure and thus, rising the durability of the sorbent in cycle period.

(2) A sorption catalyst complex containing Ni as the catalyst component and CaO as the CO_2 adsorption component has been successfully prepared. The evaluation of the sorption catalyst complex in ReSER hydrogen production showed an over 95% (v/v) H_2 yield at much lower temperature and thus, realized the enhanced function both in fixed bed reactor and fixed fluid bed reactor under the operation condition of T 600℃, atmosphere pressure, steam/carbon mol ratio 4, W/F=0.16g min ml^{-1h}.

(3) In the view point of saving energy, the Reactive Sorption
Enhanced Reforming (ReSER) process for hydrogen production by using a fluidized reactor and regenerator cycle system like FCC in petroleum refining would have an attractive potential to be developed in the future.

References

1 Carvil, B.T., Hufton, J.R., and Sircar, S.,(1996),"Sorption-enhanced reaction process", *AICHE Journal*,42,2765-2772

2 Harrison, D. P., (2008),"Sorption-Enhanced Hydrogen Production: A Review". *Ind. Eng. Chem. Res.*, 47(17): 6486-6501

3 Wang, Y.N. and Rodrigues A. E.,(2005)," Hydrogen production from steam methane reforming coupled with in situ CO_2 capture: Conceptual parametric study". *Fuel*, 84(14-15): 1778-1789

4 Ding, Y. and Alpay E. (2000). "Adsorption-enhanced steam-methane reforming." *Chem. Eng. Sci.* 55(18): 3929-3940

5 Xiu, G., P. Li, et al..(2003),"Adsorption-enhanced steam-methane reforming with intraparticle-diffusion limitations". *Chem. Eng. Journal.*,95(1-3): 83-93

6 Feng, B., W. Liu, et al. (2006),"Overcoming the Problem of Loss-in-Capacity of Calcium Oxide in CO2 Capture", *Energy Fuels*, 20(6): 2417-2420

7 Grasa, G. S. and J. C. Abanades. (2006),"CO_2 Capture Capacity of CaO in Long Series of Carbonation/Calcination Cycles". *Ind. Eng. Chem. Res.* 45(26): 8846-8851

8 Hughes, R. W., D. Lu, et al. (2004),"Improved Long-Term Conversion of Limestone-Derived Sorbents for In Situ Capture of CO_2 in a Fluidized Bed Combustor". *Ind. Eng. Chem. Res.* 43(18): 5529-5539

9 Kuramoto, K., S. Fujimoto, et al. (2003),"Repetitive Carbonation-Calcination Reactions of Ca-Based Sorbents for Efficient CO2 Sorption at Elevated Temperatures and Pressures". *Ind. Eng. Chem. Res.* 42(5): 975-981

10 Fan, L. S., R. Agnihotri, (2000),"Activation of calcium oxide for use as sorbent", *US 4226839*

11 Li,Z.S.;Cai, N.S; Huang,Y.Y.; Han, H.J. Synthesis, (2005),"Experimental Studies, and Analysis of a New Calcium-Based Carbon Dioxide Absorbent". *Energy & Fuels*, 19, 1447-1452

12 Silaban, A., M. Narcida, et al. (1992),"Calcium acetate as a sorbent precursor for the removal of carbon dioxide from gas streams at high temperature". *Resources, Conservation and Recycling*,7(1-3): 139-153

13 S. F. Wu, T. H. Beum, J. I. Yang, and J.N. Kim,(2007)," Properties of Ca-Base CO_2 Sorbent Using $Ca(OH)_2$ as Precursor", *Ind. Eng. Chem. Res.* 46:7896-7899

14 Lei Shao., Jianfeng Chen,(2005),"Synthesis and application of nanoparticles by a high gravity method", *China Particuology*, 3(1)134-135

15 Su. F. Wu, Qing. H. Li, Jong. N. Kim ,Kwang. B. Yi, (2008),"Properties of Nano CaO/Al_2O_3 CO_2 Sorbent", *Ind. Eng. Chem. Res.*47(1) 180-184

16 Su. F. Wu, He Jun, Xie. Q. Wang, (2006),"A sorption catalyst preparation and application in methane steam reforming process for hydrogen producing", *CN1903431*

Application of Novel Compressor/Driver Configuration in the Optimized Cascade® Process

Tat Tsang, Chevron Global Technology Services Company
Mike Larkin, Chevron Global Technology Services Company
Greg Robbins, Chevron Global Technology Services Company
Higino Dos Santos, Chevron Global Technology Services Company
Ruzbeh Homji, Chevron Global Technology Services Company

AIChE Spring Meeting, April 2009
9th Topical Conference on Natural Gas Utilization
Tampa, Florida, April 26-30, 2009

Introduction

The concept of the Angola LNG project first conceived in late 1997 is to eliminate gas flaring through the development of a profitable, competitive and integrated gas solution allowing the booking of gas reserves, timely oil development and contributions to the social and economic development of Angola. The LNG facility is being constructed on the northern shore of Kwanda Base near the town of Soyo in the Zaire Province. The project is designed to produce a nominal 5.2 million metric tonnes per annum (MMTPA) of LNG and related LPG products (Propane and Butane), and condensate. It is also expected to supply up to 125 million standard cubic feet per day of domestic gas for Angolan consumption. Figure 1 shows an artist's rendition of the LNG plant.

Figure 1: Angola LNG Plant

After extensive studies and a Front End Engineering and Design (FEED) competition, the investors made the Final Investment Decision (FID) in December, 2007 to move the project towards the construction phase.

The outcome of the FEED competition was the selection of Bechtel as the EPC contractor for the construction of the plant and the employment of the ConocoPhillips Optimized Cascade® Process as the licensed technology for the Angola LNG Project. The Angola LNG Project will also introduce several new firsts in the evolution of the ConocoPhillips Optimized Cascade® Process.

Angola LNG Plant Design

Angola LNG Project is completing detailed engineering using the ConocoPhillips Optimized Cascade® Process for the first gas liquefaction facility to be built in Angola. The Optimized Cascade® Process is a well proven liquefaction technology with nearly forty years of operational experience. The technology was first used in ConocoPhillips' Kenai LNG Plant in Alaska and most recently at Equatorial Guinea. Most liquefaction facilities using the Optimized Cascade® Process have so far utilized GE Frame 5 series gas turbines as compressor drivers, while the Darwin LNG facility in Australia utilized the similarly sized LM2500+ aeroderivative gas turbines. The Angola LNG Project is the first application of the Optimized Cascade® Process that utilizes GE Frame 6 and Frame 7 gas turbine as mechanical drives. This implementation has allowed leveraging the cost improvements of 5+ MMTPA plants while maintaining the high reliability of the two-trains-in-one concept. Figure 2 below shows a simplified process schematic of the Angola LNG Plant.

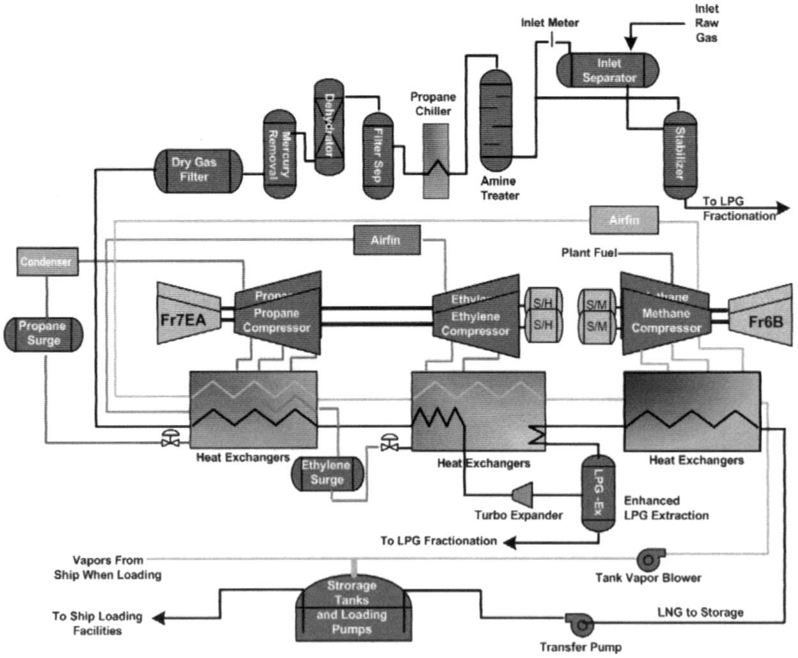

Figure 2: Simplified Angola LNG Plant Process Schematic – ConocoPhillips Optimized Cascade® Process

Refrigeration Compressor/Driver Configuration

The Angola LNG compressor/driver configuration is composed of two parallel GE Frame 7 gas turbines, each driving Propane and Ethylene refrigeration compressors on the same turbine shaft and two parallel GE Frame 6 gas turbines, each driving multiple Methane refrigeration compressors. Each Frame 7 gas turbine train has a variable speed drive starter/helper motor, while each of the Frame 6 gas turbine train has a variable speed drive starter motor. The configuration does not require the use of gear boxes. Figures 3 and 4 show the Angola LNG compressor/driver configurations.

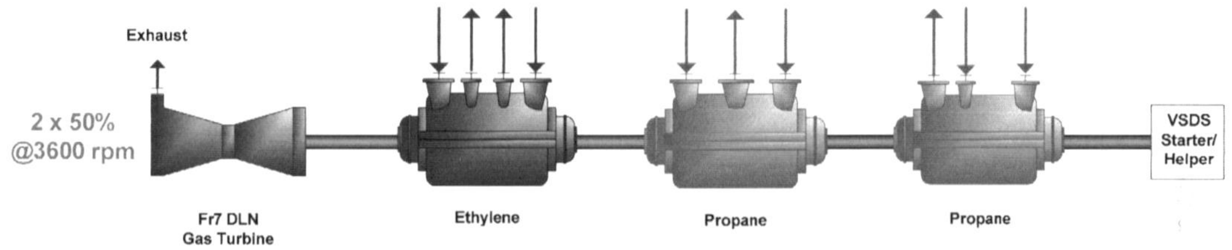

Figure 3: Frame 7 Gas Turbines with Propane and Ethylene Compressors in ConocoPhillips Optimized Cascade® Process

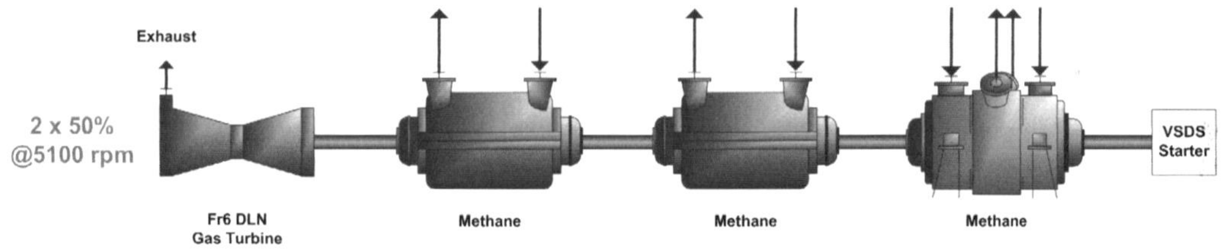

Figure 4: Frame 6 Gas Turbines with Methane Compressors in ConocoPhillips Optimized Cascade® Process

Single shaft Frame 6 and Frame 7 gas turbine drivers have been used for many years at several locations with other liquefaction technologies; single shaft Frame 5 machines have been in service for nearly four decades in the Cascade® Process at the Kenai liquefaction facility in Alaska. The Angola LNG Project has applied lessons learned associated with this well proven application in an innovative way. This results in a first-of-its-kind refrigeration compressor/driver configuration in the Optimized Cascade® Process. The novel refrigeration compressor/driver configuration of the Angola LNG Project enables the plant to use only four drivers to produce a nominal 5.2 MMTPA of LNG and related LPG and condensate products and still maintain the operating advantages of the two-trains-in-one configuration. The majority of Optimized Cascade® applications employ six drivers and produce between 3 to 3.7 MMTPA. To date, the only other 5 MMTPA Optimized Cascade® LNG plant employs 8 drivers.

Design Development in FEED and EPC Phases

As always, there were trade-offs with the new compressor/driver configurations. Operational flexibility is a potential issue due to the fact that single shaft gas turbines have a very limited speed range. The controllable speed range with the single shaft gas turbines in Angola LNG is only around 6% which is much less than that of the dual shaft gas turbines. On the Propane-Ethylene trains, controllability is further complicated by having two different services on the same shaft. This requires the implementation of advanced control strategies to provide adequate operational flexibility and ensure stable operation. The single shaft gas turbines are also very limited in start-up torque and each of Angola LNG's Frame 6 and 7 trains requires a large starter motor. In order to avoid the use of an even larger starter motor, partial depressurization of the compressor casing is required before start-up to limit the starting torque requirements. Also for the Frame 7 train, the starter motor is also used as a helper motor to augment the power provided by the gas turbine during normal operation. The starter motor for the Frame 6 train is only used for starting.

Dynamic simulations studies were performed during FEED by ConocoPhillips and the contractor to determine design requirements for compressor transient operations, shutdown, start-up, and compressor surge control. The results were used to ensure that adequate instrumentation and controls were provided to allow stable compressor operation during transients, shutdown, and start-up. The starter motor sizes were optimized based on compressor casing depressurization simulations balancing the objective to minimize refrigerant loss and establishing sizing criteria of a recovery system. The dynamic simulations were updated during EPC with as-tested compressor performance curves, actual equipment design data, and more accurate piping volumes.

The use of a configuration with multiple compressor bodies, a gas turbine and a variable speed drive motor on the same string introduced significant complexity with regards to mechanical design and rotordynamics. The overall length of the Frame 7 and Frame 6 strings are about 50 meters and 43 meters respectively. Individual compressor lateral and stability analyses as well as train torsional analyses were performed by GE Nuovo Pignone during the engineering phase. Bechtel also contracted third party rotordynamic experts to perform independent studies on some of the more complex machines. The analysis was followed by extensive testing that included a compressor model test, mechanical running tests for every compressor rotor including spares, ASME PTC-10 performance tests on each compressor class, motor type testing, control system integration testing and a full load string test for each compressor/driver type.

Full Load String Test

Full load, full speed string tests are very costly and time consuming, but the benefits far outweigh the expense. The API 617 mechanical running test, the

ASME PTC-10, Type 2 performance test, together with the full load, full speed string test can verify that the string components including auxiliaries are designed and manufactured correctly and will deliver the desired performance. Every effort was made to match the volume reduction per impeller at the full load string test with that of the guarantee point.

Data from the test provided validation of the following:

- The rotor and train model, the lateral, stability and torsional calculations
- Mechanical integrity without harmful aerodynamically or fluid induced vibrations.
- Potential interface issues between the driver and driven equipment
- Noise data
- Control and support system tuning and interfaces

The full load string test for Angola LNG Project Frame 6 and Frame 7 strings were conducted in vendor's test facilities in the third and fourth quarters of 2008. Figures 5 and 6 show the string test setup in various phase of installation.

Frame 6 String Test

The following actual "contract" components were used in the Frame 6 full load string test arrangement:

- Gas turbine MS6001
- LP Methane Compressor
- MP Methane Compressor
- HP Methane Compressor
- Dry gas seals and seal gas panels
- Torquemeter
- Couplings
- VSDS System (Starter Motor, Transformer, VFD, Jacking oil system)
- Gas turbine auxiliary baseplate
- MARK VI Control Panel
- Lube Oil Console
- Inlet Duct and Structures
- Filter House and Structures
- Exhaust Frame Cooling Fan
- DLN Skid
- Oil Vapor Separator

The compressors were loaded in closed loops in the string configuration. The test gas used in the closed loops was adjusted to load the compressors similar to the guarantee point.

The test sequence for the full load string test was as following:

- Start-up the unit
- Increase speed to maximum continuous speed and loading of compressors
- Maintain speed for four hours at specified conditions. Take complete set of readings every half hour. Each complete set of readings will consist of thermodynamic readings with the addition of mechanical parameters such as vibration and bearing parameters. Perform noise measurement.
- During the final part of the full load string test, the lube oil temperature was increased from 50°C to 55°C
- Shutdown turbine with normal sequence

Results from the string test were satisfactory and in accordance with expected values. Bearings and seals were checked and were found in good condition.

Frame 7 String Test

The following actual "contract" components were used in the Frame 7 full load string test arrangement:

- Gas turbine MS7001
- Ethylene Compressor
- LP Propane Compressor
- MP/HP Propane Compressor
- Dry gas seals and seal gas panels
- Torquemeter
- Couplings
- VSDS System (Starter/Helper Motor, Transformer, VFD, Jacking oil system)
- Gas turbine auxiliary baseplate
- MARK VI Control Panel
- Lube Oil Console
- Inlet Duct and Structures
- Filter House and Structures
- Exhaust Frame Cooling Fan
- DLN Skid
- Oil Vapor Separator

The compressor aerodynamic performance test was performed in accordance with ASME PTC-10 1997, Type 2 and API 617. The compressors were tested in closed loops in the string configuration. The PTC-10 performance tests for the Frame 7 compressors had to be performed in the string configuration because the shop drivers were too small. The test gas used in the compressor closed loops was a mixture of natural gas and Suva HFC-134A (CH2FCF3) to allow the

PTC-10 requirements of similarity, Reynolds number, Mach number, and volume ratio to be met.

The test sequence for the Frame 7 full load string test was essentially the same as for the Frame 6 train.

Results from the compressor performance test and string test were satisfactory and met the requirements of ASME PTC-10. There was an additional two hour test during which the large starter/helper motor was fully loaded. However, there was a diode failure associated with the starter/helper motor during the tests. Testing was completed after replacing all the diodes. Root cause analysis conducted by the motor vendor identified the potential cause. A temporary solution was developed and later tested on a similar type of motor for validation. The revised design will be incorporated into the starter/helper motor provided for Angola LNG Frame 7 strings. The Frame 7 string test also enabled the tuning of the variable speed drive system.

Summary

For many LNG projects, compressor/driver configuration remains to be a critical area for study. This is expected since the compressors and drivers represent a large percentage of the project cost. Unscheduled downtime from this equipment has a major effect on the facility capability. Proper selection of the compressor/driver configuration is a key to the success of an LNG project and future plant operation.

With the rapid growth of the LNG industry and the desire to drive a lower unit cost per LNG production, complex configuration of compressors and drivers are being used for LNG refrigeration services. Although advanced technology and tools are routinely used in the turbomachinery design process, a full load and full speed string test should be included in the vendor scope of work. The ability to perform a full load mechanical run test using the contract compressors, gas turbines, VSD system, dry gas seal system, and lube oil system while monitoring key equipment parameters will identify many design deficiencies, reduce field start-up problems, and avoid future operational issues.

Figure 5: Frame 6 String Test Setup in Assembly

Figure 6: Frame 7 String Test Setup in Assembly

References

1. API 617, 2002, "Axial and Centrifugal Compressors and Expander-Compressors for Petroleum, Chemical and Gas Industry Services", Seventh Edition, American Petroleum Institute, Washington D.C.

2. ASME PTC-10, 1997, "Performance Test Code on Compressors and Exhausters", American Society of Mechanical Engineers, New York

3. A. Eaton, R. Hernandez, A. Risley, P. Hunter, A. Avidan, and J. Duty "Lowering LNG Unit Costs Through Large and Efficient LNG Liquefaction Trains – What is the Optimal Train Size?", AIChe Spring National Meeting 2004, New Orleans

4. C.B. Meher-Homji, T. Matthews, A. Pelagotti, H.P. Weyermann, "Gas Turbines and Turbocompressors for LNG Service", Proceedings of the Thirty-Sixth Turbomachinery Symposium, Texas A&M University's Turbomachinery Laboratory, College Station, Texas

5. W.R Qualls and P. Hunter, "A Focus on Balance – A Novel Approach Taking the Phillips Optimized Cascade LNG Process into the Future", AIChe Spring National Meeting 2003, New Orleans

Biography of Speaker

Mike Larkin is a Project Engineer–Turbo Machinery on the Angola LNG Project. He has worked for Chevron (and previously Texaco) for 33 years. He has been involved with upstream and mid-stream compression and gas handling projects and facility operations. He was also part of early corporate wide reliability and efficiency efforts. He attended Tulane University in New Orleans and received his BS and MS degrees in mechanical engineering. He is an ASME member and a registered professional engineer in Louisiana and Texas.

7b Process Understanding of Acid Gas Removal Unit Instabilities: Lessons Learnt from Malaysia LNG TIGA

Siti Rafidah Moslim, Technical Services, Malaysia LNG Sdn Bhd (Subsidiary of PETRONAS), PO Box 89, 97007,, Tanjung Kidurong, Bintulu,, Sarawak, Malaysia

In a remarkably competitive Oil and Gas industry, sustaining plant operational stability is key in ensuring optimum business performance. Acid Gas Removal Unit (AGRU) instabilities due to foaming and carryover are typical critical operational issue that impacts plant availability, reliability and utilization. With comprehensive understanding of factors that promotes foaming, AGRU instabilities can be avoided and sufficient designs considerations can be made from Project phase in ensuring long term plant operational stability. Malaysia's PETRONAS LNG complex in Bintulu, Sarawak, comprises of Malaysia LNG Sdn Bhd, Malaysia LNG DUA Sdn Bhd and Malaysia LNG TIGA Sdn Bhd plants with total LNG deliverable capacity of about 23 million tonnes per annum (mtpa). The three LNG plants are operated as one integrated site, all being part of the Gas Sector of PETRONAS. The first plant, MLNG with three modules (Modules 1, 2 and 3) began operation in 1983. This was followed by the second plant, MLNG DUA, which started its operations in 1994 also with three modules (Modules 4, 5 and 6). The third plant, MLNG TIGA with two modules (Modules 7 and 8) was later integrated with the existing site and come into operations in 2003. This paper enlightens the process knowledge learnt from the operational experience of MLNG TIGA AGRU for better design of future installation. It highlights key factors to be considered in ensuring AGRU robustness towards foaming promoters especially heavy hydrocarbons (HC) contamination; particularly in preventing HC ingress into AGRU amine solvent and effectively removing HC from accumulating in amine system.

Cold Box Optimization Program

Dr. Mona Setoodeh
LNG Consultant, Process Engineering

Timothy P. Gushanas
LNG Consultant, Process Engineering

CH·IV International, Houston, Texas

AIChE Spring National Meeting 2009
9th Topical Conference on Natural Gas Utilization
Tampa, FL April 26-30, 2009

Abstract

Cold Box Optimization Program (CBOP) represents the migration of standard high-value strategic technologies and services in the petrochemical process industries to the world of LNG Peakshaving. The program enables the owners and operators to modify the process variables such as the pressure and composition of the refrigerant in order to maximize their LNG production rate within the system constraints. Increased LNG production of 5% to 10% without increasing energy consumption have been proven.

Two major factors contribute to the decline in the production rate of the peakshaving facilities as they age. First is the decrease in the equipment efficiencies over time. Most peakshavers are at least 20 years old, some almost 40, and the consequent drop in the efficiencies of the key equipment such as the refrigerant compressors is inevitable and will result in production losses. Second is that almost every plant has a different feed gas composition or battery limit condition today than it did when it was originally designed and started up. Since the design of the refrigeration loop is directly based on the feed gas composition, temperature and pressure, any variation in the feed gas condition can negatively affect the production rate.

Realizing the demand in the peakshaving industry to maintain their LNG production rates, CH·IV has launched the Cold Box Optimization Program, whereby owners and operators of natural gas liquefaction facilities can realize substantial benefits during the liquefaction season. The first activity in CBOP is to perform an independent diagnostic of the entire liquefaction section. Once the Diagnostic Phase has been completed, the owner/operator can implement the recommendations put forth by the CBOP. CBOP then moves to the Monitoring Phase where regular rigorous model-based performance monitoring and off-line optimization is performed to assure that the liquefaction plant continues to operate at peak efficiencies and production rates. Operating at the optimum set points and mixed refrigerant compositions can also prolong the life of equipment

Introduction

LNG peak-shaving facilities are used for storing surplus natural gas that is to be used to meet the requirements of peak consumption. Each peak-shaving facility has a regasification unit, but may or may not have a liquefaction unit. These facilities without a liquefaction unit depend upon tank trucks to bring LNG from other nearby sources to them. From the approximately 113 peakshaving facilities in North America, 57 have liquefaction capabilities.

Two major factors contribute to the decline in the LNG liquefaction rate of the peakshaving facilities as they age. First is the decrease in the equipment efficiencies over time. Forty-seven (82%) of U.S. peak shavers [with liquefaction capabilities] were built during the 1960s and 1970s. Only five were built in the 1980s and five in the 1990s[1]. The consequent drop in the efficiencies of the key equipment such as the refrigerant compressors is inevitable and will result in production losses.

The second and sometimes more significant factor is that almost every plant has a different feed gas composition and/or inlet conditions today than it did when it was originally designed and put into service. In addition the feed gas composition changes can also occur over the course of a single liquefaction season due to natural events such as hurricanes reducing gas from Gulf Coast sources and also the shift of the natural gas supply to Coal Bed Methane and Shale Gas sources. Since the design of the refrigeration loop is directly based on the feed gas composition and pressure, any variation in the feed gas can negatively affect the production rate.

The Cold Box Optimization Program (CBOP) provides the facilities with a powerful tool to modify the process variables such as the pressure and composition of the refrigerant in order to maximize their LNG production rate within the system constraints.

Optimization Process

Generally speaking, by changing a process variable which in turn affects the shape or profile of the exchanger cooling curve (duty versus temperature), the liquefaction rate can be increased. This increase can be achieved by optimizing the mixed refrigerant composition and the compressor profile, i.e., suction discharge and intermediate temperatures and pressures. The optimization is performed in three main phases:

In *Phase I*, CH·IV reviews the relevant documents such as the PFD and the original heat and material balance of the facility, and creates a base case model using Pro/II® simulation software and validates the thermodata. The preliminary modeling is normally followed by a site visit of the facility, based on which the model is "tuned" against actual plant performance (compressor efficiencies, actual pressure drops, calculated UAs, etc).

In *Phase II*, CH·IV simulates and optimizes the process with an objective of maximizing production or reducing the energy requirement. The optimization objective is achieved by manipulating system variables within the system constraints.

In *Phase III* CH·IV reviews the hydraulics of the existing systems, including the heavy knock-out and tail gas systems if applicable, and makes recommendations for modifications required in order for the facility to be capable of handling its current feed gas safely, and for the facility to operate at the highest feasible efficiency. In addition to presenting the performance analysis of the facility, CH·IV also identifies constraints and bottlenecks and provides recommendations on equipment and line sizing, or revamp options and cold box replacement projects if applicable.

- ***Optimization Objective***

  The objective of optimization for peakshaving facilities is to either maximize LNG production or minimize energy consumption, subject to system constraints.

---

1 http://www.atp.nist.gov/eao/ccar/section_3.htm

- ***Process Optimization Variables***

  Process optimization variables are set points that can be independently controlled or manipulated by the plant operators. For Peakshaving facilities typical optimization variables are as below:

  - Feed gas heavies knock-out temperatures
  - Refrigeration compressor suction and discharge pressures
  - Refrigerant condensing temperatures
  - Mixed refrigerant composition

- ***System Constraints***

  The optimization process is performed within actual system constraints. Typical main system constraints are as below:

  - Refrigeration capacity and power limits:

    The feed gas is liquefied ultimately by the flow of the nitrogen or the mixed refrigerant (MR) through the coldbox. The refrigeration capacity is directly proportional to the nitrogen or the mixed refrigerant flow, which is powered by the nitrogen/ MR cycle compressor. The compressor performance is the ultimate limiting factor in the overall liquefaction capacity of the facility.

  - LNG exchanger heat conductance (UA):

    The product UA, known as heat exchanger conductance, is a proportionality constant that relates the heat flow rate or duty (Q) to the driving force or temperature difference (CMTD[2]) by equation $Q=UA \cdot CMTD$. This heat exchanger performance characteristic is calculated using actual plant data and defines the limiting capacity of the heat exchanger battery in the optimization process.

  - LNG density differential

    The density of the LNG generated is normally maintained within ± 2% of the design density. It is important to control the density of the LNG produced, because one of the most effective means of preventing rollover in the LNG storage tank is to avoid stratification, which could occur if LNGs of differing densities are stored in the same tank. Therefore rigorous density control is essential if any feed gas variations are anticipated.

  - Heavies knock-out tail gas system design

    Higher non-methane hydrocarbon feed gas streams will generate more heavy liquids in the process versus the lower case. There is a limit to the amount of additional heavy liquids that each facility can handle, and a limit to dealing with these heavy liquid hydrocarbons in terms of revaporizing them and sending them out as a tail gas stream.

---

2 Corrected Mean Temperature Difference

Optimization Case Study

Presented is the process optimization performed by CH·IV for an operating peakshaver. CH·IV studied the liquefaction sensitivity to feed gas pressure and optimized the process for natural gas feed pressure increase of 50 psi.

A schematic of the coldbox for the subject plant is shown in Figure 1. Base case feed gas condition for the subject peakshaver is shown in Table 1.

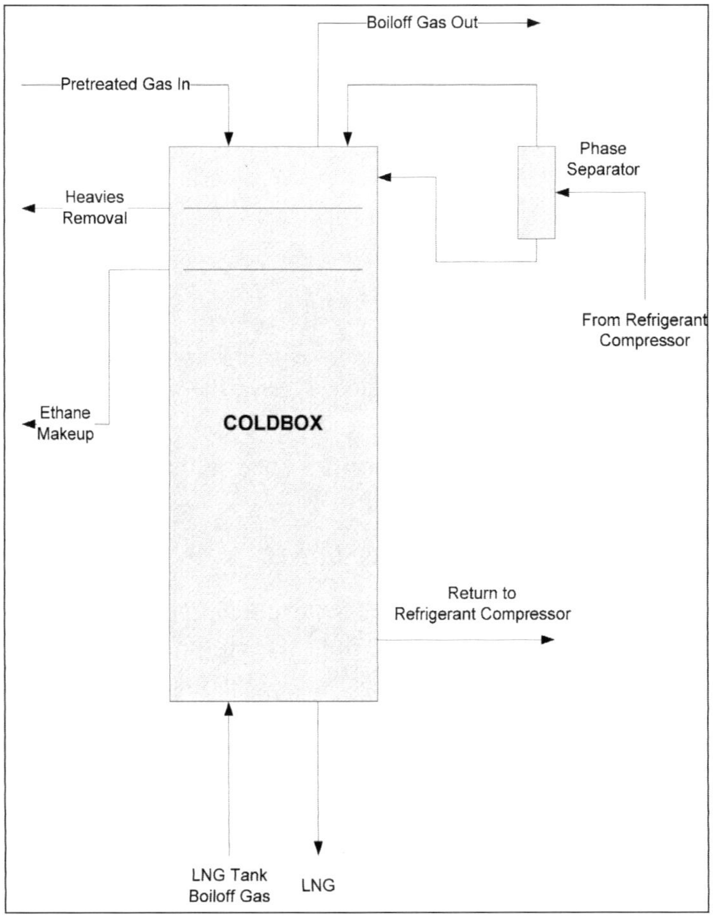

Figure 1: Cold Box Diagram

Table 1: Base Case Feed Gas Condition

Composition	Design Case
C1 % mol	96.68
C2 % mol	1.81
C3 % mol	0.50
nC4 % mol	0.10
iC4 % mol	0.10
N2 % mol	0.80
Temperature °C	70 ° F
Pressure psig	500 psig

- ***Optimization Objective:***

  To maximize LNG production at the new higher feed gas pressure of 550 psig.

- ***Optimization Variables:***
  - Refrigerant compressor suction pressure
  - Mixed refrigerant composition (N_2, C_1, C_2, C_4)

- ***System Constraints:***
  - LNG exchanger heat conductance (UA)
  - LNG exchanger Minimum Internal Temperature Approach (MITA)
  - Mixed refrigerant compressor horsepower(s)
  - Mixed refrigerant dew point(s) at interstage pressure(s)

Optimization Outcome

When the feed pressure is increased by 50 psi the natural gas cooling curve shifts to a warmer level, reducing the latent heat and the total refrigeration requirement. To take advantage of the higher feed gas pressure a "warmer" or heavier refrigerant mix should be used and the refrigerant compressor suction pressure should be lowered. Maximum advantage is gained when the system is optimized. Comparison of mixed refrigerant composition for feed gas pressure of 500 psig and 550 psig is shown in Table 2. Note that the "heavier mix" corresponds to the lower suction pressure at the MR compressor.

The LNG heat exchanger cooling curve for feed gas pressure of 550 psig with optimized MR composition is shown in Figure 2. The resulting duty requirement for the optimized case is lower. The temperature approach between the "hot" and "cold" sides are closer and resulting MTD is lower for the optimized case. As a result the LNG production rate for the given system can be increased by 8.9%. The summary of the optimization outcome is shown in Table 3.

Table 2: Comparison of Mixed Refrigerant Composition, 500 psig vs. 550 psig Feed Gas Pressure

Feed Gas Pressure	PSIG	500	550
Refrigerant Conditions			
Stream Description		MR TOTAL	MR TOTAL
Phase		Vapor	Vapor
Temperature	F	55.00	55.01
Suction Pressure	PSIG	52.73	50.07
Flow Rate	Lbmol/hr	1754.56	1753.21
Composition			
N_2		10.22%	9.86%
C_1		34.37%	32.47%
C_2		38.97%	40.66%
C_3		0.00%	0.00%
nC_4		16.44%	17.01%

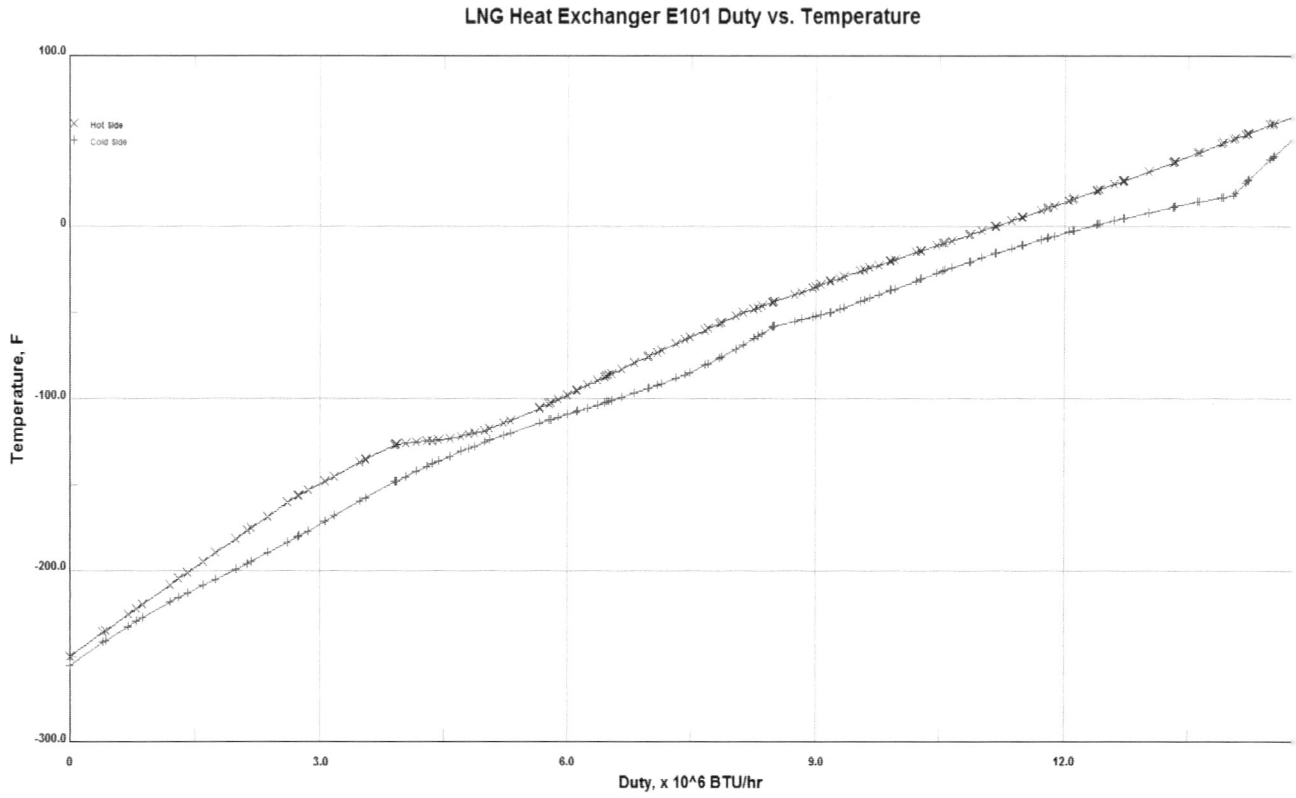

Figure 2: LNG Heat Exchanger Cooling Curve for 550 psig Feed Gas and Optimized MR Composition

Table 3: Summary of Optimization Outcome

Case	NG Feed (mmscfd)	Gross LNG (gpm)	NET LNG (gpm)	Production Rate Change
Base Case	4.54	38.4	33.6	**0.0%**
Increase Feed Pressure by 50 psi	4.57	38.6	33.7	**0.3%**
Increase Feed Pressure by 50 psi - & Optimize	4.92	41.6	36.6	**8.9%**

Conclusions

The Cold Box Optimization Program represents a differentiating technology solution for peakshaving facilities. Improvements of 5% to 10% are achievable through high fidelity first principle modeling and optimization. The requirements are technology "Know-How", real world operating experience, and the right tools.

Biography of Speaker

Dr. Mona Setoodeh received her Ph.D. in Chemical Engineering from University of Houston. She is currently an LNG Consultant for CH·IV International in Houston, Texas. She has been involved in providing consulting engineering services to clients with regards to land-based and offshore liquefaction of natural gas, storage, handling and re-gasification of Liquefied Natural Gas.

Maximized Values of LNG Specifications in the LNG Industry: Producer and Buyer Perspectives

7e

Joseph H. Cho, Ph.D., P.E.
Director of Houston Tech. Center
Jaeyoung Yang, Ph.D.
Senior Process Engineer
D. Y. Kim, Ph.D.
Process Engineer
Hong-Chul Kim, P.E.
Sr. Consultant of Process Engineering
SK Engineering & Construction

AIChE Spring Meeting, April 2009
9th Topical Conference on Gas Utilization
Tampa, Florida, April 26-30, 2009

ABSTRACT

In the past, the major LNG buyers in Far East Asian countries prefer to import LNG with high heating value and to meet this specific requirement of the consumer, LNG industry started producing high heating value LNG. However, there is a new trend with the potential buyers of LNG to have some flexibility in the ranges of natural gas properties they were able to choose from. As the growing demand increases for lean LNG (low calorific value), especially in the US and the UK, LNG producers are forced to address new gas quality issues.

The challenge is to understand the requirements of the potential new markets and to plan a flexible way to meet the requirements of both the LNG exporters and importers. This paper discusses the flexibility and allowable ranges of LNG specifications which satisfies both producers as well as importers with reference to maximizing the LNG production flexibility and importers issues on interchangeability. These specifications are compared in detail to those of LNG being produced along with the options and problems for conditioning the natural gas on either end of the LNG.

INTRODUCTION

Producing LNG from the natural gas is a simple phase change process without any chemical reactions involved. However, LNG production plant is composed of many other facilities whose purpose is to either protect equipment and/ or to meet the LNG product specifications. For instance, heavy hydrocarbon removal, dehydration, and mercury removal processes could be the first purpose oriented processes, while NGL extraction processes could be the second purpose oriented processes. When it comes to acid gas removal process, it is built for both purposes.

Although in general terms, all the LNG production facilities look alike, but on a closer look, every LNG production plant is unique in its nature and characteristic. The facility would be only designed to cater the needs of its captive consumers. It is interesting to note that unlike other industries, traditionally, the conception of a LNG project used to start by completing the LNG chain; abundant natural gas supply, and a consumer with specific quality and quantity requirements of LNG which is compatible with the natural gas resources. Usually long term Sales and Purchase Agreement (SPA), take or pay type of contracts used to support and justify the construction of LNG project. However, this traditional LNG industry is now facing a challenge to supplement new gas demand and meet the gas quality at the same time.

WORLDWIDE LNG PRODUCDER AND LNG QUALITY

LNG is produced worldwide in about 20 LNG production facilities. Depending on the geographical region, LNG products have different qualities which are basically attributed to the raw natural gas composition. In order to operate the plant in a safe manner and to maintain LNG quality, there are several processing steps involved such as Acid gas removal, dehydration, and heavy hydrocarbon removal. In Acid gas removal section, poisonous and corrosive gases such as H2S and CO2 are removed to the maximum extent.

In a separate unit, heavy hydrocarbons and aromatics are removed prior to cryogenic process to prevent freezing in the main cryogenic heat exchanger. The other components including nitrogen, methane, ethane, propane and heavier hydrocarbon are major constituents of LNG and their composition is very important since their composition determine the LNG product markets in which the LNG products are marketable. Therefore, LNG producers should consider over the LNG product quality, as well as natural gas feed quality, which will be produced in their LNG production plant.

In a table 1, LNG product composition and quality is listed for the representative LNG products worldwide [1].

Table 1: Typical LNG composition and calorific value

	Alaska	Brunei	Malaysia	Indonesia	Australia	Oman	LNG A	Qatar
Composition,mol%								
Methane	99.57	89.40	91.15	89.18	87.82	87.66	92.63	89.87
Ethane	0.16	6.30	4.28	8.58	8.30	9.71	6.89	6.50
Propane	0.09	2.80	2.87	1.67	2.98	2.04	0.35	2.25
Butane	0.04	1.30	1.36	0.51	0.87	0.58	0.06	1.04
Pentane	0.01	0.05	0.01	0.02	0.01		0.06	0.01
Nitrogen		0.05	0.32	0.03	0.01	0.24	0.01	0.34
HHV, MJ/scm	37.7	42.0	41.8	41.5	42.5	42.1	39.9	41.6
Wobbe Index	50.6	53.1	52.7	52.6	53.3	53.0	51.8	52.6

LNG MARKET AND GAS SPECIFICATION

LNG Market

It is considered that Asia, Europe, and US are some of the major LNG markets worldwide. Traditionally, Asia is the largest market for LNG producers. Asian countries import rich LNG which contains much ethane and heavier hydrocarbons. On the other hand, the US and UK import lean LNG which contains less ethane and heavier hydrocarbons. This is due to their pipeline gas specifications that LNG importer should meet.

Before LNG consumption is increased, the US and most European countries depend on their natural gas supply through compressed pipeline natural gas. However, worldwide energy deficiency has forced the countries to import LNG more than ever. However, as their original pipeline network requires lean natural gas, conventional rich LNG product does not comply with their pipeline gas specification. Under this market environment, the new emerging lean LNG product markets are very attractive to LNG producers since their energy consumption is remarkable.

In addition to the new lean natural gas market, spot market became a new market trend which deals short term base or cargo based contracts. This market is believed to be very crucial in the near future because LNG demand is seasonal and variable, although LNG production rate cannot be responded easily to variable market demands.

Pipeline Gas Specifications

Among many other gas specifications, calorific value is believed to be one of the most important specifications. Calorific value is generally expressed by high heating value (HHV) or by Wobbe index. Worldwide HHV allowable range in the market is shown in figure 1. Many LNG producers are producing rich LNG for their existing markets and are not in a position to change their product to cater the ever increasing demand for lean LNG.

However, calorific value specification is not only the driver for marketability. The other components such as ethane, butane and oxygen and CO_2 specifications are not uniform around the global market. For example CO_2 content limit is 2~3% in Europe. In California, ethane and CO_2 content are limited at 6% and 3% respectively. Maximum butane content is 2% and inert should be less than 4% in the US. For Oxygen, Europe allows only 0.01%, while Canada and California allows 0.4% and 0.2%, respectively. Australia LNG can not be marketable in US and UK because its HHV is too high to be allowed in their pipeline, while it is marketable in Asian counties and continental Europe. For an another example, LNG A in table 1 could be distributed in the US market in view of its calorific value but not in California due to its component based specification which need less than 6 % ethane content.

As a result, it is clear that LNG product has a very limited and specific marketable area and do not enjoy a global market like other hydrocarbons. Due to this inevitable linkage between LNG product quality and LNG market, the LNG producer can not have flexibility in marketing.

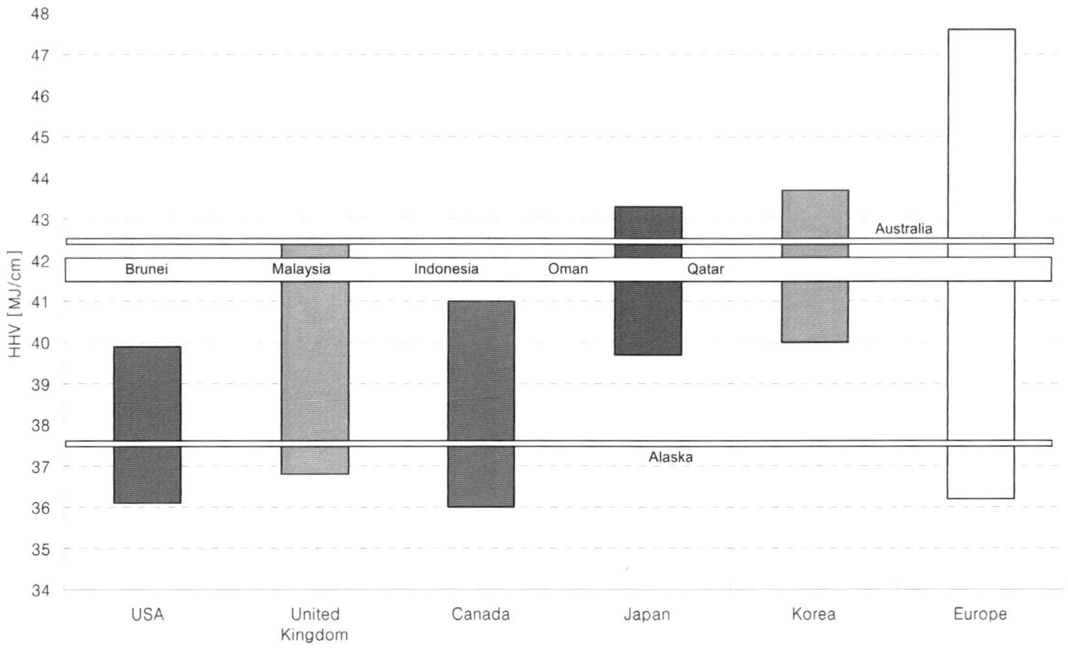

Fig. 1 - Worldwide HHV Specifications

GAS INTERCHANGEABILITY

In the LNG industry, expanding gas demand and emerging new markets are opportunities for both LNG producers and LNG importers. Since LNG produced is not compatible with every existing pipeline gas specification worldwide, LNG importers do not have many choices in selecting LNG on the market since natural gas source is inevitably restricted to some specific LNG producers. In order to get some bargaining power in the LNG market, and to meet the increasing gas demand, LNG importers need to buy any LNG and should be in a position to adjust its specification to meet the local pipeline specification.

Therefore, the issue of interchangeability dealt with quality specification and its suitability for end user including appliances, gas turbines, process applications and vehicular fuels [2]. Therefore it refers to a measure degree to which the combustion characteristics of vaporized LNG are compatible with those of conventional pipeline gas [3]. It is a new imminent challenge for LNG importers to adjust the imported LNG to be interchangeable with existing pipeline specification. At the same time, from the LNG production side, LNG producers need to have some facilities to adjust their produced

LNG quality in accordance with LNG market demand.

CALORIFIC VALUE CONTROL TECHNOLOGY

Since calorific valve is important specification, new technologies for adjusting calorific value have been developed to pursue the new market trend. These available technologies help LNG importers decision making for any imported LNG interchangeability projects. The calorific value control technologies are classified into two groups: decreasing calorific value and increasing calorific value.

- Decreasing calorific value: N_2/CO_2/Air/Fuel gas injection, NGL extraction
- Increasing calorific value: LPG injection, N_2/CH_4 rejection

In addition to these technologies, dual calorific value production system has been developed to produce lean gas and rich gas without blending LPG and injecting inert [4]. The calorific value control technologies and their basic characteristics are described briefly below.

Technology for decreasing calorific value

Inert Injecting Process

Membrane process, pressure swing adsorption (PSA) or cryogenic air separation process are utilized to produce N_2 for injection, depending on the required production capacity. It is generally believed that cryogenic air separation process is more economical than membrane process for high capacity plant. For air injection, it only needs to compress air to pipeline pressure, but accompanying O_2 with air might not be acceptable for most pipeline networks. Flue gas injection is another very attractive option as it is considered less cost, but inherently more difficult to accept because of water vapor and excess O_2.

The two representative calorific value indices (HHV and Wobbe Index) show different sensitivity on calorific value control technologies. As the degree of N_2/CO_2 injection increases, both HHV and Wobbe index decrease, but with different amount. As Wobbe index has specific density in the denominator, the molecular weight of the injection component is another factor which influences the Wobbe index as well as the amount of injection. Wobbe index is more sensitive to the amount of N_2/CO_2 injected. In fact, the variation of HHV is irrelevant to the material injected if they have no heating value as shown in Fig. 2. HHV is dependent only on the amount of inert material.

However, as can be seen in Figure 2, Wobbe index is more sensitive on the injection amount when CO_2 is injected rather than when N_2 is injected because CO_2 has higher relative density than N_2. If Wobbe index is crucial to meet specification and needs to be reduced more than HHV, CO_2 injection is more efficient method of calorific value control in technical point of view without considering environmental issue of emitting more CO_2. However, when inert injection is considered to decrease HHV, it needs to be checked out if inert composition violates pipeline specification of inert maximum content.

When air is injected instead of nitrogen or carbon dioxide, O_2 in air is a controlling factor. If O_2 content is limited to maximum 0.2% as in California, air should be injected so as not to violate the maximum limit. With this low O_2 limit, HHV can not be regulated much by injecting air, even though this method is cost effective compared to N_2 injection.

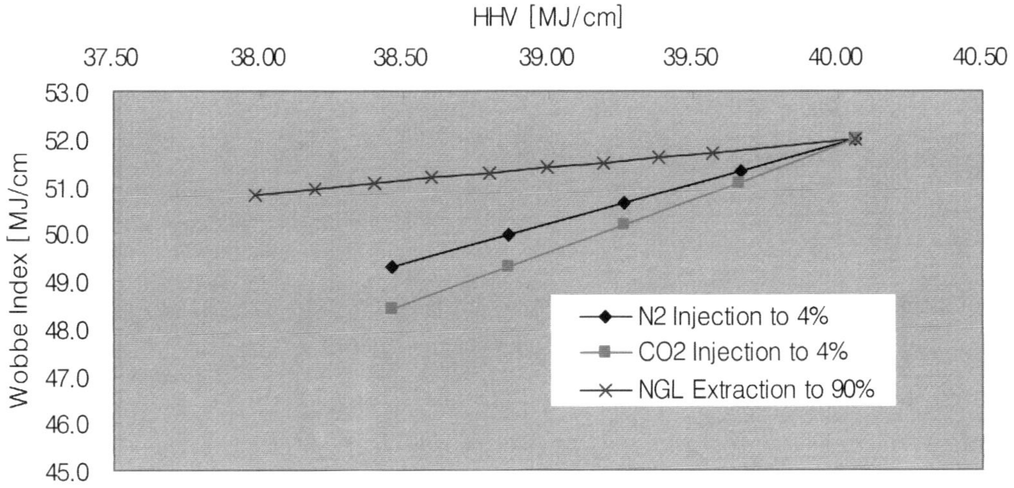

Fig. 2 - The Effect of Inert Injection on Calorific Value

NGL Extraction Process

NGL extraction is most effective and easily applicable method to control the calorific value of LNG product in LNG production plants. Up to now, it is believed that it is economically attractive or worth of capital expenditure when the NGL extracted is marketable without high transport cost. If NGL extraction is done only for calorific value control, the facility does not need to have distillation columns as many as multi-products LNG production plant has, in which pure ethane, propane, butane and C_5+ are produced simultaneously. In general, by modulating ethane recovery into methane-rich main stream, LNG product calorific value can be controlled.

Although NGL extraction is used for calorific value control in LNG terminals, NGL extraction is more common concept in the liquefaction plants to reduce heavy component which might cause freezing problem in cryogenic heat exchangers. For the purpose of preventing freezing, a simple scrubber can be utilized to remove heavy hydrocarbons, but it is not efficient to lower the calorific value as there are not freezing-inducing components in natural gas as much to lower the calorific value. For the purpose of calorific value control, a hydrocarbon extraction facility is installed with a series of distillation columns including a deethanizer as well as a demethanizer with cold reflux. NGL extraction processes are well investigated in some paper and patent [5, 6, 7, and 8].

A method commonly employed in a gas processing plant and liquefaction plant is expander-compressor combination with some heat recovery heat exchangers. Ethane content in the extracted NGL varies depending on the operating condition of demethanizer, especially on how much reflux rate is supplied to the column. As the demethanizer should run below the critical pressure at low temperature to make separation possible, the feed gas should be let down to a lower pressure by expander or J-T valve as well as to get some liquid traffic in a demethanizer (Fig. 3)

Contrary to the NGL extraction process in the baseload LNG plant, some heat exchangers are necessary to heat up and vaporize LNG partially to accomplish separation in a distillation column for the LNG terminal NGL extraction facility. The general NGL extraction process from LNG is illustrated in figure 4. In general, LNG is pumped to a distillation column operation pressure. A cold LNG feed stream can be used for cooling medium to supply reflux stream to the top of the column, while some heat exchangers are used to provide energy into the column using a feed gas heater, a side reboiler, and a bottom reboiler.

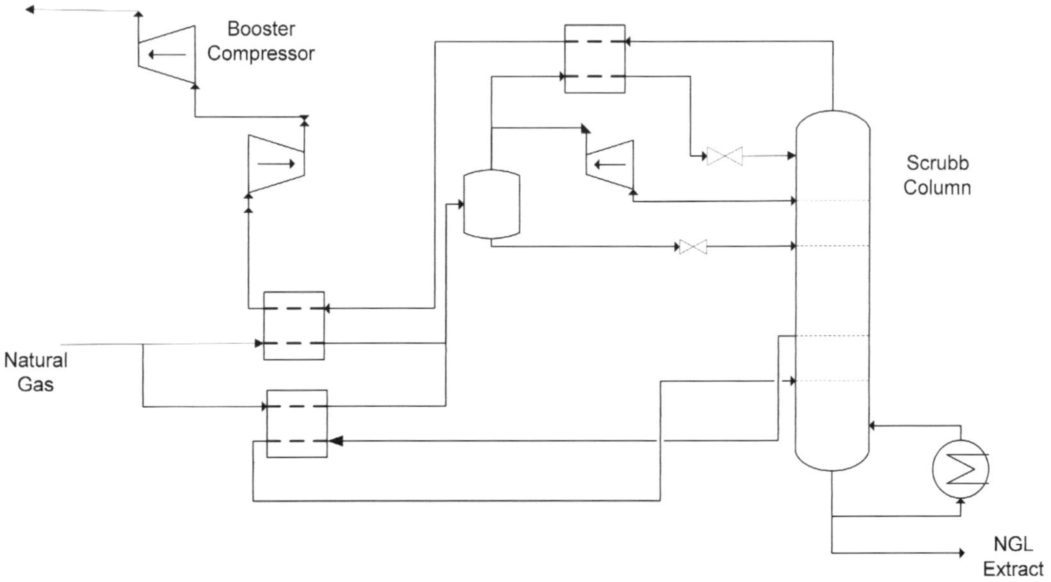

Fig. 3 - NGL Extraction Process in a Liquefaction Plant (Expander-compressor process)

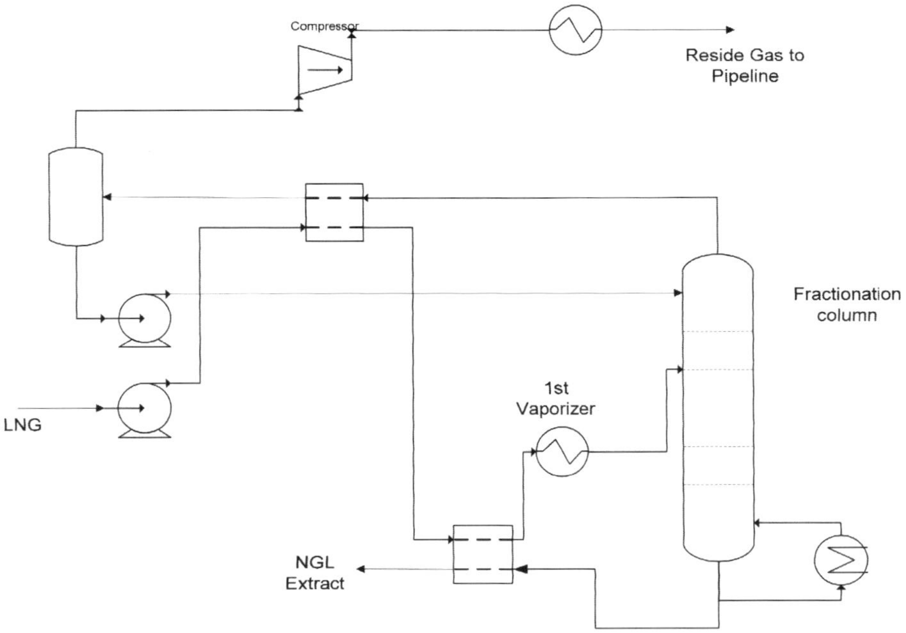

Fig. 4 - NGL Extraction Process in a LNG Terminal

Technology for Increasing Calorific Value

As inert injection can decrease calorific value, removing inert is a way to increase calorific value. However, calorific value control by rejecting nitrogen is not easy as nitrogen injection, since the effect of nitrogen rejection on the calorific value is prominent only when raw natural gas stream contains high nitrogen content. On top of that, N2 rejection should be performed in accordance with fuel balance of the LNG plant and requires another cryogenic distillation column to reduce the content enough to achieve satisfactory decrease of the calorific value.

LPG injection is another way to increase calorific value of natural gas, although it is costly and damages the image of LNG which is known to be a clean energy [4]. HHV increases, in general, about 1.8% with 1% increase of LPG fraction by injecting LPG. Where high HHV is required as in Japan and Korea, LPG injection is a simple and effective way to enhance calorific value of imported lean LNG.

ECONOMIC EVALUATION OF CALORIFIC VALUE CONTROL TECHNOLOGY

Economic evaluation for the two representative calorific value control technologies has been performed. This analysis was based on 750 mmcfd send-out capacity and 15 years life cycle. Discount rate was assumed as 10%, while power cost is 5 cents/kwh. As a result, NGL extraction facility needs large capital cost comparing to the N2 injection facilities. However, the facility can produce hydrocarbon products such as ethane, propane, butane, etc, while there is shrinkage of natural gas send-out rate due to rejection of the hydrocarbons. The net operation profit is remarkable which makes the facility attractive in spite of large capital cost. However, N_2 injection does not create any economic value unless the rich LNG is less expansive than lean LNG so that net profit is produced from the purchasing cost.

Relative Costs	NGL Extraction	N2 Injection
CAPEX	100%	15 – 20%
OPEX	100%	180 – 220%
Net Operating Profit	100%	-
NPV	100%	-100%

Unit: Thousand USD

TAILORING LNG PRODUCTION AND DISTRIBUTION

It is certain that LNG producers can get better position in LNG market, if they have facilities that can control calorific value of their products. It would be more appropriate if their facility can control individual component content along with calorific value. If LNG storage facilities allow not-less-than two calorific value LNG products, LNG producer can produce two different LNG products for different markets. The production schedule is optimized, that is, the production quantity for each product is optimized, depending on the market situation.

If it is more profitable to sell their product in lean LNG market, they can increase lean LNG production rate by operating NGL extraction facility. The emerging of LNG spot market is their new opportunity to get more profit from their LNG production plant since they can optimize their production schedule as per spot market demand. Theoretically, LNG producer portfolio should be variable and optimized on the product rate for lean and rich LNG. An LNG producer can produce a long-term base LNG product, while allowing remaining LNG production capacity to be flexible to make maximum profit by optimizing production schedule/rate.

Asian countries need more natural gas in winter season for residential heating and they need to import more LNG from market. In case the Asian country can not acquire rich LNG as much as they require from their conventional supplier, they usually import lean LNG and inject LPG to meet the pipeline calorific value requirements. Figure 5 shows the example where Alaska LNG with low HHV which is marketable only in the lean gas market and Europe is imported to Asian countries and the LNG can meet the pipeline specification by injecting LPG.

On the other hand, LNG producer's tailoring to market demand can be illustrated as in figure 6. As an example, Angola natural gas after treatment has a very high HHV of 44.5 MJ/cm. If the natural gas is liquefied into LNG, it can be marketable only in continental Europe. However, LNG importers in Asian countries who have N2 injection facility can import the LNG and turn it into pipeline compatible gas. As an another method of HHV control, LNG producer with NGL extraction facility can decrease HHV to 38.2 MJ/cm, by running NGL extraction process shown in Fig. 3, which is marketable in lean LNG gas markets of the UK, Canada, and UK. NGL extraction at LNG terminal using a process shown in Fig. 4 can also reduce the HHV which is also marketable in the lean gas markets mentioned above.

As shown in Fig. 7, LNG which is exported to continental Europe depends highly on gas demands of Europe. Since Europe has very wide range of HHV specification, European countries can import most of LNG produced worldwide. If European countries are more interested in buying other LNG with lower HHV, the high HHV LNG falls into very unfortunate situation, by loosing market. This situation can overcome by using calorific value control facilities which make the LNG marketable to Asian countries, US, UK, etc.

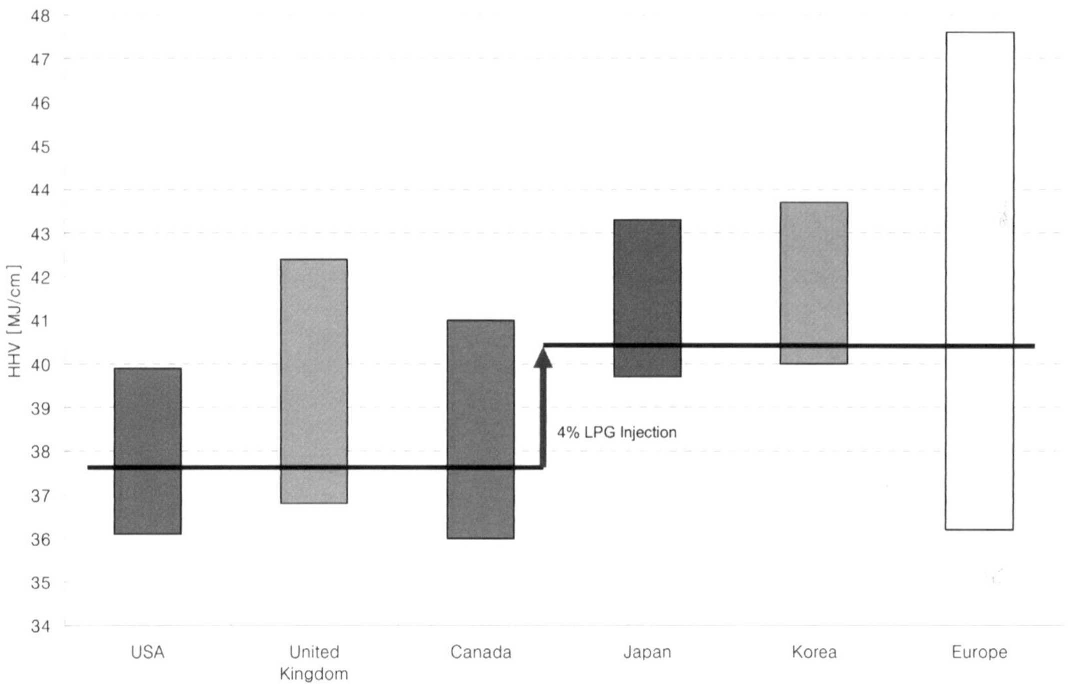

Fig. 5 - Meeting Calorific Value by Increasing HHV

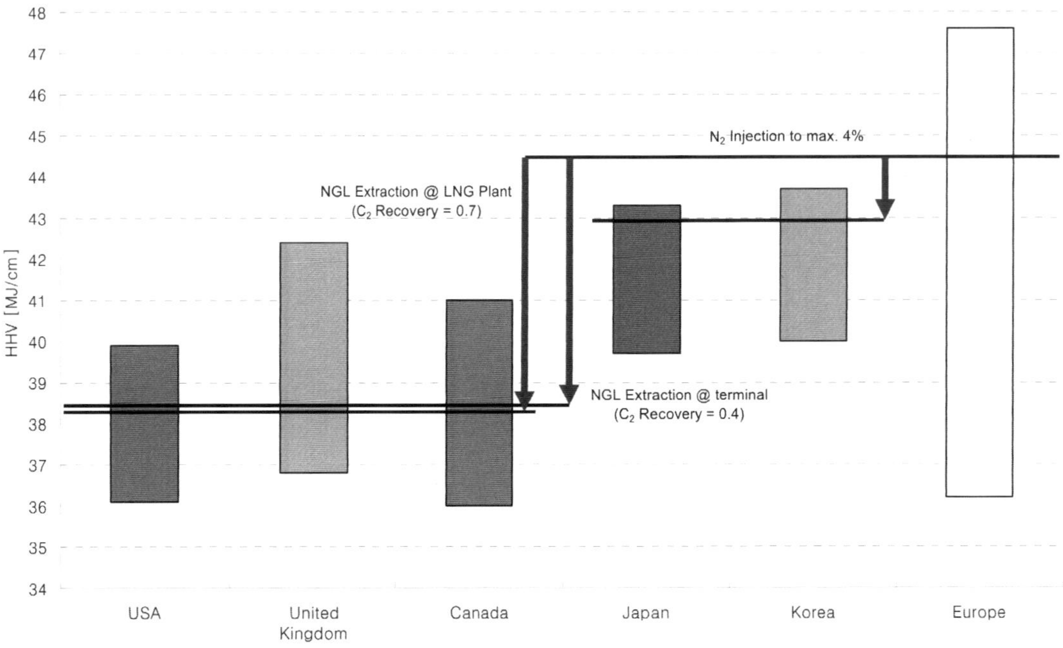

Fig. 6 - Extending a Market by Decreasing HHV

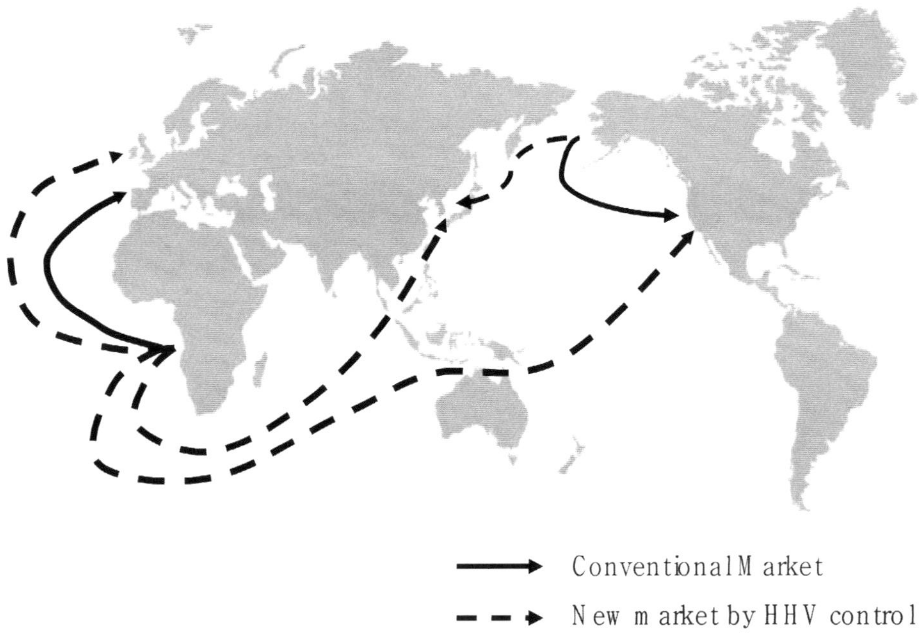

Fig. 7 - Diversification of LNG Market by Calorific Value Control

CONCLUSION

Current LNG market is faced with new challenges as well as opportunities due to increasing demand for natural gas. LNG producers and LNG importers are responding to new market environment by employing several calorific value adjustment technologies. It is no doubt that NGL extraction facility in LNG production plants as well as LNG terminals is a reliable and efficient method to reduce HHV. Inert injection and LPG injection in a LNG terminal can also control the calorific value. But, they have some component base limitations such as maximum inert and/or butane content.

By selecting appropriate method for calorific value control, the calorific value control facility provides LNG importers and LNG producers with more flexible and economical advantage in a LNG market. By installing calorific value control facility, natural gas can have wide range of HHVs, allowing LNG producers to enter new markets and LNG importers to import less expansive LNG. The facility makes them have more bargaining power in LNG market and maintain stable send-out rate to pipelines.

REFERENCE

1. D. McCartmey, Gas Conditioning for Imported LNG, 82nd Annual convention, Gas Processors Association, 2002.
2. D.J. Hawkins, Uncertainty about gas quality could delay US LNG imports, Sept. 20, 2004, Oil & Gas Journal.
3. D.R. Rogers, Gas 'Interchangeability' and Its Effects On U.S. Imports Plans, August 2003, Pipeline & Gas Journal.
4. J Cho, S.N. Shah, J.T. Kwan, and G-H. Lim, Development of a practical process for a natural gas calorific value adjustment system without blending rich gas.
5. J. Cho, Cost Effective and Practical NGL Stripping Processes, April 2008, AIChE Meeting
6. D. Coyle, F. Vega, and C. Durr, Natural gas specification challenges in the LNG industry, LNG 15, 2007
7. J. Cho, H. Kotzot, F. Vega, and C Durr, Cost Effective Processes for Heating Value adjustment, GasTech 2006
8. J.D. Wilkinson and H.M. Hudson, Liquefied Natural Gas Processing, US Patent 7155931B2
9. S. Huang, H. Kotzot, F. Vega, and C. Durr, Selecting the optimal scheme for N2 injection in an LNG terminal. 84th Annual convention, Gas Processors Association, 2005

Evaluating the Debottleneck Options for Existing LNG plants using Simulation Models

7f

Jaleel Valappil, Ram Tekumalla and Larry Hall
Bechtel Oil, Gas and Chemicals
Houston, Texas

Kent Anderson and Dale Ortego
ConocoPhillips
Houston, Texas

AIChE Spring Meeting, April 2009
9th Topical Conference on Natural Gas Utilization
Tampa, Florida, April 26-30, 2009

Introduction

There is a significant emphasis on utilizing the existing natural gas assets to their maximum potential. The first step in achieving this is operating the existing LNG plants at their maximum capacity [1]. Beyond that, there are economic incentives in debottlenecking the existing LNG plants to increase their capacity. There are various options available to perform the debottlenecking of existing plants. The evaluation and selection of the best options to increase the plant capacity are important for optimum return on the investment. This paper discusses the methodology utilized to debottleneck LNG plants based on ConocoPhillips Optimized Cascade® Process*. An overall plantwide simulation model is used as one tool to perform the evaluation and optimization of the various debottleneck options.

The dynamic model is used here as a tool to screen the various debottleneck options and to identify the most economical one. Once the appropriate debottleneck option is selected, the steady state model is used to design the plant based on the selected option. Detailed engineering design and construction is based on the steady state model simulation. The dynamic model allows efficient screening of various options before the detailed engineering design. The steady state and dynamic simulation models herein after will be referred to as simulation models.

It is important to identify the existing plant constraints before performing any debottleneck evaluations. These constraints vary with the operating conditions like ambient temperature, feed composition etc. One of the main constraints that limits production in LNG plants is the refrigeration compressor drivers. Also, various other equipment limitations and hydraulic constraints may be present in the plant. In addition to steady state model, plantwide dynamic model provides a suitable tool to identify these existing plant constraints over the range of operating conditions.

(*Optimized Cascade services are provided by ConocoPhillips Company, Phillips LNG Technology Services Company and Bechtel Corporation via a collaborative relationship with ConocoPhillips Company. Optimized Cascade, the Optimized Cascade logo, ConocoPhillips and its logo are trademarks of ConocoPhillips Company. Bechtel and its logos are trademarks of Bechtel Group Inc.)

The simulation models are built during the engineering and construction phase of the plant to evaluate and predict the plant performance prior to initial operation. This initial model is updated to incorporate actual tested equipment performance data and baselined to reflect the actual plant behavior. The model parameters are tuned with the actual plant data that is reflective of the normal plant operation over time.

Once the existing constraints are identified, various options to debottleneck the LNG plant are studied. These include upgrades or modifications to existing equipment, replacing or adding modifications to existing equipment or changes to the process configuration. Of particular interest are plant changes that can be made with minimum downtime or within the downtime required for other purposes. For example, gas turbine drivers can be upgraded to those with higher power ratings.

The simulation models can be run with the actual field conditions to identify the plant performance with these debottleneck modifications individually or in selected combinations. The incremental LNG production is estimated at various times of the year to obtain an annualized production increase. This is important to optimize the selection of LNG plant debottleneck modifications that can be used in estimating the return on capital investment.

LNG Process Description

The studies presented in this paper are based on the ConocoPhillips **Optimized Cascade®** process (Fig 1). The feed gas is first processed in the feed-treatment section of the plant. Treated gas from the amine system is fed to the molecular sieve dehydrators and processed through activated carbon beds to remove any mercury. The treated gas is fed to the liquefaction unit where it is cooled in stages and condensed prior to entering the LNG tanks. The liquefaction process includes three refrigeration circuits consisting of predominantly pure component refrigerants, propane, ethylene and methane. Although not required, each refrigeration circuit typically utilizes parallel compressors (up to two or three per refrigerant service) combined with common process equipment. The feed gas passes through multiple stages of chilling in the propane, ethylene and open loop methane circuits. Each successive stage is at a lower temperature and pressure. The resulting LNG product is pumped to storage where it is stored at near atmospheric pressure and –161°C.

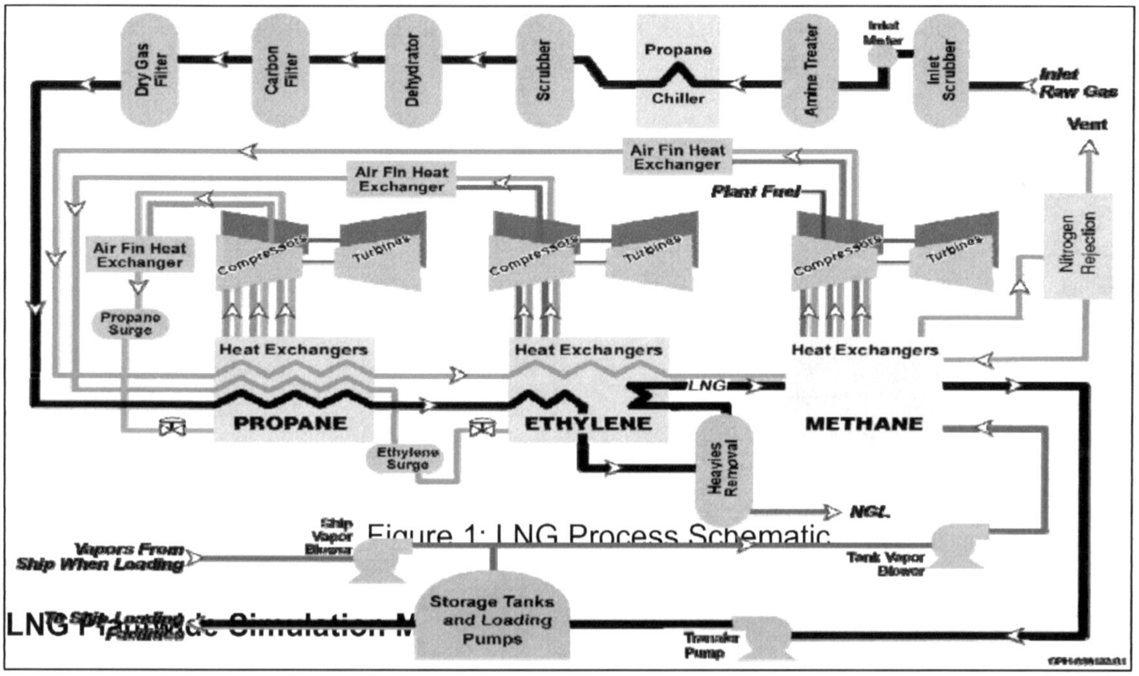

Figure 1: LNG Process Schematic

LNG Plantwide Simulation Model

The plantwide simulation model used here is fundamental in nature and based on rigorous thermodynamics [2]. Some of the main components in the LNG process are turbine driven refrigeration system compressors and the heat exchangers. The compressors are modeled with the as-tested performance curves. The governor and turbine dynamics and turbine power output are computed based on a model derived from vendor provided data. The anti-surge control system and the performance control system are modeled in the simulation environment. The Brazed Aluminum Heat Exchanger is another important piece of equipment in this process. Detailed equipment design information is used to model the brazed aluminum exchanger in the simulation environment. The application of the pressure profile to the simulation model is based on the isometric information wherever available.

Baselining LNG Plant Simulation Model

The simulation model was first baselined with the operating data from the operating LNG plant. This is essential to establish a valid baseline for further debottleneck simulations. The plant model is based on as-tested compressor curves and equipment data from the plant construction stage. Some of the model parameters are adjusted to obtain a better match with the plant operating data. The main focus of the baselining exercise was to validate the performance of the refrigeration systems (especially the compressors and gas turbines) against the plant operation.

The philosophy utilized in model baselining is that of matching the overall plant operation, rather than looking at specific variables. The attempt here is to look at the match between the plant and the model from an overall process viewpoint, rather than select variables. The model is first baselined against the data at a particular ambient temperature. The key model parameters are adjusted to fit this operation. The model is then validated to see how well it matches the plant data at a higher ambient temperature without making any further model changes. This is helpful in establishing the validity of the model at conditions different from the baselining one used for model tuning.

Evaluating Debottlenecking Options for LNG Plant

There are significant economic benefits in debottlenecking the existing LNG plants by increasing their capacity or availability. There are various options to perform the debottlenecking of existing plants. The main objective is to identify and remove the capacity limiting constraints in the plant. Once the existing constraints are identified, various options are considered to debottleneck the LNG plant and are studied with the simulation models. These studies include:

- upgrades or modifications to existing equipment,
- replacing existing equipment with ones that provide higher capacity, or
- changes in process configuration or conversion to alternate technologies.

These options were considered for the plant under consideration. The upgrade that provided maximum economic benefits with minimum plant downtime was identified to be the power augmentation of the refrigeration compressor drivers.

Identification of Capacity Limiting Constraints

The LNG plant under consideration had various constraints active under various operating conditions. At low ambient temperature conditions, the refrigeration compressor was the limiting factor, operating at its highest allowable speed. At higher ambient conditions, the refrigeration compressor turbine power was the limiting factor. Also, other capacity limiting constraints like equipment limitations were examined and identified during the simulations.

Various debottleneck options were examined to increase the capacity of the plant. This included upgrading the refrigeration compressor drivers to turbines with higher ratings, increasing air cooled condenser areas and inlet air chilling for the gas turbines. A description of power augmentation options studied is provided below.

Power Augmentation Options

This study considered power augmentation of the existing refrigeration turbines in the LNG plant and the addition of propane condenser bays as options to enhance LNG production. The power augmentation simulations included turbine inlet air chilling and conversion of the existing gas turbines to ones with higher power rating.

1. Refrigerated Inlet Air Cooling: Two refrigerated inlet air cooling cases were evaluated in this study. These involve chilling the turbine inlet air to two different temperatures. Since the gain in turbine output is a function of the reduction in inlet air temperature, the greater the reduction the higher is the potential performance gain.

2. Conversion of existing refrigeration compressor gas turbines to ones with higher power ratings: In this case, the gain in turbine output provides additional refrigeration capacity, and hence higher LNG production.

3. Conversion of existing gas turbines to ones with higher efficiency: This upgrade helps to increase the efficiency of the LNG plant along with nominal increase in production.

Simulation Results

The production rate of the LNG plant at various ambient temperatures was calculated using simulation. Figure 2 presents a comparison of the different study options giving the estimated LNG production in relation to the ambient temperature. The existing operation is used as a reference to compare the various debottleneck options. It is found that below a particular temperature, the various options do not add any incremental benefit. This was due to the fact that the constraint changed from the driver power to another equipment limitation (like the refrigeration compressor) or a process operating parameter.

The simulation model was also used to identify the incremental production increase during various months of the year. Further, the effect of process modifications like changing the fuel gas source and increasing the area of air-cooled condensers were evaluated. The insights obtained from these simulations were valuable in selecting the optimal plant debottlenecking strategy.

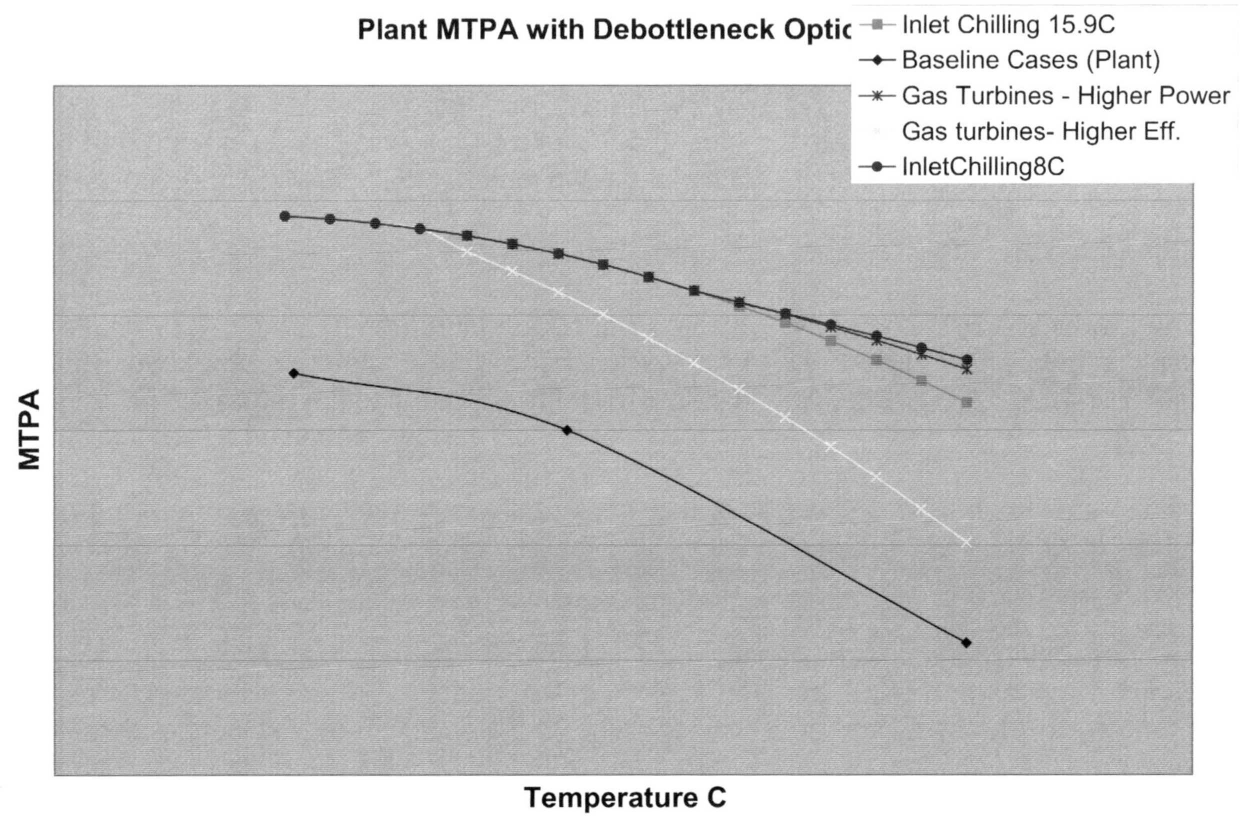

Figure 2: Plant MTPA vs. Ambient Temperature with various Debottleneck Options.

References

1. J Valappil, V Mehrotra, and D. Messersmith, "LNG Lifecycle Simulation", Hydrocarbon Engineering, October 2005.
2. J Valappil, V Mehrotra, D Messersmith, and P Bruner, "Virtual Simulation of LNG Plants", LNG Journal, January/February 2004.

Biography of Speaker

Jaleel Valappil is a senior engineering specialist with Bechtel Oil, Gas and Chemicals Advanced Simulation Group in Houston, Texas. His expertise covers dynamic simulation, real-time optimization, operator training simulation, and advanced process control. He currently applies advanced simulation and optimization technology to improve the design and operation of plants being built by Bechtel. Before joining Bechtel, Dr. Valappil was a senior consulting engineer with the Advanced Control Services Group of Aspen Technology, Inc. Dr. Valappil holds BS and PhD degrees in Chemical Engineering from the Indian Institute of Technology and Lehigh University in Pennsylvania, respectively.

jvalappi@bechtel.com

Design Considerations of Domestic Gas Plant for LNG Baseload Plant

7g

J.C. Kuo, P.E.
Senior Advisor
Chevron Energy Technology Company

AIChE Spring Meeting, April 2009
9th Topical Conference on Natural Gas Utilization
Tampa, Florida, April 26-30, 2009

For many based load LNG Projects around the world, there is a requirement to install a Domestic Gas (DOMGAS) train for each based load LNG liquefaction train. This is to ensure the domestic energy demands are fulfilled before sending the LNG to other place for sale.

The requirements to design the DOMGAS Plant are to
(1) Meet export pipeline delivery pressures and
(2) Domestic gas specifications, which include acid gas specification, removal of water and heavy hydrocarbons to meet export pipeline moisture and hydrocarbon dewpoint specifications.

1.0 Typical Domestic Gas Specifications

The following table lists the typical specifications.

Table 1. DOMGAS specifications

Product Characteristic	Units (note 1)	DBNGP Specification
Higher Heating Value	MJ/Sm^3	37.0 – 42.3
Wobbe Index	MJ/Sm^3	46.5 – 51.0
Gas HC Dew point	°C	< 0 °C @ 2.5 to 8.72 MPa (note 3)
Maximum Inert Gases	Mol%	7.0
Maximum Carbon Dioxide	Mol%	4.0
Maximum Oxygen	Mol%	0.2
Maximum Water	mg/Sm^3	48
Maximum Total Sulfur		
Un-odorized	mg/Sm^3	10
Odorized	mg/Sm^3	20
Maximum Hydrogen Sulfide	mg/Sm^3	2
Minimum Extractable LPG	t/TJ	0
Maximum Radioactive Components	Bq/Sm^3	600

Notes:

1) Sm^3 means 1 cubic meter of dry gas at 15°C and an absolute pressure of 101.325 kPa.
2) MAOP (maximum allowable operating pressure) is the maximum pressure at which a pipeline may be operated. MAOP for pipeline is 120.0 Barg.

3) MPa is an absolute pressure

2.0 Acid Gas Removal Unit (AGRU)

The objective of the Acid Gas Removal Unit is to remove acid gas components. Removal of CO_2 and H_2S prevents them from freezing out hydrate at low temperatures in the pipeline and also make the treated gas to meet product specifications include H2S, total inerts (CO2 and Nitrogen).

Typical Acid Gas Removal Unit specifications are

H_2S exiting Gas Treating Unit (to meet sales specifications)	< 3.5 ppmv (2 mg/Sm[3])
CO_2 exiting Gas Treating Unit (to meet total inert and CO2 specifications)	depending on Nitrogen %
Total Sulphur (H_2S + COS + Organic Sulfur) exiting Gas Treating Unit (to meet sales specifications)	< 20 mg/ Sm3

Acid-Gas Treating processes have been broadly classified into chemical absorption, physical absorption, adsorption, physical separation, and hybrid processes. Selection of an acid-gas removal systems generally start with a preliminary screening and conclude with economic studies comparing the final choices. Considerations that can be used for screening are

- *Simplicity.* A simple, single-step removal process based on an inexpensive solvent should always be considered.
- *Purity requirements.* Amines can reduce CO_2 and H_2S to low levels with little co-absorbed hydrocarbon losses.
- *Arrival conditions of the gas to be treated.* High pressure gas with high acid-gas partial pressures or concentrations may favour absorption processes if co-absorption hydrocarbon losses do not render these processes unattractive.
- *Operating Pressure and temperatures.* Operating at high pressures reduces recompression costs for transfers to other units. Some solvents have lower freezing points and can therefore operate at lower temperatures.
- *Amounts of contaminants.* Some amines form non-regenerable products with COS and CS_2 which accumulate and degrade the amine solution. Large amounts of mercaptans may favour processes that have mercaptan removal capabilities.
- *Selectivity of H_2S and CO_2 removal.* Some processes remove H_2S but allow some CO_2 to slip. This lowers costs because less treatment and less regeneration heat duty are required.
- *Level of acid gas components in feed gas.* High levels of CO_2 or H_2S may favour a pairing of two processes with bulk removal at the first stage and purification at the second.

- *Amount of hydrocarbons*. Physical absorption can be attractive when heavy hydrocarbons are low.
- *Solvent capacity*. Processes that use solvents with high acid-gas loading capacity have lower circulation and regeneration costs.

Discussion of the five groups of acid-gas removal processes are given below.

2.1 Chemical absorption

In these processes, H_2S and/or CO_2 are chemically bound to the active ingredient in the treating solution; therefore, the residue gas can be treated to very low levels for these contaminants in most of the processes. Another common advantage of these processes is low co-absorbed hydrocarbon losses. Current commercialized chemical solvent processes use weak bases like alkanolamines, alkali salt solutions, potassium carbonate, or a chelate solution.

2.1.1 Amine Based Processes

The amine processes in commercial use: including
Monoethanol Amine (MEA),
Diethanol Amine (DEA),
Diglycol Amine (DGA),
Methyl Diethanol Amine (MDEA)
Formulated Amines
There are many proprietary MDEA based formulated solvents that are marketed by different companies. These formulated solvents not only retain the benefits of MDEA but also enhance them for specific applications by the addition of proprietary additives. Some formulations allow slipping of a larger fraction of the inlet gas CO_2 while reducing H_2S to pipeline specification of 4 ppmv.
Hindered Amines
An example of this technology is ExxonMobil's FLEXSORB solvents.

2.1.2 Non-Amine Based Processes

In these processes, either an alkaline salt or a chelate solution is used to remove the acid gas components:

Hot Potassium Carbonate Process
Redox Process
The two commercial iron redox processes are Lo-Cat and Sulferox.
Biochemical Processes
In the Shell-Paques process, which uses liquid phase biological oxidation, H_2S is converted to elemental sulfur

2.2 Physical absorption

In these processes, the solvent does not chemically react with the acid gas components. The physical solvent has a higher affinity for absorbing acid gases compared to lighter hydrocarbons like methane and ethane (the key components of natural gas). In other words, the physical solvent preferentially absorbs acid gases contained in a hydrocarbon stream. In some cases, physical solvents can also remove COS, CS_2, mercaptans and water in addition to H_2S.

Examples of Physical Solvent Processes are:

Selexol
Marketed by Dow Chemical, this solvent has higher solubility for H_2S versus CO_2 making it possible to slip some CO_2 into the residue gas. In addition to H_2S and CO_2, the Selexol solvent can also remove some COS, CS_2 and mercaptans without degradation of the solvent

Propylene Carbonate (Fluor)

- The advantage of this process over amine processes is that it has high acid gas loadings and thus a corresponding decrease in capital costs. Regeneration of the loaded solvent is with successive pressure reductions and requires no regenerative heat.

Rectisol (Lurgi and Linde)
In this process methanol is used as the physical solvent. Because methanol has high vapor pressure at ambient temperature, this process is operated at temperatures of -30 to -100 F. This process is best suited for bulk removal of H_2S and CO_2 when the ethane and heavier content in the gas is low

2.3 Adsorption

Adsorption processes are usually only cost effective at low feed concentrations and volumes. In these processes, a solid bed preferentially adsorbs the acid gas components as it flows through the bed. The acid gas components are then stripped from the bed by flowing a high temperature regeneration gas through the bed. Some examples of solid bed adsorption processes are:

Molecular Sieve (UOP, Zeochem and others)
This process involves the adsorption of acid gases on molecular sieves. It can be designed to remove H_2S and some CO_2 or remove both H_2S and CO_2 as well as COS, CS_2 and mercaptans.

Activated Carbon
H_2S and other sulfur compounds are physically adsorbed. Treated gas purity of 0.0 – 0.2 ppmv sulfur can be obtained. During regeneration with steam, the adsorbed sulfur compounds are desorbed.

2.4 Physical separation

Physical separation processes are usually only considered for bulk removal of acid gases. Commercially available physical separation technologies for acid-gas removal are membranes and cryogenic distillation.

Cryogenic Distillation
Distillation is a capital and energy-efficient means of treating gases with high levels of CO_2. Conventional distillation can only remove CO_2 to a level of about 15% without CO_2 freezing. Cryogenic fractionation uses special internals and intermediate paraffins to modify the CO_2 solid-liquid equilibrium to allow removal of acid gases without CO_2 freezing problems. The processes are dry and have low corrosion rates but require alloyed steel and low pressure. These processes are reported to be attractive when removing large amounts of CO_2 from gases that contain low concentrations of hydrocarbons heavier than methane.

Membranes

In membrane processes, a polymeric film (membrane) is used to separate the acid gas components for the gas being treated. While the acid gas components permeate through the membrane, the hydrocarbons permeate to a much lesser extent.

2.5 Mixed/Hybrids

The better processes use an aqueous solution of both a chemical solvent and a physical solvent thereby combining the benefits of both chemical and physical solvent processes. The gases can be treated to high purity, with high acid gas removal capacity and regeneration energy requirements that are between chemical and physical solvent processes.

Following are some examples of these processes:

Sulfinol (Shell)

- This process uses an aqueous solution of Sulfolane with either Diisopropanol Amine (DIPA) or MDEA. Sulfinol with DIPA is referred to as Sulfinol-D while that with MDEA instead is referred to as Sulfinol-M.
- Other examples of mixed solvent processes are Optisol (Natco), Amisol (Lurgi) and Flexsorb PS (ExxonMobil).

3.0 Considerations of Dehydration Technologies

After AGTU, the gas stream is saturated with water. When the gas stream is cooled below its hydrate temperature, either due to pipeline transportation, pressure reduction or cooling, free water condensation or hydrate formation will occur. Such problems can be avoided by dehydration to remove the water from the hydrocarbons prior to cooling or injection of a hydrate inhibitor into the hydrocarbons during cooling.

TEG (triethylene glycol) contactor and MEG (monoethylene glycol) or other Glycol such as EG (ethylene glycol) injection is considered. Dry-bed dehydrator using solid desiccants generally costs more to purchase and operate than glycol units, and is only needed for applications such as high H_2S content gases, very low water dew point requirement, or processes where cryogenic temperatures are encountered. Since the DOMGAS plant does not have these situations, dry-bed dehydration is not included here. Descriptions of the TEG Dehydration and MEG injection technologies are given below.

3.1 TEG Dehydration

The TEG dehydration system removes the water from the hydrocarbons by absorption in a contactor, or an absorber tower. Lean (concentrated) TEG enters the tower above the gas-liquid contacting section (trays or packing) and absorbs the water from the inlet gas stream. It leaves the bottom of the tower as rich TEG. Since a small portion of the hydrocarbons, primarily methane, is also absorbed into the TEG, the pressure of the rich TEG is reduced through a flash tank to remove most of these hydrocarbons. The TEG is then filtered to remove solids, heated in an exchanger (usually lean/rich TEG exchanger), and fed to the regenerator. The regenerator's reboiler may be heated by steam, hot oil, or direct firing. The TEG is re-concentrated by removing most of the water in the regenerator.

The leaner the TEG concentration, the lower the water content of the dehydrated gas leaving the contactor. To achieve a higher TEG concentration than 98.6%, the reboiler needs to be operated at a temperature higher than 400°F. However, the TEG will begin to decompose or degrade above 400°F. The most common method to achieve the high TEG concentration while also avoid exceeding the degradation temperature has been by use of stripping gas or by means of vacuum in the reboiler. Most TEG units requiring high glycol concentration use stripping gas. The stripping gas is fed to the bottom of a separate stripper where it flows counter-currently to the lean TEG for final water removal. Figure 1 illustrates a TEG dehydration unit with stripping gas.

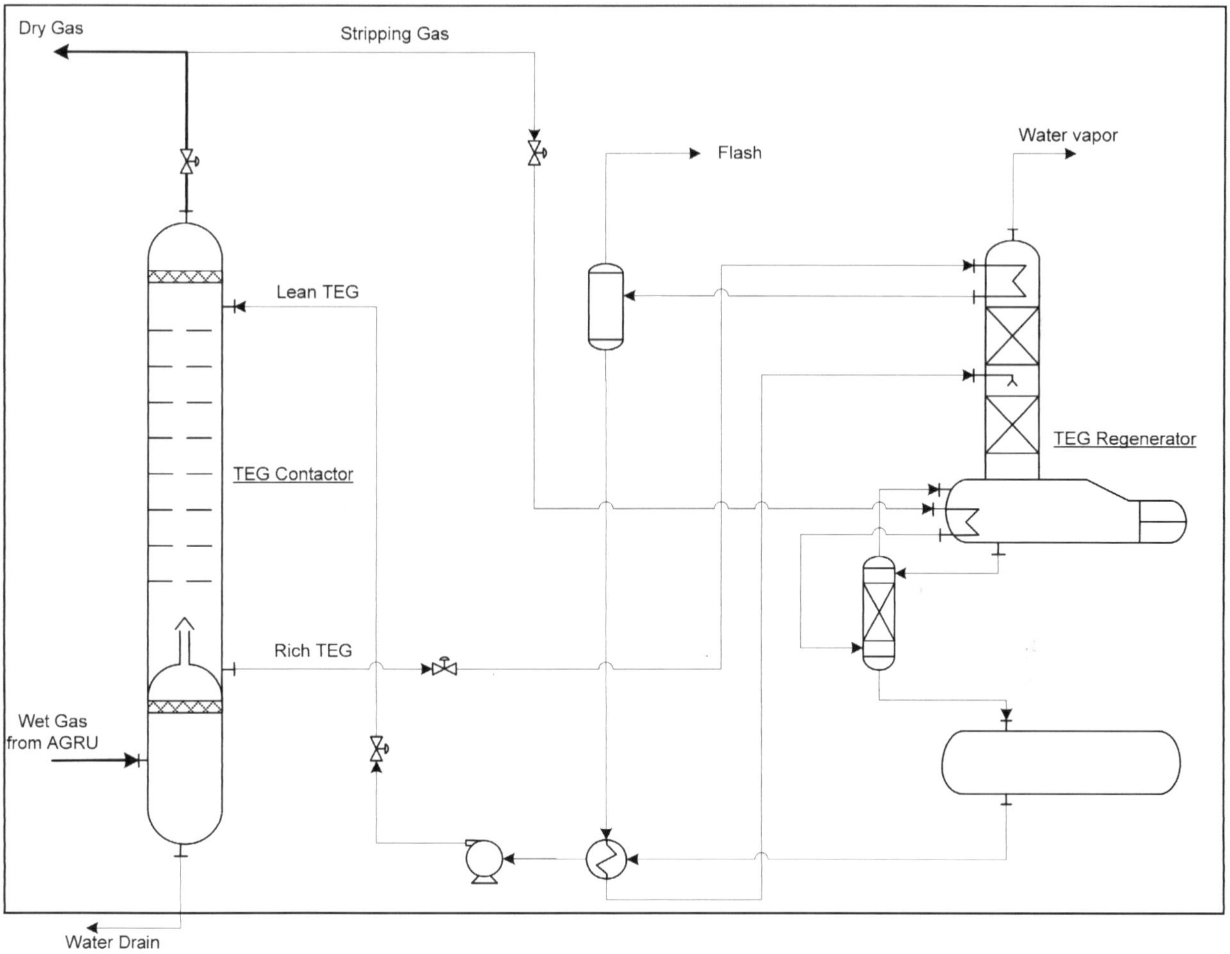

Figure 1. TEG dehydration unit with Stripping Gas.

3.2 MEG Injection

In glycol injection systems, the glycol provides some dehydration. But its primary function is to act as an "antifreeze" agent in suppressing the formation of solid hydrates. Hydrate inhibition utilizes injection of the glycol into a process stream where it can combine with the condensed hydrocarbon phase to lower the hydrate formation temperature at a given pressure. Although ethylene glycol (EG), diethylene glycol (DEG), and triethylene glycol (TEG) could all be used for hydrate inhibition, the most popular one has been MEG because of its lower cost, lower viscosity, and lower solubility in liquid hydrocarbon.

To minimize the possibility of solid-phase formation in the glycol solution, compositions near the eutectic (i.e., 60 to 80 wt% glycol in water) are commonly employed, compared to 95% or higher concentrations used in more conventional dehydrator designs. The dilute aqueous solutions have the further advantage of low solubility in liquid hydrocarbons.

Glycol is regenerated by vaporizing the extra water in the reboiler. Heat is supplied to the reboiler by a fired tube that burns fuel gas, a hot oil loop, or by steam. The lean glycol is then pumped from the surge tank, through the lean/rich glycol exchanger, and again injected into the inlet gas. Figure 2 illustrates an MEG injection unit.

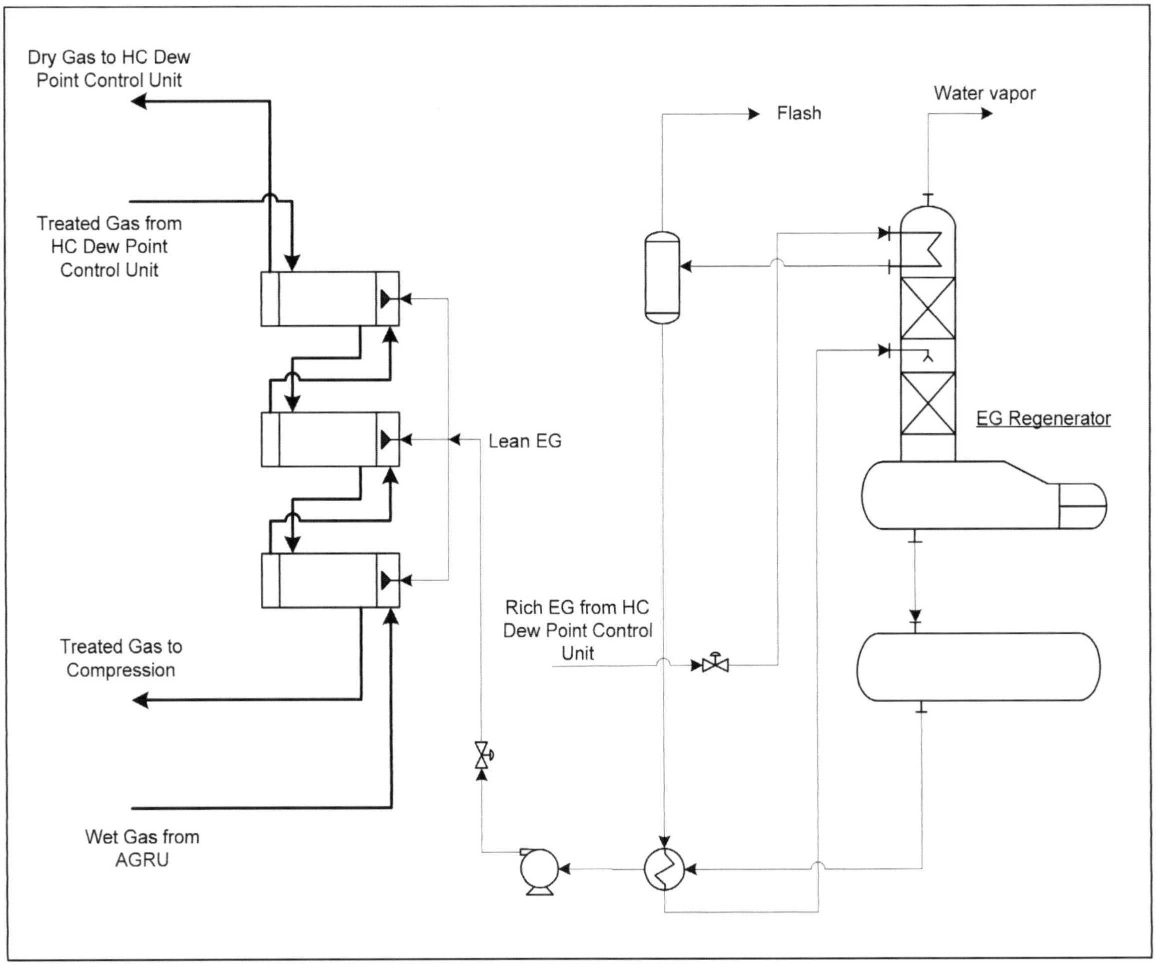

Figure 2. MEG injection unit.

4.0 Considerations of Hydrocarbon Dew point Control Technologies

To prevent the formation of liquids when gas is transported in pipelines, it is necessary to control the hydrocarbon dew point below the pipeline operating conditions. Condensation of liquids can cause problems such as liquid slugging, formation of two-phase flow, metering accuracy and safety. It is common to have a pipeline specification on hydrocarbon dew point. A large number of technologies are available to control the hydrocarbon dew point. Six dew point control technologies are listed for evaluation

In evaluating the six dew point control technologies, many factors are considered. The major factors are performance, economics, availability, precedent, and number of suppliers. Other factors include efficiency, operability, size, and simplicity.

4.1 LTS (Low Temperature Separation with TEG/JT)

The use of the Joule-Thompson (J-T) effect to remove heavy hydrocarbons is an attractive method mainly due to its simplicity of design and operation. The concept is to chill the gas by expanding the gas across a J-T valve. The cooling causes the heavy hydrocarbons to condense and separate from the gas. The degree of dew point control is directly related to the pressure reduction across the J-T valve, and therefore the increase in the power of gas compression. This cooling effect can be amplified by using the cooled gas to pre-cool the inlet gas in a gas heat exchanger. The efficiency of a J-T unit is therefore directly related to the efficiency of the gas heat exchange involved in the process.

To prevent hydrate formation, the gas needs to be dehydrated with TEG contactor prior to the J-T valve or injected with MEG. A LTS (TEG/JT) process scheme is given in Figure 3.

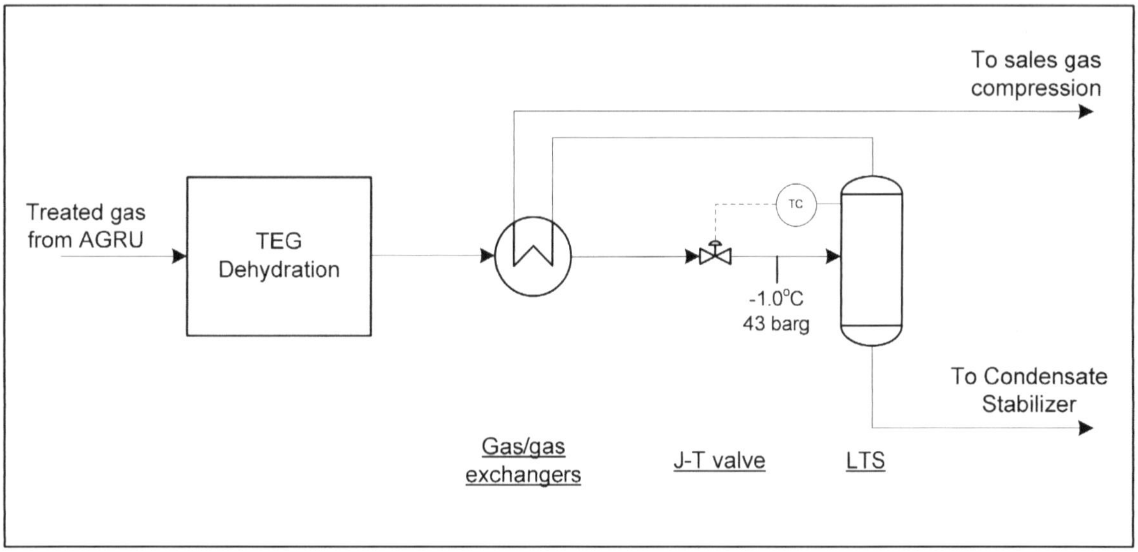

Figure 3. LTS (TEG/JT) Process Scheme.

4.2 C3-Refrigeration/ TEG

Propane refrigeration systems potentially may be the most economic approach to dew pointing because of simplicity of design and better energy efficiency. Refrigeration of a gas stream causes condensation of both water and hydrocarbons and therefore can be used to reduce the dew point of the gas. However at high pressure and low temperature liquid water form hydrates and may plug valves and exchangers of the plant. In order to prevent hydrates formation it is necessary either to previously dehydrate the gas stream or to inject hydrate preventers like methanol, ethylene glycol or diethylene glycol. Methanol is used only in emergency or at start-up because it is mostly lost with the gas stream. Glycols are preferred since they can be easily regenerated and recycled. A Propane Refrigeration scheme is given in Figure 4.

In Figure 4, the gas pressure is essentially maintained through the process allowing for pressure drops across equipments. The gas is heat exchanged and then cooled by the refrigeration chiller to a specific temperature. Liquid is separated in a knock-out drum, and the cooled gas is reheated in heat exchangers before entering the compressor. Hydrate formation is prevented by either TEG contactor upstream of the refrigeration unit or by MEG injection to the gas stream.

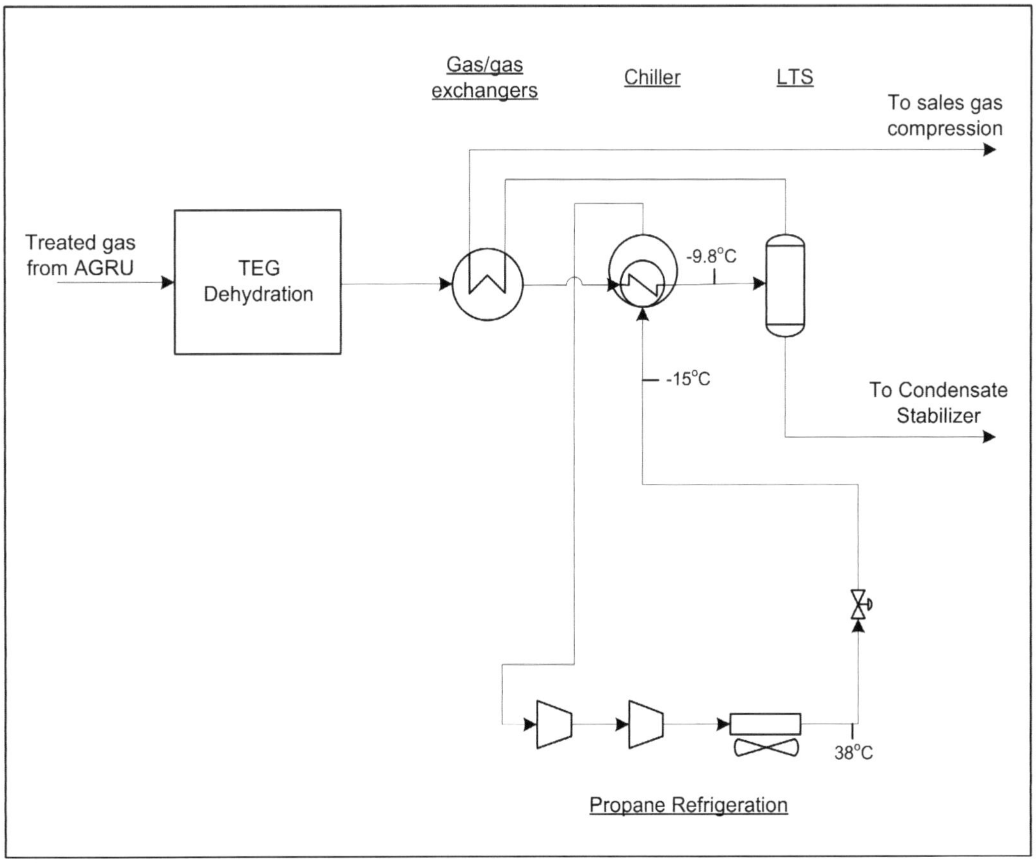

Figure 4. A Propane Refrigeration Scheme.

4.3 Oil Absorption

Although many old plants that were built based on the lean oil process are still in use today, most lean oil plants have been shut down or replaced with modern straight refrigeration or turbo expander processes (GPSA, p.16-8).

Absorption is the process in which liquid is used to remove some of the components from a gas stream. A liquid hydrocarbon solvent (lean oil) retains the heavy hydrocarbon components in an absorber column. The spent or diluted oil (rich oil) is sent to a distillation column to get rid of the heavy hydrocarbons, re-concentrate the oil, and is recycled back to the absorber column to complete the cycle. In order to improve separation in the absorber, lean oil and natural gas are chilled before entering the absorber. A process scheme is given below.

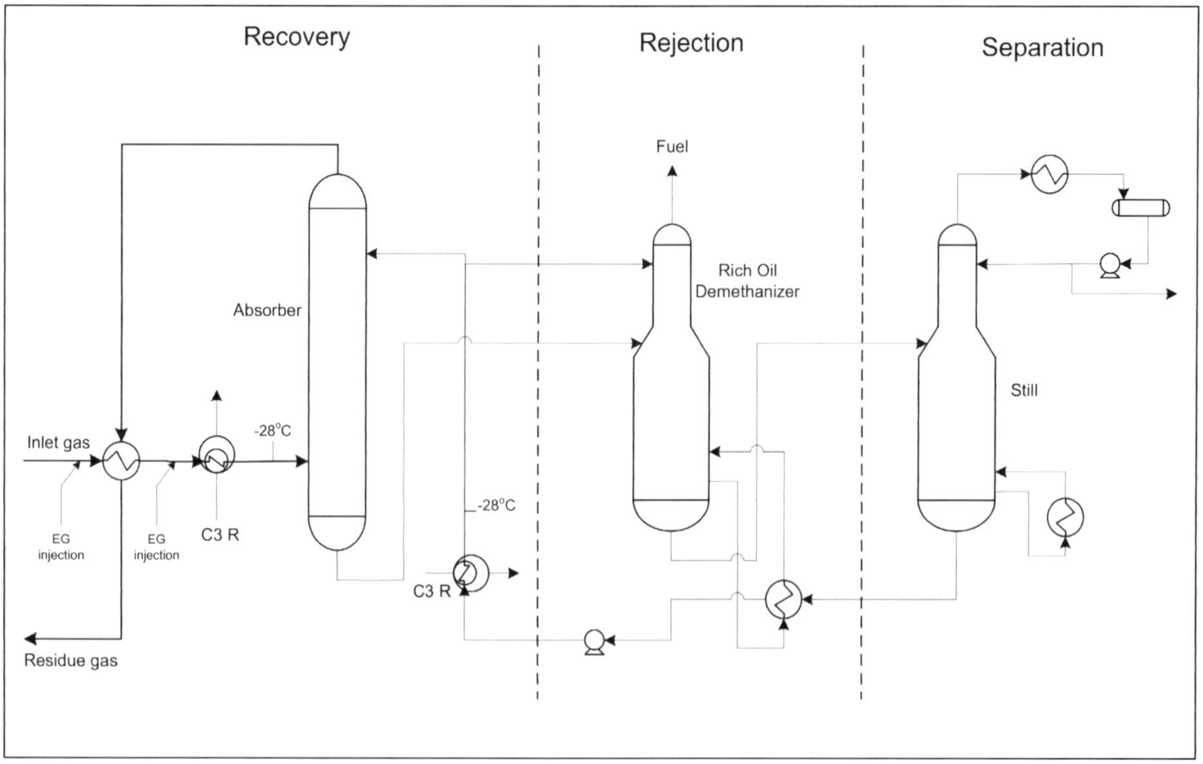

Figure 5. Lean Oil Absorption Process Scheme.

4.4 Twister™

The Twister technology is expected to be competitive under the NPV analysis when applied to DOMGAS dewpointing. However, it is not as well proven or accepted as propane refrigeration and TEG/JT LTS.

Twister takes advantage of gas expansion for liquid separation. It uses static equipment to achieve the desired separation and focuses on replacing Joule-Thomson expansion valves and/or turbo-expanders (GPSA, p.16-6). The process starts with the inlet stream expanding and passing across vanes which swirl the stream. The swirling gas then passes through a supersonic nozzle in which the pressure is reduced and liquid is formed. The centrifugal motion forces the liquid in the supersonic gas stream to the wall where it is drained from the apparatus. The vapor is then expanded in a diffuser nozzle and recovers 70-80% of the initial pressure. This technology is focused on hydrocarbon dew point control and dehydration applications in both onshore and offshore locations.

Two studies have shown that the Twister system when compared to the JT system has a similar simplicity, robustness, and order of magnitude project CAPEX. Schinkelshoek and Epsom (2008); Genesis Oil & Gas Ltd (2007) Furthermore, when the focus is on higher NGL recovery, the Twister results in a higher NPV compared to the JT system.

- Twister is a low temperature separation process, for which performance can be optimized by inlet cooling. Figure6 below shows a typical Process Flow Diagram of a Twister dewpointing system.

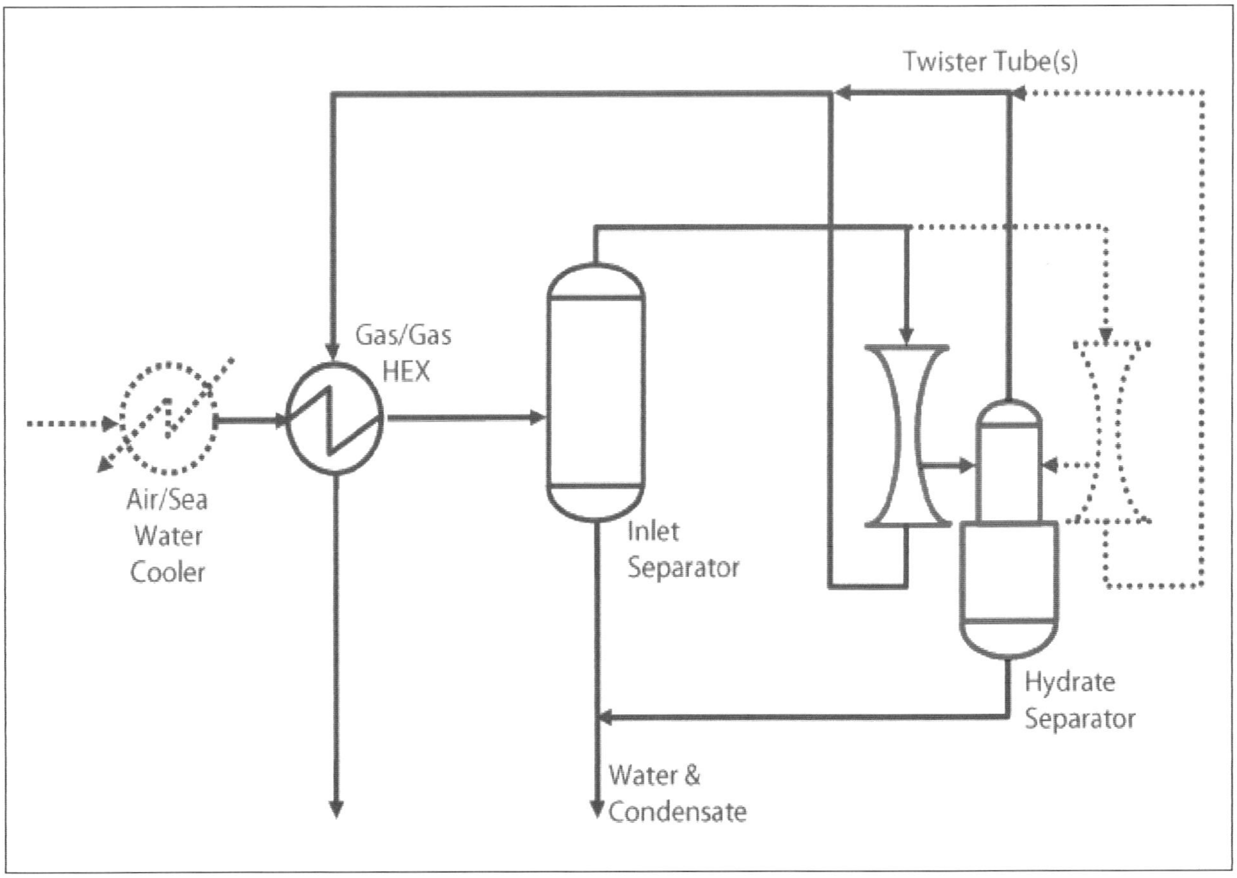

Figure 6. Twister Dew pointing System

Although the Twister technology has some success in commercial testing of offshore dehydration, similar testing in onshore fuel gas conditioning has only recently started. Therefore, commercial success in simultaneous moisture and hydrocarbon dewpointing will remain in the future.

4.5 Membranes

Membranes probably will not be competitive in DOMGAS dew point control applications with high gas flows. Membranes have been considered for natural gas dehydration and NGL removal. Baker (2002); GPSA (2004). For NGL removal, Baker (2002) judged from NGL removal performance data that membrane technology will be very competitive with propane refrigeration for flows up to 20-30 million scfd. At higher gas flows, economies of scale favor condensation by turbo expansion. For natural gas dehydration, membranes have no technical problem in removing water from the feed gas because selectivities are high. However, glycol dehydration is well accepted by the industry, and its cost is very low. To be competitive, the membrane system must minimize loss of methane with the permeate water. Dehydration by membrane permeation is normally only considered for plants that can make use of low pressure natural gas fuel. GPSA, p.20-45.

The driving force for separation of a gas component in a mixture is the difference between its partial pressure across the membrane. As pressurized feed gas flows into the metal shell of the separator, the fast gas component, such as water and CO2, permeate through the membrane. This permeate is collected at a reduced pressure, while the non-permeate stream, i.e., the dry natural gas, leaves the separator at a slightly lower pressure than the feed.

Some natural gas dehydration designs that have been considered are depicted in Figure7. In the first design shown in Figure 7 a, a small one-stage membrane system removes 90% of the water in the feed, producing a permeate gas representing 4-5% of the initial gas flow. The amount of methane and other natural gas components in the permeate stream is dependent on pressure drop and the surface area of the membranes. However, 5–10% of the feed stream is a realistic figure. Water/methane membrane selectivities are very high (>200), so the separation achieved is completely pressure-ratio-limited.

This methane loss cannot be reduced by making the membrane more selective because the process is completely pressure-ratio-limited. If the permeate gas could be used as a low-pressure fuel, this design would be economical. Unfortunately, most plants do not need this quantity of fuel gas, and the stream is too large to flare. A compressor could be used to recompress the permeate gas, as shown in Figure 7b, but the capital costs of the system would then more than double, and the process would only compete with glycol dehydration where glycol units are not well suited,.

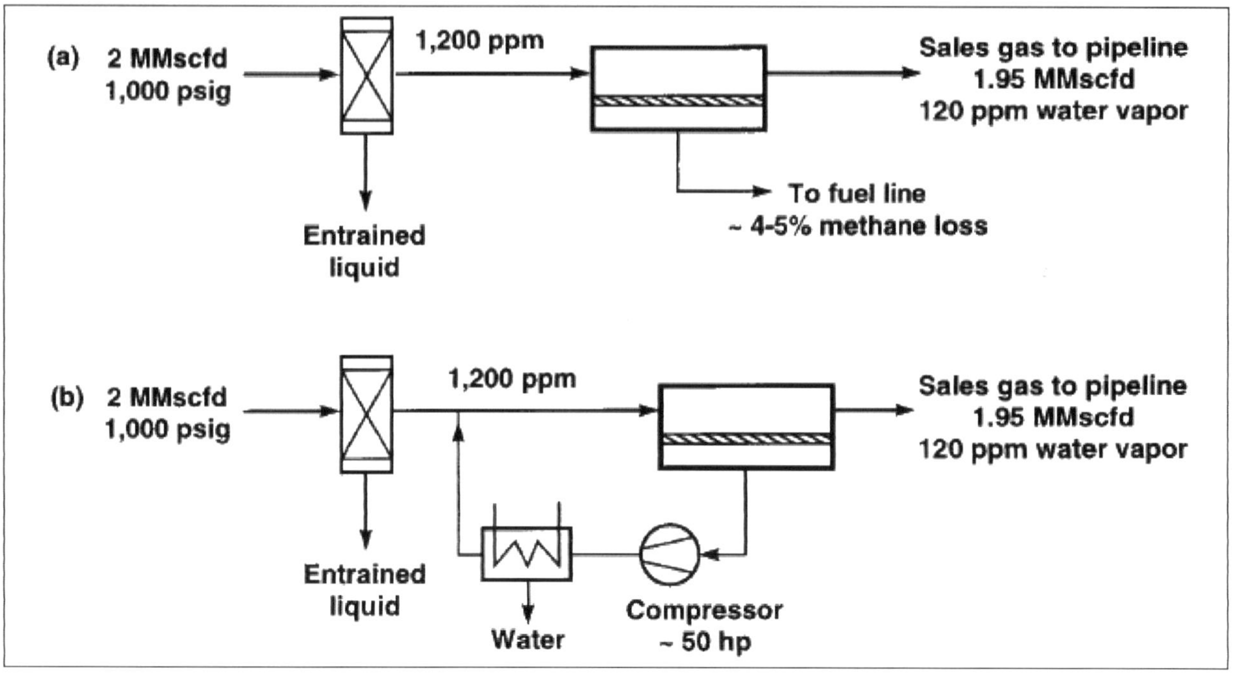

Figure 7. Two simple membrane processes for the removal of water from high-pressure natural gas.

Membranes have several advantages over other systems including:

• ease of installation - they are modular and skid mounted and there are usually only three connections necessary to connect up a unit;
• Weight and space savings - at low throughputs they are smaller than alternative processes - and can be configured to suit the available space;

• They have no moving parts (unless a compressor is used to recycle gas to increase product recoveries or improve purities) allowing them to operate as automated units and with limited maintenance crews;
• They have low environmental impact since no chemicals are circulated;
• they can work with no power or utilities but it may be beneficial to use gas cooling or heating to precondition a stream as this can increase efficiency and improve membrane life.

In 2004, GPSA, p.21-3 wrote: "rubbery membranes may have potential applications in heavy hydrocarbon dewpointing. However this technology has been tested only to a limited extent at this point of time." The U.S. Department of Energy has since then sponsored some field testing of membranes modules supplied by Membrane Technology and Research (MTR) Inc. The system startup and initial testing began in February 2005. After December 2005, the unit operated continuously and smoothly, with essentially no change in performance levels through September 2006.

4.6 Silica Gel Adsorption

The use of solid desiccant units is typically limited to applications such as high H2S content gases, very low water dew point requirements, simultaneous control of water and hydrocarbon dew points, and special cases such as oxygen containing gases, etc. because solid desiccant units generally cost more to buy and operate than glycol units (GPSA, p.20-36; p.21-3).

For simultaneous control of water and hydrocarbon dew points, silica gel is selected over molecular sieves because of lower regeneration costs and acid resistance. Alumina also has comparable costs and has even better acid resistance. However, it does not release hydrocarbons as readily as silica gel during regeneration. Silica gel beds may be used in parallel such that one bed is regenerated while the other is in service. The beds are regenerated by heating and desorbing the hydrocarbons. The heavy hydrocarbons are recovered from the regeneration gas via condensation.

The use of silica gel to recover hydrocarbons from natural gas is declining in favor of continuous processes such as cryogenic condensation and absorption (Kohl, p.1086).

Recently, Mitariten and Lind (2007) proposed using an integrated silica gel system for simultaneous removal of water and hydrocarbons from the feed gas by treating the streams before and after the acid gas removal unit (AGRU). The three steps of hydrocarbon removal, regeneration, and cooling will include an additional step of removing water from the AGRU exit stream. Only one vessel needs to be added for the additional step instead of an additional 3-step unit. This integrated system should reduce costs associated with foaming in the AGRU and treating BTEX emissions from glycol regeneration. However, this recent proposed integrated operation has not been tested.

5.0 References

1. Holub, P.E., Sheilan, M., "Fundamentals of Gas Sweetening," Proceedings of Laurance Reid Gas Conditioning Conference, Feb 27–Mar 1 2000, Norman, OK, USA.
2. Klinkenbijl, J.M., Dillon, M.L., Heyman, E.C., "Gas Pre-Treatment and their Impact on Liquefaction Processes," GPA Nashville TE Meeting March 2, 1999.
3. Kohl, A.L., Nielsen, R.B., *Gas Purification (5th Edition)*, Ch 2, Elsevier 1997.

4. Habibullah, A., "Alaska North Slope LNG project considers various CO2 removal processes," O&G Journal, **100**(22), p.46 (June 2002).

5. Stone, J.B., Jones, G.N., Denton, R.D., "Selection of an Acid-Gas Removal Process for an LNG Plant," 75th Annual GPA Convention March 11-13, 1996, Denver, CO, USA.

6. Baker, R.W., "Future Directions of Membrane Gas Separation Technology," *Ind. Eng. Chem. Res.* **2002,** *41,* 1393-1411.

7. Bikson, B., Giglia, S., and Hao, J., "Novel Composite Membranes and Process for Natural Gas Upgrading," Annual Progress Report-2002, DOE Award Number: DE-FC26-99FT40497. (pp12-15 economic evaluation of TEG v membrane)

8. Blachman, M., McHugh, T., Sour Gas Dehydration Technology and Alternatives, Paper presented at the LRGC February 27 – March 1, 2000 Norman, Oklahoma.

9. Ebeling, H.O., Lyddon, L.G., Convington, K.K., "Reduce Emissions and Operating Costs with Appropriate Glycol Selection," Proceedings of the Seventy-Seventh GPA Annual Convention. Tulsa, OK: Gas Processors Association, 1998.

10. Gas Processors Suppliers Association, *Engineering Data Book,* 12th Edition (2004).

11. Genesis Oil and Gas Consultants, "*Twister NGL Recovery Study*," December 2007.

12. Mehra, Y.R., Gaskin, T.K, "Guidelines Offered for Choosing Cryogenics or Absorption for Gas Processing", Oil & Gas Journal, March 1, 1999, Vol. 97, No. 99, p.62.

13. Membrane Technology and Research, Inc., "Field Demonstration of a Membrane Process to Recover Heavy Hydrocarbons and to Remove Water from Natural Gas," Final Report Prepared for the United States Department of Energy National Energy Technology Laboratory, August 30, 2007, DOE Award No.: DE-FC26-99FT40723.

14. Mitariten, M., Lind, W., "The Sorbead™ Quick-Cycle Process For Simultaneous Removal of Water, Heavy Hydrocarbons and Mercaptans from Natural Gas," Presented at the Laurence Reid Gas Conditioning Conference February 25-27, 2007.

15. Schinkelshoek, P., Epsom, H.D., "Supersonic Gas Conditioning-Commercialisation of Twister Technology," GPA 87th Annual convention, Grapevine, Texas, USA, March 2-5, 2008.

16. Twister BV, *Products & Services.* 22 January 2009 < http://twisterbv.com/>

BIOGRAPHY

J.C. Kuo is a veteran of the Gas Processing/LNG industry, with 35 years experience in the field. He is senior advisor for Chevron.LNG Major Capital Projects. He has served as the Process Manager /Lead for many projects, including the Wheatstone LNG, Gorgon LNG, Delta Caribe LNG, Casotte Landing and Sabine Pass LNG terminal projects. He also served as the Technical Reviewer for Angola, Olokola, Algeria and Shtokman LNG projects. Before working with Chevron, he was the Technology Manager for IPSI/Bechtel and served as the Process Manager for the Pemex Catarell offshore project, the Egyptian LNG (Idku) train 1/2 and China Shell Nan Hai Projects. He has contributed to LNG and gas processing technology improvements through a book, patent and many papers on the subject. He has a BS from Chung Yuan Christian University, Taiwan and a MS from the University of Houston, both for Chemical Engineering. He also holds a MS in Environmental Engineering from Southern Illinois University. He is a registered Professional Engineer in Texas and a member of AIChE. He is the president of the 99 Power Qi Qong Texas divisions.

Defining Community and Economic Benefits Associated with Large Energy Infrastructure Projects: LNG Case Study and Project Implementation Issues[1]

31a

Ruzanna Makaryan, Sr. Energy Analyst
Dmitry Volkov, Energy Analyst
Michelle Foss, Chief Energy Economist and Head

Center for Energy Economics
Bureau of Economic Geology
The University of Texas at Austin
Houston, Texas

AIChE Spring Meeting, April 2009
9th Topical Conference on Natural Gas Utilization
Tampa, Florida, April 26-30, 2009

Introduction

Large energy infrastructure projects—indeed, large infrastructure projects of any sort—encompass substantial risks and uncertainties. They range from initial feasibility to final investment decision, from the engineering, procurement and construction process through the operation and eventual decommission of the facility. A significant risk, and a specific uncertainty to be addressed, is the response of local communities and populations to project proposals and filings and the myriad tradeoffs among costs and benefits associated with project development.

Uncovering and reaching consensus on the net benefits associated with large energy infrastructure projects is a complex task that involves a diversity of issues (dimensions) and stakeholders. Stakeholder perceptions of value associated with infrastructure projects vary with each issue and over time.

The CEE-UT research team undertook a study effort to develop a methodology and approach that could address these complexities and help improve identification of benefits and costs as well as associated issues. Our study was centered on liquefied natural gas (LNG) import receiving and regasification terminals and associated infrastructure in North America. We also evaluated projects other than LNG import terminals to assess applicability of our approach and findings to other types of energy infrastructure. We also considered applicability of our approach and findings to energy projects being undertaken outside of North America.

We undertook this research effort by devising a study protocol for data collection from a sample of projects using the US federal regulatory process as our framework (US

---

[1] This paper is based on CEE's 2008 special project: *Defining Community and Economic Benefits Associated with Large Energy Infrastructure Projects: LNG Case Study*. The effort was undertaken within the scope of CEE's public research and education consortium on LNG, which has produced a number of papers comprising the *Guide to LNG in North America. The Guide* provides extensive information to those interested in U.S. energy trends and security; LNG industry and market developments. The CEE web site, www.beg.utexas.edu/energyecon/lng provides links to industry, government and public information sources.

Federal Energy Regulatory Commission, FERC, for onshore licenses and US Maritime Administration/US Coast Guard (USCG), MARAD/USCG, for offshore licenses). We disaggregated stakeholders into distinct groupings based on proximity to, and likelihood of being affected by, LNG import terminal facilities and examined stakeholder positions across key issue dimensions identified from the regulatory process and based on pre-study surveys and analysis, including input from direct observation and outside sources. Findings and conclusions were derived from comparative analysis across the sample of projects and a set of summary hypotheses that were tested once data collection was complete.

Methodology/Study Protocol

In this study, 20 LNG projects have been analyzed (see Table 1). These included import receiving and regasification terminals and associated infrastructure in five geographic regions of North America. Stakeholders in each project were grouped using the following groupings: Immediate Site-Host Community; Waterway Community; Local Community (city/county); Greater Community (state/city or province/city); National Community (federal); External Stakeholders.

Ten issue dimensions were defined: safety/security, wetlands, fisheries, energy costs, roads, taxes, employment, air emissions, property value and other/intangibles. The regulatory process was used as a study timeline. Information from dockets and other public domain sources was collected to build data templates. Summary diagrams were drawn from data aggregated by geographic region, and hypotheses developed and tested. We also evaluated projects other than LNG import terminals to assess applicability of our approach and findings to other types of energy infrastructure. And we considered applicability of our approach and findings to energy projects being undertaken outside of North America.

Summary Results

Across all projects, on average, energy costs emerged as one of the main perceived benefits for the greater community or state stakeholder group. Local stakeholder groups tend to perceive benefits associated with taxes and employment. Also on average for all projects, the strongest tendencies to perceive costs are for the fisheries and safety/security dimensions, with a weak tendency to perceive potential costs with respect to property value. When perceived benefits are isolated, the indication is that for one or more projects energy costs are attractive to multiple stakeholder groups, including the local community (city/county) as well as the greater community (state/city or province/city), national groups and external interest groups. Taxes represented a strong convergence of perceived benefits among the waterway, local, greater, and national groups. Air emissions represented the third strong convergence, with local (in particular), waterway, greater and national groups perceiving benefits. Of note is that for all projects in all locations the immediate site communities mainly are indifferent.

Table 1. Profiled LNG Import and Regasification Projects

Region	Project	Location	Type	License
Pacific Northwest	Jordan Cove Energy	Coos Bay, OR	Onshore	FERC
	Kitimat LNG	Kitimat, BC, Canada	Onshore	BCEAO/EC
California	Cabrillo Port	Southern CA (Ventura-Los Angeles counties)	Offshore	MARAD/USCG
	Clearwater Port	Southern CA (Ventura County)	Offshore	MARAD/USCG
	Esperanza Energy	Southern CA (Los Angeles-Orange counties)	Offshore	MARAD/USCG
	Sound Energy Solutions	Long Beach, CA	Onshore	FERC
Central/Western Gulf Coast	Sabine Pass	Sabine, LA	Onshore	FERC
	Cameron LNG (Hackberry)	Hackberry, LA	Onshore	FERC
	Main Pass Energy	Offshore LA (Plaquemines-St. Bernard parishes)	Offshore	MARAD/USCG
	Casotte Landing LNG	Pascagoula, MS	Onshore	FERC
	Compass Port	Offshore AL (Mobile-Baldwin counties)	Offshore	MARAD/USCG
Florida	Port Dolphin LNG	Western FL (Manatee County)	Offshore	MARAD/USCG
	Ocean Express*	Southern FL (Broward County pipeline landing)	Offshore	FERC
	Calypso LNG	Southern FL (Broward County pipeline landing)	Offshore	MARAD/USCG
Northeast	Crown Landing	Logan Township, NJ	Onshore	FERC
	Broadwater Energy	Long Island Sound, NY	Offshore	FERC
	Northeast Gateway	Boston, MA	Offshore	MARAD/USCG
	Weaver's Cove Energy	Fall River, MA	Onshore	FERC
	Canaport	St. John, NB, Canada	Onshore	NBELG/EC
	Project Rabaska	Ville Guay/Beaumont, Québec, Canada	Onshore	BAPE/EC

Notes: *FERC is US Federal Energy Regulatory Commission; MARAD/USCG is US Maritime Administration and US Coast Guard; BCEAO is British Columbia Environment Assessment Office; EC is Environment Canada; NBELG is New Brunswick Environment and Local Government; BAPE is Bureau d'audiences publiques sur l'environnement. * Indicates US pipeline only.*

When perceived costs are isolated, fisheries, safety and security, property value and air emissions stand out as the primary concerns. Wetlands are a strong, and overlapping, dimension. Our results illustrate the complexity of stakeholder management. For example, local community groups everywhere do not view air emissions in the same way. And where immediate site communities may be indifferent with regard to benefits, they show strong tendencies to perceive costs associated with safety and security, fisheries and property value.

In the process of reviewing the projects, we found sufficient regional differentiation across all projects in our sample to enable broad aggregation of data into our five geographic areas (Pacific Northwest, California, Central/Western Gulf Coast, Florida, and Northeast) with synthesis and summary conclusions on that basis. Summary conclusions within each region are provided in the following chapters.

Pacific Northwest Projects

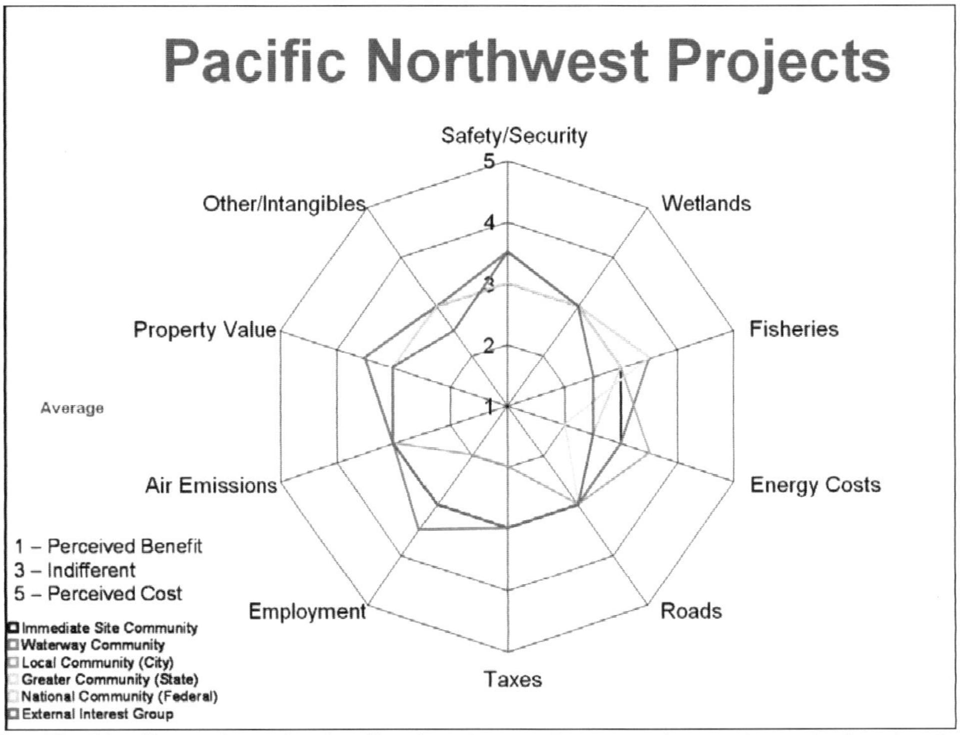

For the Pacific Northwest projects, the tendency to perceive net costs is charted across five dimensions (safety/security, fisheries, energy costs, employment and property value). Of interest is receptiveness of the local communities (cities) to jobs and taxes, the greater community (state) to energy costs and the tendency toward net positive positions on fisheries by external interest groups (most likely related to displacement of dams and restoration of salmon runs). Overall, the Pacific Northwest projects appear to trigger stronger negative reactions, but care must be taken as the stakeholder characterization scores are averaged across a smaller number of projects than for other regions. That said, much more variability across stakeholder groups is registered than for Florida, which also has a small number of projects represented in the sample.

Safety/security

- Immediate site community, waterway community and local community have expressed concerns about safety and security for the projects proposed in the region. Some of these related to increased LNG tanker traffic. Other concerns were related to emergency responses and hazard mitigation capabilities (e.g. firefighting). External interest groups that opposed LNG projects in other locations had interactions with local groups in the region in relation to safety and security concerns.

Fisheries

- Both of studied projects in the Pacific Northwest had indications of major concerns on the effect of the increased tanker traffic and required dredging on the impacts on marine habitats, commercial and recreational fishing. There were also concerns on perceived impact on the access to the marine environment. However, in the case of the Port of Coos Bay, the Port will actually receive benefits from a full utilization due to tanker traffic, which will help stabilize port operations and bring additional revenue.

Energy costs

- In projects proposed in the Pacific Northwest, natural gas, from the imported LNG, is perceived as going to other states, such as California. Local and state stakeholders perceive limited benefits, which is becoming a more important issue and requires the identification of community benefits that target these stakeholders. As more projects are proposed in the Pacific Northwest, the perception of Oregon being a transit State is leading stakeholder groups to ask for a regional planning effort in LNG regasification siting. In the case of projects in Canada, the perceived benefits outweigh the transit nature of the natural gas.

Taxes

- Local community stakeholders perceived a tax benefit from the projects as communities would receive tax revenues from the LNG project

Employment

- Positive impact of additional employment has been identified by the local and greater (state) communities. However, concerns were also raised by the waterway community that there may be an adverse effect on employment in commercial fishing due to increase in tanker traffic.

Property value

- There are concerns about property values along the waterway from immediate site community as result of increased activities in the waterway.

Other/intangibles

- Among other issues for this region was the fact that industrialization is not necessarily seen in a positive light. In some cases in the Pacific Northwest, continued use of brownfields leads some stakeholders to raise issues of environmental justice. Despite being in an industrial setting, waterway and local community has expressed concern with LNG tanker traffic and LNG storage.

- We have observed that projects that have been proposed for locations in the Pacific Northwest that are extremely low in population density have shown fewer tendencies towards objections.
- Stakeholder groups that did not perceive benefits from the project and had the opportunity to exert pressure or create an obstacle for the project did so. Hypothesis not observed in Pacific Northwest cases studied.

No concerns have been expressed in relation to the roads, wetlands and air emissions dimensions.

California Projects

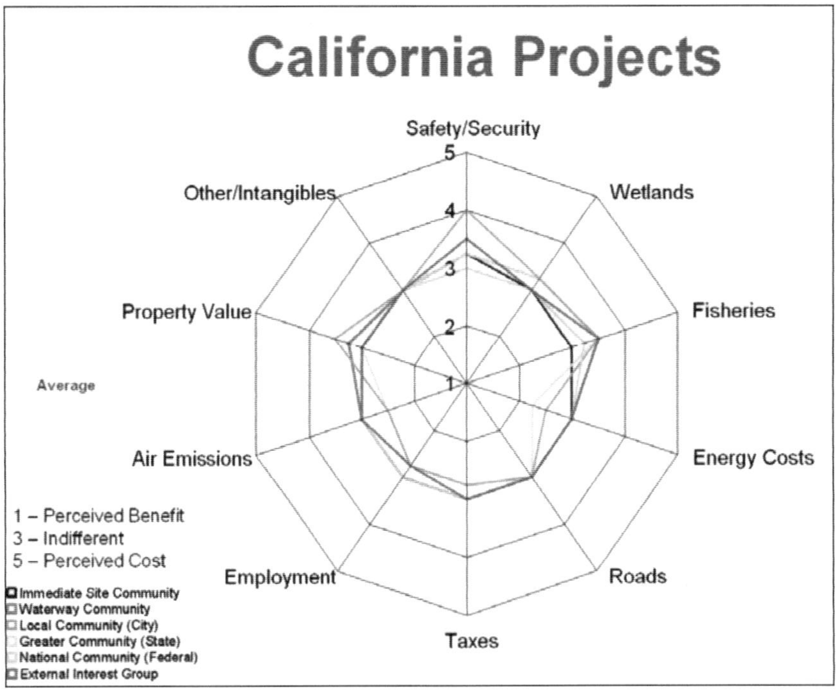

In California, initial support for new LNG infrastructure has given way to diverging views, one that continues to support (albeit with considerable constraints) LNG as a source of energy for the State and another that rejects projects outright. Projects face increasing hurdles as new issues are incorporated into the public debate (safety, NOx emissions, CO_2 emissions, pipeline safety and environmental impact).

Despite being in an industrial setting, waterway and local community has expressed concern with LNG tanker traffic and LNG storage. Increased LNG tanker activity in ports and waterways of California has not been accepted by some stakeholders despite already high tanker traffic. Projects are moving towards addressing vaporization emissions and pipeline footprint. Opposition to LNG projects in California from local communities and state agencies have resulted in rejection or denial of land or coastal lease agreements. However, local communities removed from the host community (in the case of California's local schools in the Port of Long Beach) tend to support (or at least not reject) the projects and do not show general tendencies towards a perception of negative benefit.

Safety/security

- Safety and security concerns are the most outstanding for offshore facilities with storage proposed in California. Safety and security concerns have been expressed by most levels of stakeholders (from immediate site community to national community). These issues have been more difficult to address. Fears include potential terrorist attacks, effect of the facility on the schools and roads, safety of the beaches, and others.

Wetlands

- Construction of pipelines that would deliver vaporized LNG to shores raised concerns about its impacts on the wetlands for most of the proposed projects in California. In the case of the Port Esperanza project, this issue was addressed by proposing an underground “utility conduit” that would avoid coastal impacts.

Fisheries

- Local communities, greater state communities were all concerned about the potential effect of tanker traffic and other activities on the marine life, although projects were proposed in areas with already high activity or in a very close proximity to the existing ports. This concern in Cabrillo project was addressed with a location already in the heavily transited shipping lanes and the lack of a permanently moored ship.

Energy costs

- In California, benefits in energy costs from natural gas imports in the form of LNG are not perceived by many stakeholders leading stakeholders to oppose the project since they only perceive negative impacts.

Taxes and Employment

- These two dimensions have been accepted by the local and state communities as major potential benefits. For example, impact from the Sound Energy Solutions proposal is estimated at 1000 of union jobs in construction with an economic impact of $401 million.

Air emissions

- Perceptions surrounding air emissions and their federal permits generated elements that were used to create opposition to LNG projects developed in California. Implementation of state of the art NOx reduction technologies in California has not proven to be enough for local and state level stakeholders. Project proponents have incorporated technical solutions into their proposals that address the main observations from opposition groups: no new emissions, no on-site storage, and no surface shore crossing.

Property value

- Waterway and local communities expressed concern on property values due to safety issues and visible impact from the presence of the floating storage and regasification unit.

Central/Western Gulf Coast Projects

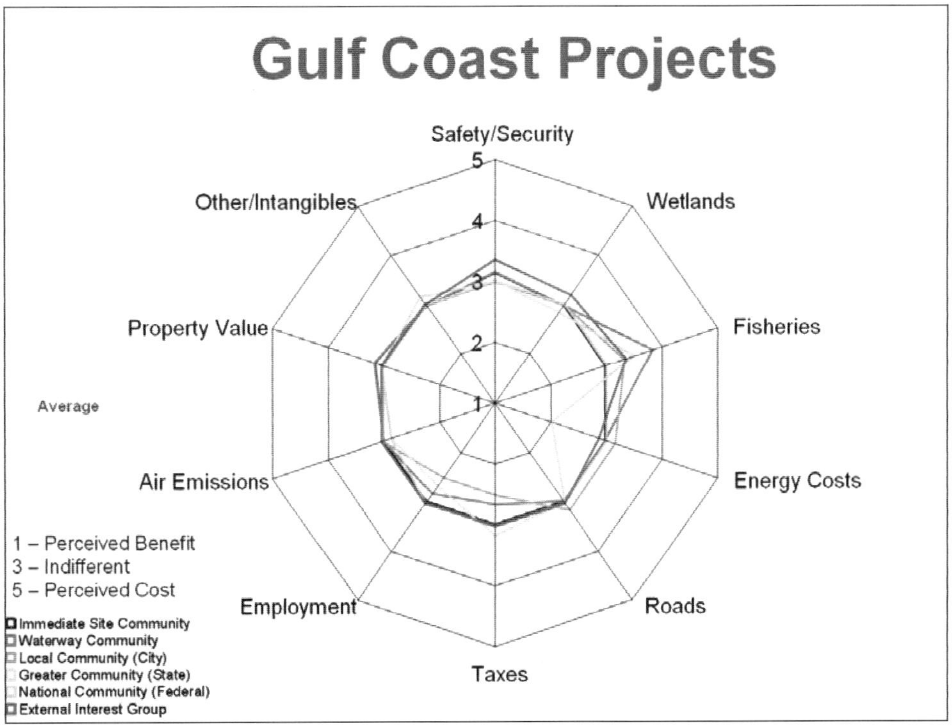

Environmental impacts from LNG terminal operations are not completely understood by some stakeholder groups in the Gulf Coast leading to higher levels of concern than would be expected given similar infrastructure already in use in power plants. In spite of having expressed environmental concerns related to fisheries, the Gulf Coast region maintains interest in keeping construction jobs and tax base by host community.

Fisheries

- Gulf Coast proposals that included open loop vaporizers were perceived as generating a perceived possible loss of value associated with activities associated with fisheries which led to rejection by State stakeholders (guarantee of Veto). The concerns were raised late in the process and studies that covered the related environmental issues have not changed the joint position of Gulf Coast Governors in supporting vetoes to projects that considered open loop vaporizers. The issue has been addressed by dropping open loop vaporizers by most project proponents.

Energy costs

- Local benefits in most cases of the Gulf Coast projects were clearly identified. In cases that they were not (e.g. Alabama) the regasified LNG was perceived to benefit other States and to only transit through the State. This posturing led to the offer of a purchase option for a portion of the delivered natural gas.

Employment, roads, taxes

- Local and greater (state) communities are positive about increased revenue based for local taxes and support the projects early due to increases in jobs and tax revenues.

In the Gulf Coast region, no significant concerns have been raised on the rest of the dimensions (*safety/security, wetlands, roads, air emissions, property value, and other/intangibles*). Other minor concerns included bird migration sites, dredging impacts on wetlands. In the case of Casotte Landing LNG, thermal integration was added as an additional energy cost benefit. The process would allow to use the refinery heat to heat LNG vaporizers, bringing the savings over a billion dollars over the life of the projects in energy costs.

Florida Projects

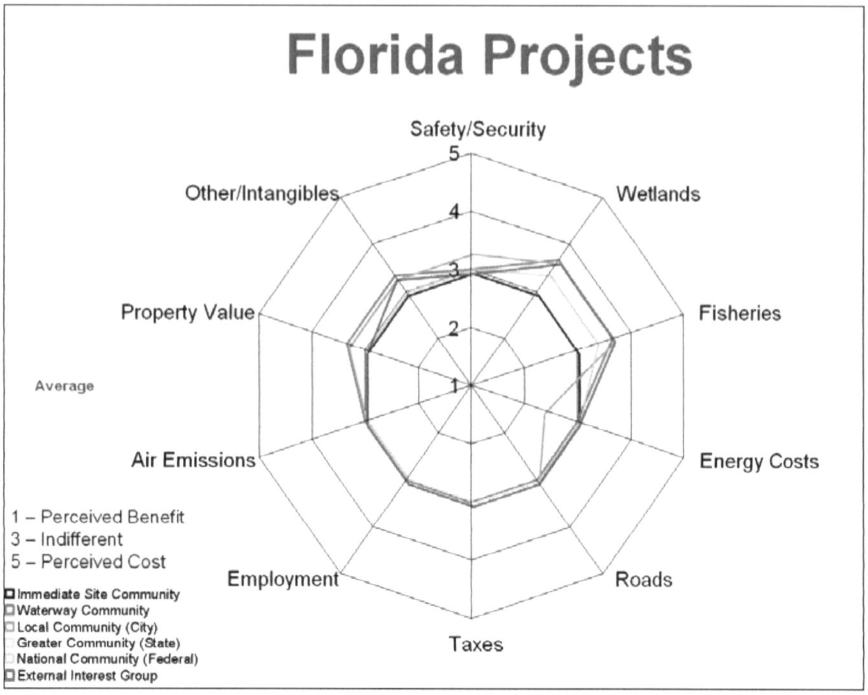

Florida's local and stakeholder opposition to the new energy infrastructure projects is mostly based on perceived environmental risks and safety concerns that have led to roadblocks with local permitting despite considerable lobbying efforts at the federal and state level. In Florida, fear of environmental impact creates a barrier to entry to infrastructure projects despite projections for increased consumption of natural gas in the region.

Fisheries and wetlands

- Effect of the proposed projects on fisheries and wetlands issues in Florida has been the major issue for the local stakeholders. Opposition to projects due to environmental concerns and previous environmental impacts has led to project designs that minimize the projects environmental footprint (pipelines). Pipelines still have to meet special environmental requirements due to fragile ecosystem. Implementation of novel technologies for construction of pipelines (horizontal direction drilling) in areas of environmental interest increased concerns instead of providing stakeholder acceptable mitigation. Offshore construction technology and environmental impacts remain an open issue. Local communities may still prevent projects from moving forward by denying local permits for pipelines.

Energy costs

- Florida's local stakeholders do not explicitly express a perceived benefit from the availability of natural gas to other more removed stakeholders (main customers are perceived to be other parts of Florida) and not directly the interested waterway communities.

Other/intangibles

- Concerns regarding the possible impacts on Florida's tourism from the perceived negative environmental impacts from pipeline construction were originally addressed but not overcome leading to eventual revoking of pipeline rights of way and license for one project.

Other dimensions (roads, taxes, employment, air emissions, and property value) have not been as outstanding as the three above. Minor concerns included impact of onshore activities on the existing transportation facilities, such as for example, state highways and an international airport runway. Another concern was that port right-of-way capacity is already at its maximum and would not allow other pipelines to be added.

Northeast Projects

Despite familiarity with an LNG import facility in the Northeast, and abundant LNG storage throughout the region, LNG terminals still generate concerns in many dimensions. Negative perceptions have been greatest in Northeast projects that have considerable waterway communities that also act as host and immediate site communities. Offshore projects in federal waters remove the projects from host communities though may create concerns for some members of the waterway community (fishermen). Regional perceived benefits have not outweighed perceived negative benefits by local and waterway communities in the Northeast.

Safety/security

- Stakeholders on all levels have expressed great concerns about safety and security for most of the projects proposed in the Northeast. Marine safety was given as one of the reasons. Communities expressed the need for an exclusions zone and other measures. Increased awareness and concerns have generated efforts for a regional approach to siting (MA). As a mitigation approach for one of the project, the project proponent has committed to use US crews on the tankers that would help reduce fears of potential safety/security threats.

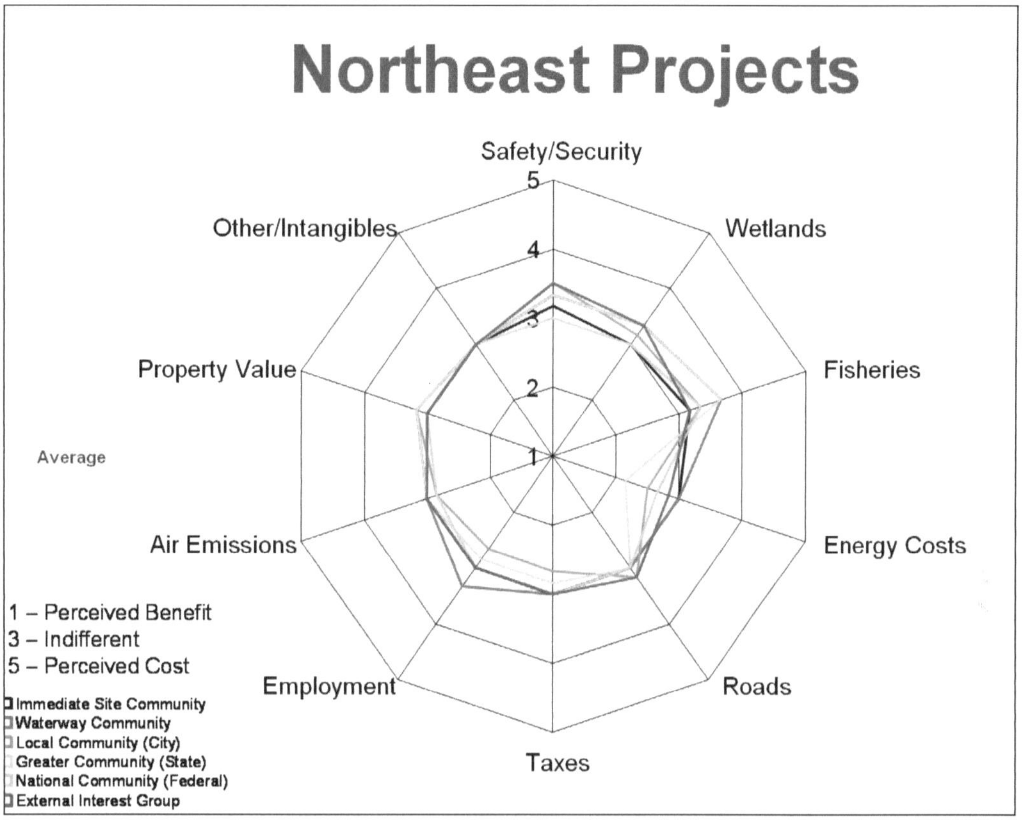

Wetlands and Fisheries

- There is a perceived loss of value by local stakeholder groups (fishermen, lobstermen).
- Dredging that would be required in the construction of the terminals has generated environmental concerns regarding disposal of the dredged material and the overall impact on the wetlands and fish habitat.
- There are also concerns for large LNG tanker traffic interfering with smaller craft traffic of fishermen.
- Mitigation approaches for environmental impacts of the activities include grants for gathering and monitoring of the coastal and seafloor environment.

Energy costs

- There is a clear recognition of benefit on energy costs and fuel choice by local communities on all levels (local through state) and external interest groups.

Roads

- For those projects where roads and bridges are in close proximity to the proposed sites, this has been a great concern for the local and waterway communities as the LNG tankers would be in transit and represents a potential safety issue. Also of concern are disruptions for bridge operations and road congestions. One project has proposed a grant for improvements of public transportation and public access ramps for boats.

Taxes and Employment

- Local communities and other stakeholders clearly identify a potential positive impact on job creation and tax revenue for all projects in the region.

Air emissions

- Although in one project, the state commented that the project did not meet the state's ozone standards, other stakeholders recognize that the greater use of natural gas over other fossil fuels will bring environmental economic benefits based on fewer pollutant emissions.

Property value

- Local community perceived a loss of value (property value, "quality of life") due to the presence of a regasification facility (an "industrial site") which was not compensated through other means led to the rejection of a land lease agreement in the Northeast project. In another project, local and waterway stakeholder groups have sought means to influence the regulatory process by providing obstacles to the project (e.g. a bridge that blocks large tanker traffic).

Other/intangibles

- Overall, in the Northeast region, there are considerable concerns from site and waterway communities regarding safety and impacts on livelihood and lifestyle. Only solution has been to go offshore, address specific concerns early and provide early compensation to "affected" industries and reduce safety concerns by committing to all US crews.

Observations Applying to All Regions

- The study found that for the projects studied, stakeholder groups that were indifferent on some dimensions and perceived costs on the rest (no tradeoffs) oppose the project. The stakeholder group that did not perceive a benefit from the project was usually either the waterway or the immediate-site community. This was observed in all regions.
- We observed that when safety and security concerns are present, stakeholders tend to oppose the project implying that stakeholders place a very large valuation on this dimension. When other tradeoffs are present, stakeholder response varies. This was observed for all regions.
- Concerns regarding safety and security need to be addressed quickly and comprehensively since their relative valuation outweigh other issues.
- National benefits were not shown to be valued by local and state stakeholders. In addition, if local and state stakeholder did not perceive additional benefits, the issue of environmental and economic justice was raised (e.g. letter from Louisiana Governor Blanco to MARAD vetoing Main Pass Energy Hub).

Project Management Implications – Working with the Stakeholders

Successful large infrastructure projects can be the engine of economic development in the immediate project community and outside of it. In addition to benefits for the project developers, these have impact on education systems, job security, increased revenue flows and numerous other benefits to the communities. However, to get the project to start providing these benefits, it requires success in the entire chain of steps taken to its completion. Financial aspects, skilled manpower, transparent working procedures, complex technology, familiarity with the approval process are important components. However, technical reasonableness only is not sufficient. The project has to be perceived reasonable by its stakeholders, including local citizens and communities, regional groups and others. Classical approach dominated in the industry in the 1980's and 1990's and implied focus on the technical aspects of the infrastructure projects.[2] The industry was operating under false premise that "societal groups" did not have power and impact on the fate of the projects. Negligence of the public opinion was causing conflicts between the stakeholders of the project. The industry slowly came to realization of the fact that project planning must include a rigorous social "due diligence" process. It would typically include analysis of the major interest groups, "development of external communication procedures, negotiations for finding support, the anticipation on information needs".[3] It is also important to look into future possible various options for different stakeholders in terms of legal/illegal means to resist a project and plan negotiation strategies in advance. This new approach could be categorized as and interdisciplinary one where engineers, public relation professionals and social scientists work together on the planning problems.

There are a number of examples in the non-LNG industry of the projects had issues with planning and implementation. In 1999, there were more than 1,500 large projects being developed across the world (Miller and Lessard 2000).[4] Many found cost overrun of over 40%.The Suez Canal construction cost 1,900% more to build than planners originally estimated. The Sydney Opera House cost 1,400% more.[5] The cost of the construction of the Channel Tunnel (Chunnel) exceeded estimates by 80%. Flyvbjerg's findings from a study on megaprojects in 20 countries around the world between 1927 and 1998 are that 90% of the projects had cost overruns, consistently across all 20 countries and constant, not improving in the 70-year period. [6]

These cost overruns are attributed to many factors, including additional safety considerations, environmental concerts, etc. Yet, in many projects, delays happen

---

[2] Rob E. C. M. vander Heijden, "Planning large infrastructure projects: seeking a new balance between engineering and societal support", Dokumente und Informationen zur Schweizerischen Orts-, Regional- und Landesplanung (DISP), jrg. 32 (1996), pp. 18-25

[3] Rob E. C. M. vander Heijden, "Planning large infrastructure projects: seeking a new balance between engineering and societal support", Dokumente und Informationen zur Schweizerischen Orts-, Regional- und Landesplanung (DISP), jrg. 32 (1996), p. 19

[4] The Management of Large Engineering (Physical Infrastructure) Projects: Debating a Research Agenda, Nuno Gil, AoM 2007 Philadelphia, http://www.personal.mbs.ac.uk/ngil/documents/IntroPDWPhiladelphia.ppt

[5] Megaprojects and Megamistakes, by Cindy Wagner http://www.wfs.org/trend3nd03.htm

[6] Megaproject Policy and Planning: Problems, Causes, Cures, Bent Flyvbjerg, 2007 http://www.plan.aau.dk/GetAsset.action?contentId=3592306&assetId=3614859

because of rejection of a project by local communities. By far, the key to the local acceptance of any project (regardless of the scope and size) is the communication with the project stakeholders, timely and efficient, as discussed below. This applies not just to LNG project or energy projects in general, but is true for many infrastructure projects such as highways, dams, airports, bridges, and many more. As we look back at success stories and failures, we can draw many lessons for all involved in the projects.

Planning and Organization

The first step of any project is planning. In addition to typical financial and technical (engineering, design) considerations, it includes initiation of the project; sets out the project; and plans what, when and how the project will be implemented. Planning addresses resource allocation, personnel, milestones and timing of the project. A carefully planned project has clear, tangible objectives that are targeted to be reached at a certain point. Project plans would also outline the steps in the regulatory process. Considering potential impact, a sound planning process would also include evaluation of local concerned groups and other opposition, and would outline approaches which project developers would use to mitigate and address the issues raised by the opposition. Anecdotal evidence suggests that projects where authorities commit to a certain option or interest of a specific stakeholder group in early stages of decision-making process, often fail. Public authorities have to have an independent comprehensive view on a particular project and try to be as objective and as long as it is possible.

Inclusiveness and Involvement

Participation of stakeholders is one of the keys in identifying risks and uncertainty of projects. "The Facility Siting Credo" calls to "institute a broad based participatory process"[7] where representatives of all groups affected by the siting of LULUs ("locally unwanted land uses" such as waste storage nuclear facilities, for example) are invited to partake in the siting process through interviews and surveys of stakeholders, or through advisory committees. Early involvement of stakeholders is beneficial to both local communities and the project developers. For project developers, it would give opportunities to communicate about issues, explain costs and benefits of the project to the community and would prevent negative, quick conclusions about the project without a deep familiarity with these issues. For stakeholders, an advisory group would allow the community to voice their concerns in an organized manner. Communities will feel represented and understand that their concerns are taken into account.

There are, however, drawbacks to having a citizens' advisory group. As outlined by P. T. Allen[8], these include the following:

---

[7] Facilities Siting Credo: guidelines for successful facilities siting, Article by Howard Kunreuther and Lawrence E. Susskind in *Environmental Impact Assessment Review*. Publication Services, University of Pennsylvania, 1991. http://web.mit.edu/publicdisputes/practice/credo.pdf

[8]Public Participation in Resolving Environmental Disputes and the Problem of Representativeness Peter T. Allen, Risk: Health, Safety & Environment 297 [Fall 1998]

- Legitimacy of non-expert input, where citizen panels cannot provide value to the planning process due to the lack of technical expertise.
- Lack of accountability of such citizen groups, which led to the fact that decisions promoted were not necessarily feasible technically or financially.
- There are also issue of legitimacy of group and representativeness. How close to the community such advisory groups stay without losing touch with their constituencies and whether representative elected to these groups represent the education level and income similar to those of the community.

Regardless of these drawbacks outlined above, recognition, understanding and involvement of all stakeholders are necessary.

Timing

Timely involvement of stakeholders in the process can bring a benefit of eliminating negative reactions to the project Stakeholders need to be engaged in the process as early as possible. If their concerns are addressed and issues are mitigated at the early stages of the project development, there are more chances of getting stronger support of the project by these communities. Even after the approval for the project siting has been received and construction has started, there needs to be a mechanism for continuous engagement of the communities through the project life.

Public resistance to a project development could significantly lengthen its implementation, since it generally increases time needed for clearing necessary regulatory approvals among other things. A recent study[9] that analyzed 258 projects around the world concluded that longer sluggishness in project implementation causes significant cost overruns, as also stated above. For every year of project delay construction cost overruns of an average 4.64% were expected. Additionally, there are financing costs which are also usually very sensitive to project delays.

Communication

This is probably the most important aspect of stakeholder management in any project. Transparent, open, complex and extensive communication needs to happen among various parties involved in the project, project developer, stakeholders, authorities, and media.

The first step, for example, taken by the World Bank team in the Bumbuna Hydroelectric Project[10] in Sierra Leone was to use "communication-based assessment" (CBA) to identify political, social, and cultural environment of the project. CBA was also used to assess the position of stakeholders including their level of information, attitudes, perceptions and concerns, practices, behaviors and interests. The tasks of the CBA included:

---

[9] Bent Flyvbjerg, Mette K. Skamris Holm and Soren L. Buhl, "What causes cost overrun in transport infrastructure projects?", Transport Reviews, Vol. 24, No. 1, 3-18, January 2004

[10] The Role of Communication in Large Infrastructure, The Bumbuna Hydroelectric Project in Post-Conflict Sierra Leone, *Leonardo Mazzei, Gianmarco Scuppa, World Bank Working Paper No. 84*

1. Understanding the history of the project.
2. Evaluating the political, social and cultural environment around the project.
3. Identifying stakeholders and assessing their level of information, perceptions, interests and concerns.
4. Identifying communication problems to be addressed and related objectives.
5. Assessing the government counterpart communication capacity.
6. Analyzing the available media (mass and proximity media) and communication professionals.
7. Designing a communication strategy and creating the conditions for its successful implementation.

It is crucial that the party, leading communication process with the stakeholder had sufficient experience in such line of work and negotiation expertise. Quite often both project developers and authorities use a top-down approach in planning and implementation, which result in emergence of public resistance to the project and formation of opposition alliances of various stakeholder groups (environmentalists, local authorities, citizen groups and various non-governmental organizations). Understanding of key interests of these groups is an important step in planning of communication campaign.

Information dissemination to all stakeholder groups at all stages of project planning and implementation is essential to its success. It is also desirable to try to overcome a popular tendency to overstate potential benefits of the project, especially public benefits as early in the project planning process as possible.

A World Bank 2004 study, conducted jointly with Hydro Quebec analyzed 16 large hydropower projects. They found that the "main factors were directly or indirectly related to communication. [11] Nearly all participants in this study brought up early involvement of local communities and potentially affected populations in the project planning as key to its successful implementation: *Even if the public participation process may add time and costs to a project, this is nothing compared to the cost of canceling a project or of struggling on for several years . . .The balance of past experience is clearly in favor of finding ways to associate the public in project planning and design, especially with respect to the planning and design of project related environmental and social mitigation and compensation requirements."*

Conclusion

Dealing with stakeholder issues is necessity. This may not always get listed among typical risks for infrastructure projects; yet it is a very practical aspect of any project development, and especially large infrastructure projects with potential risks and societal changes involved. Stakeholders expect transparency, accountability, participation and an understanding of their needs in the process of developing large

[11] The Role of Communication in Large Infrastructure, The Bumbuna Hydroelectric Project in Post-Conflict Sierra Leone, *Leonardo Mazzei, Gianmarco Scuppa, World Bank Working Paper No. 84*

infrastructure project. These are necessary for the stakeholders to feel ownership in the project and accept the project in their community.

In LNG siting process, early identification of key stakeholder groups that influence public acceptance is required. Concerns regarding safety and security need to be addressed quickly and comprehensively since their relative valuation outweigh other issues. Also, to avoid the impact of external influences, immediate and local communities need to perceive definitive benefits from the project. It is up to the project developer to timely communicate such benefits in a transparent matter, along with clear identification of risks and/or perceived loss of value. While there are commonalities in stakeholder issues, each project and each community will require a unique approach. Thus, project siting and other project development strategies need to be adapted to different locations. Project developers need to observe how other project discussions are evolving since they may result in new issues surfacing in their own discussions. Fast response is usually required to avoid additional problems from the issues.

CEE will continue its research on issues related to the siting and implementation of large infrastructure development projects.

References

1. P. T. Allen, "Public Participation in Resolving Environmental Disputes and the Problem of Representativeness Risk: Health, Safety & Environment 297, Fall 1998.
2. Center for Energy Economics, "*Defining Community and Economic Benefits Associated with Large Energy Infrastructure Projects: LNG Case Study,"* 2007-2009, Publication pending.
3. B. Flyvbjerg, "Megaproject Policy and Planning: Problems, Causes, Cures," 2007
4. B. Flyvbjerg, M. K. Skamris Holm and S. L. Buhl, "What causes cost overrun in transport infrastructure projects?", Transport Reviews, Vol. 24, No. 1, 3-18, January 2004.
5. N. Gil, presentation "The Management of Large Engineering (Physical Infrastructure) Projects: Debating a Research Agenda," Philadelphia, AoM 2007.
6. R. E. C. M. vander Heijden, "Planning large infrastructure projects: seeking a new balance between engineering and societal support", Dokumente und Informationen zur Schweizerischen Orts-, Regional- und Landesplanung (DISP), 1996
7. H. Kunreuther and L. E. Susskind, "Facilities Siting Credo: guidelines for successful facilities siting," Environmental Impact Assessment Review. Publication Services, University of Pennsylvania, 1991.
8. L. Mazzei, G. Scuppa, "The Role of Communication in Large Infrastructure, The Bumbuna Hydroelectric Project in Post-Conflict Sierra Leone," World Bank Working Paper No. 84, 2006.
9. C. Wagner, "Megaprojects and Megamistakes," The Futurist Magazine, 2003

Biography of Speaker

Ruzanna Makaryan is a Senior Energy Analyst at the Center for Energy Economics. She has over 12 years of experience in energy research, international development assistance programs, management consulting, financial management, and project/ program management. She participates in CEE's research on oil and gas sector in CIS; research and public outreach for CEE's consortium on Commercial Frameworks for LNG in North America; and research on national oil companies. She also manages CEE's training programs and facilitates public information and outreach for CEE activities. She previously worked in Turkmenistan with PA Consulting Group, implementing USAID-sponsored technical assistance programs for the Turkmen government agencies in the oil and gas sector. Ms. Makaryan holds an MBA from the University of Houston's Bauer School of Business and a B.A. in Linguistics from the Turkmen State University. ruzanna.makaryan@beg.utexas.edu

EN 1473 and other EN LNG Standards – The European Norms and Standards for the LNG Industry

Eginhard Berger, Head of Division Marketing & Sales 31b
IMPaC Offshore Engineering, Hamburg, Germany

AIChE Spring Meeting, April 2009,
9th Topical Conference on Natural Gas Utilization
April 26-30, 2009 Tampa, Florida

Introduction

In 1992 LNG experts from 12 countries in Europe initiated the development of modern European Norms (EN) for the growing LNG industry. The target was to provide information on this exciting technology with temperatures as low as -160°C and give guidance to potential users for the safe handling of LNG. In the meantime these EN LNG standards have been extended into most of the LNG units of this industry covering the aspects of health, safety and environment.

The tenor of the EN standards becomes evident in one of the preambles: "The objective of the European Standard is to give functional guidelines for LNG installations. It recommends procedures and practices that will result in safe and environmentally acceptable design, construction and operation of LNG plants. It need not be applied retrospectively, but application is recommended, when major modifications of existing installations are being considered."

There is a principal difference to several other standards: the EN´s for LNG mostly do not prescribe physical dimensions or properties of LNG related equipment and systems. Mainly points of special concern, safety targets and risks are defined. How these targets are achieved and implemented is subject to the technical and scientific approach. This leaves sufficient freedom for the introduction of advanced and innovative solutions. And it is up to the operators and authorities to require the verification of the EN requirements for the proposed LNG installations.

Survey of the European LNG Standards

A summary of the existing EN standards is listed below:

- EN 1160, Installations and equipment for liquefied natural gas — General characteristics of liquefied natural gas
- EN 1473, Installation and equipment for liquefied natural gas - Design of onshore installations
- EN 1474, Installation and equipment for liquefied natural gas — Part 1: Design and testing of loading/unloading arms; Part 2: ... of LNG hoses; Part 3: ... offshore LNG transfer systems
- EN 1532, Installations and equipment for liquefied natural gas – Ship to shore interface
- EN 12065, Testing of foam concentrates designed for generation of medium & high expansion foam and of extinguishing powders used on LNG fire
- EN 12066, Testing of insulating linings for LNG impounding areas
- EN 12308, Suitability testing of gaskets designed for flanged joints used on LNG piping
- EN 12838, Suitability testing of LNG sampling systems

- EN 13645, Design of onshore installations with a storage capacity between 5 t and 200 t

Some features of these EN´s are highlighted below:

EN 1160, Installations and equipment for liquefied natural gas — General characteristics of liquefied natural gas

The general characteristics of LNG are described including the thermodynamic and other physical properties, evaporation, spillage, ignition and containment. In addition some particular phenomena of LNG are amplified like BLEVE (boiling liquid expanding vapor explosion) or roll-over in LNG tanks.

A comprehensive list of materials used in the LNG industry together with instructions for the handling of thermal stresses is included. Important topics are health and safety related issues including exposure to cold, exposure to gas, fire precautions and protection, odor and odorisation.

EN 1473, Installation and equipment for liquefied natural gas - Design of onshore installations

This is the most comprehensive EN standard to be used for large onshore LNG installations. Purpose and application of this EN standard is to give guidelines for the design, construction and operation of all onshore liquefied natural gas (LNG) installations including those for the liquefaction, storage, vaporisation, transfer and handling of LNG.
This European Standard is valid for the following plant types:

- LNG export installations, between the designated gas inlet boundary limit, and the ship manifold;
- LNG receiving installations, between the ship manifold and the designated gas outlet boundary limit;
- Peak-shaving plants, between designated gas inlet and outlet boundary limits.

The scope of this Standard includes the following topics: normative references, terms and definitions, safety and environment, jetties and marine facilities, storage and retention systems, LNG pumps, vaporisation of LNG, Pipe-work, reception / send out of natural gas, boil-off gas recovery and treatment plants, auxiliary circuits and buildings, hazard management, control and monitoring systems, construction, commissioning and turnaround, preservation and corrosion protection, training for operations, pre-operational marine training,
Important details and definitions, specific equipment and systems for LNG installations are provided in the annexes, like: thermal radiation threshold values, definitions of reference flow rates, seismic classification, specific requirements for LNG pumps, specific requirements for LNG vaporizers, criteria for the design of pipes, description of the different types of onshore LNG installations, definition of different types of LNG tanks, frequency ranges, classes of consequence, levels of risk, typical process steps of liquefaction, odorant systems

In addition to the other LNG EN´s indicated in the list above, the following references to other relevant EN standards are recommended:

- EN 287-1, Qualification test of welders — Fusion welding — Part 1: Steels
- EN 473, Non destructive testing — Qualification and certification of NDT personnel — General principles
- EN 571-1, Non destructive testing — Penetrant testing — Part 1: General principles
- EN 809, Pumps and pump units for liquids — Common safety requirements
- EN 970, Non-destructive examination of fusion welds — Visual examination

- EN 1092-1, Flanges and their joints — Circular flanges for pipes, valves, fittings and accessories, PN designated Part 1: Steel flanges
- EN 1127-1, Explosive atmospheres — Explosion prevention and protection — Part 1: Basic concepts and methodology
- EN 1435, Non-destructive examination of welds — Radiographic examination of welded joints
- EN 1514-1, Flanges and their joints — Dimensions of gaskets for PN-designated flanges — Part 1: Nonmetallic flat gaskets with or without inserts
- EN 1714, Non-destructive examination of welds — Ultrasonic examination of welded joints

EN 1474-1, Installation and equipment for liquefied natural gas — Design and testing of marine transfer systems - Part 1: Design and testing of transfer arms

This first part is valid for conventional LNG loading arms and it contains the following topics: design of the arms, safety precautions, connection with the ship, operating and control design, inspection and tests.

Included are also definitions of ship movements, operating envelope with connection, operating and drift zone, alarm areas with first and second step alarm, mechanical thresholds and limits.

EN 1474-2, Installation and equipment for liquefied natural gas — Design and testing of marine transfer systems - Part 2: Design and testing of transfer hoses

This draft standard includes descriptions of typical hose designs and auxiliary devices with flexible leak tight inner hose, reinforced layers for structural support, leak tight outer hose (depending on the design), end connection pieces, identification rings, leak detection system, insulation system to minimize ice formation, intermediate layers to prevent leakages, bending stiffeners or limitations, handling devices, saddles etc.

A leak tight layer of a hose made of composite material comprises several single layers, which form a labyrinth seal. The hose system should include a QCDC (quick connect / disconnect coupling). The hose system should also be equipped with an ERS (emergency release system) with valves and an ERC (emergency release coupling).

The LNG hose may carry piggyback a hydraulic or pneumatic hose and electrical cables for power transfer to the ERS and QCDC systems. It should be equipped with a leak detection system.

There are three principle types of hoses:

- hoses made of a variety of composite layers
- hoses made of composite layers reinforced by flexible metallic winding
- pipes made of corrugated steel

Combinations of these types can also be considered as feasible.

The corrugated steel pipes allow

- vacuum super-insulation
- leak detection
- water tightness from outside
- sufficient self stiffness

Flexible hoses or pipes for LNG transfer systems require special design features including design motions of the production unit and of the shuttle carrier, respective loads on the system including ambient conditions, hose or pipe length to be determined by dynamic load calculations, life cycle requirements due to temperature, pressure, mechanical load cycles, aging etc., typically the life cycle should exceed 5 years.

Due to the importance of the LNG transfer system within an LNG supply chain specific inspection and tests are required. As a general rule it is expected that all parameters / features of an offshore LNG transfer system have to be verified and approved. This standard includes detailed test and approval specifications to demonstrate the suitability of the proposed LNG transfer system based on detailed material selection and manufacturing.

EN 1474-3, Installation and equipment for liquefied natural gas — Design and testing of marine transfer systems - Part 3: Offshore transfer systems

This EN 1474-3 of European Standard gives guidance for the design of liquefied natural gas (LNG) transfer systems intended for use on offshore transfer facilities or on coastal weather exposed transfer facilities. The transfer facilities considered may be between floating units, or between floating and fixed units.

The following main topics are included in this draft standard: specification of the offshore LNG transfer system, description, general safety philosophy, general functional targets and requirements, design principles and risk assessment method, design principles, risk assessment, safety relevant elements, capacity standards, risk reduction.

A new offshore LNG transfer system has to be qualified by detailed technology assessment, risk assessment – failure mode identification, analysis and approval, reliability analysis and verification statements.

The following design basis and criteria for an offshore LNG transfer system shall be considered: support structure, diameter of the transfer system and product data, dimensions and motion frame, load analysis, dynamic load behavior, cyclic motion amplitudes, velocity and fatigue, special operational modes, product swivel joints and support structures, connection / disconnection devices, handling during connection, disconnection, standby position, communication, evacuation and rescue.

Accordingly, strict safety measures shall be incorporated in an offshore LNG transfer system like: communication, position control, alarm and shut-off system, ERS (emergency release system), safety interfaces, fluid transfer control, connection to LNG shuttle carrier, operation and control design, inspection and checks, quality assurance and control, safety documentation.

EN 1532:1997 Installations and equipment for liquefied natural gas - Ship to shore interface. It is intended that this standard will be superseded by an adequate ISO standard entitled:

Installations and equipment for liquefied natural gas - Ship to shore interface and harbor operations: to be applicable for conventional LNG terminals:

This standardization project has commenced early 2009 and shall include:

- Requirements for ship, terminal and harbour administration
- Safe passage of the LNG carrier through the harbor area
- Safe and efficient LNG transfer

It shall be valid for:

- pilots, operators of VTS (Vessel Traffic Service), tugs, terminals and ships
- suppliers of fuels, lubricants and other stocks and services while the LNG carrier is moored to the terminal.

It shall also contain:

- the safe passage of the LNG carrier
- safe approach, mooring and unmooring
- access from the landside to the ship
- operative communication between ship and shore
- all instruments, data and electrical interface connections, including OPS (Onshore Power Supply) where applicable
- Supply with LN2 (if available)
- Considerations on ballast water handling

EN 12308:1998 Installations and equipment for LNG. Suitability testing of gaskets designed for flanged joints used on LNG piping

This standard contains requirements concerning

- Design specifications of tested gaskets
- Technical documentation by the gasket manufacturer
- Check of compatibility between bolting load and mechanical strength of joint components
- Means and equipment required for testing
- Testing
- Acceptance criterion
- Test report

Important informative descriptions are provided in the annexes of this standard like:

- Thermo-mechanical behaviour of a flanged joint with gasket
- Test rig
- Method for verifying the bolting load
- DN (nominal diameter) and PN (nominal pressure) equivalence for tests

EN 12838:2000 Installations and equipment for liquefied natural gas - Suitability testing of LNG sampling systems

High levels of accuracy are required for the continued or discontinued measurement of the LNG respectively NG quality mainly concerning the heating value, as this is the basis for any commercial and fiscal assessment of the product transfer.

Accordingly strict requirements to the LNG sampling are provided in this standard. It includes descriptions of continuous and discontinuous LNG sampling systems, characteristics of the test rig, measurements required, general requirements of suitability test performance, operating procedures, calculations of suitability criteria for the tested sampling system

Conclusions

Nowadays it is almost unimaginable in many parts of the world to implement new LNG installations without the instructions from these EN norms. Thus the EN´s have certainly contributed to the high safety standard of this industry resulting in minimum accident statistics.
The EN´s are permanently extended and updated and reissued every 5 years. Several of these standards are under revision or amendment. A co-operation between the EN and the ISO standardization groups dedicated to the LNG industry has been initiated. This is desirable in order to harmonize the requirements for LNG installations worldwide.

Biography of Speaker

Eginhard Berger is currently head of sales and marketing division with IMPaC in Hamburg, Germany, a company providing engineering services for the oil and gas industry, both offshore and onshore. He graduated as MSc. from the Technical University in Munich. He is engaged in the process engineering industry for more than four decades after spending a few years in the aviation industry. As project and sales manager he was involved in several LNG projects in Europe and Asia. He was an early member of the European Technical Committee TC 282, which initiated and developed the LNG EN´s (European Norms) for LNG. eginhard.berger@impac.de

31c Advanced Modeling of LNG Receiving and Regasification Terminals

Ming Yan and **Jichuan George Hu**. URS Washington Division, 10550 Richmond Avenue, Houston, TX 77042

This paper presents how the front-end work process can be improved by applying an advanced model using standard commercial simulation software in a generic LNG receiving and regasification terminal. Process design engineering hours and schedule can be reduced by using this advanced universal model, which is flexible to cover all operation modes and different design configurations. It also allows quick understanding of the consequence of changing design parameters without explicitly getting into complex simulation models. In any of the engineering design phases, such as Concept Selection, FEED, and EPC Design, the work process can be improved by using the same modeling concept. The engineering work process can further be improved by integrating this model with Project Database software to generate process datasheets. This modeling concept can also be used in operations training or plant operations optimizations to fine tune certain operation parameters to improve plant performance. A recent LNG terminal project has utilized this model concept and has demonstrated its improvement in process design.

Supersized LNG Full Containment System: Is It a Dream or Practical?

31e

Eui-Seung Park, Executive Vice President, Plant Division
Hong-Sung Kim, General Manager, Plant Division
Se-Jin Jeon, Ph.D., Principal Researcher, Institute of Construction Technology
DAEWOO E&C C.P.O. Box 8269 Seoul, Korea

AIChE Spring Meeting, April 2009
9th Topical Conference on Natural Gas Utilization
Tampa, Florida, April 26-30, 2009

ABSTRACT

This paper describes the technical challenges of supersized LNG storage tank, which is a 9% Ni full containment system. This paper also discusses the development of a full containment tank beyond 200,000 m^3 storage capacity based on the current codes and its consequences of restriction in designing the tank. Economics and operation flexibility behind this supersized LNG tank is followed by the technical features of this tank design.

INTRODUCTION

The construction costs of the LNG tank occupy a major portion of the LNG terminal's total construction costs. Therefore, various efforts have been made to increase the cost efficiency of the LNG tanks, which can be largely divided into two parts: (1) There has been a clear tendency toward LNG tanks with higher capacity during the past decades. This can achieve a decrease in the relative cost per unit of stored capacity and enable a more effective land use. It also matches well with the increasing capacity of LNG ships. (2) The cost efficiency as well as the shortening of the construction period have been pursued through the technological development in material and structural aspects.

Table 1 shows the record capacity in each type of LNG tanks. For the in-ground or underground tanks, 200,000 m^3 capacity has been already operated in Korea and Japan. This article addresses the full containment type above-ground tanks, which are popular these days and the associated concerns with super sized LNG storage tanks. Such as one being built in Korea with a capacity of 200,000 m^3 capacity.

Table 1: LNG Storage Tanks with Largest Capacity

Type		Capacity (m^3)	Location
Above-ground	W/ concrete roof	200,000	Tongyeong / Pyeongtaek (Korea)
	W/O concrete roof	180,000	Senboku / Himeji (Japan)
In-ground		200,000	Negishi (Japan)
		200,000	Inchon (Korea)
Underground		200,000	Ohgisima (Japan)

As the capacity increases, special attention should be paid to the existing code provisions and the efficient procedure should be in place for designing LNG tanks with structural and cost efficiency as primary requirements. Encouraged by the above-mentioned trend and with proven experience of LNG tank construction over 20 years (Table 2), Daewoo E&C has been investigating an efficient design procedure of above-ground prestressed concrete LNG tanks with 200,000 m^3 capacity or more. Some noteworthy points related to the large above-ground LNG tanks with concrete roof are reviewed in this article, mainly focusing on the design aspects.

Table 2: Daewoo E&C's Experience of LNG Tank Construction

Location	Type	Capacity (m^3)
Incheon	Above-ground	100,000×3
	In-ground	200,000×6
Tongyeong	Above-ground	140,000×6 200,000×2
Pyeongtaek	Above-ground	200,000×2
Yemen	Above-ground	140,000×2
Total		3,420,000

Fig. 1 - Above-ground LNG Tanks(200,000 m^3) under Construction at Tongyeong, Korea

FEATURES OF SUPERSIZED LNG TANK DESIGN

Design of 9%Ni Inner Tank

Recent documents have constantly reported that some revisions of the codes are required, especially for the inner tank, to design full containment type above-ground LNG tanks of higher capacity .

The most commonly referenced codes for the design of inner tanks have been API 620 and BS 7777. Also, many of the domestic constructions in Japan have been based on Japanese own codes. Recently, BS 7777 has been superseded by BS EN 14620 that reflects several important points required to deign more reasonable dimensions of an inner tank with higher capacity. There are three points in BS EN 14620, i.e. an increased maximum design pressure up to 500 mbar, an increased maximum thickness up to 50 mm of the inner tank plate and partial height hydrostatic testing are noteworthy. These substantially relax some of the very conservative provisions specified in BS 7777, i.e. maximum thickness of 30 mm and full height hydrostatic testing, by which it is expected that the way toward higher capacities will be wide open. The revised code actually reflects the state-of-the-art such as advanced manufacturing and welding techniques of thicker 9% Ni plates.

The height to diameter ratio of the inner tank has a primary significance since the overall proportion of the LNG tank is determined from the inner tank rather than outer tank in many cases. The following aspects can be discussed in relation to the sizing problem. As for the inner tank, provided that the strength of the inner tank plate is given, the only load-resisting factor is the thickness of the plate. On the other hand, there are some more load-resisting factors in the outer tank including concrete strength, thickness, and the amounts of reinforcing bars and prestressing tendons. When we consider that both the inner and outer tanks should possess the strength to independently resist design loadings from a design concept of the full containment type tank, it is possible that the overall proportion of the tank is more dependent on the size of the inner tank, which has less redundancies in the load-resisting mechanisms.

Thus, it is recommended that in a preliminary design stage, several alternatives of the inner tank dimensions (height, diameter, thickness, etc.) are compared together to select a candidate that minimizes the material cost of the outer tank as well as the inner tank.

Design of Dome Roof

Many of the existing concrete roof domes of above-ground LNG tanks have a radius of curvature equal to d (diameter of outer wall). It corresponds to a rise to diameter ratio of 1/8, which is often recommended as an efficient structural shape for the roof domes where self-weight or externally distributed load is dominant. On the other hand, a higher rise is advantageous when the roof is subjected to internal pressure, which is one of the main design loadings in LNG tanks, since the higher curvature can endure the internal pressure with less tensile stresses.

Therefore, it can be mentioned that the conventional rise of the above-ground LNG tank domes is not optimal, at least from the structural aspects. When designing large LNG tanks with increased dome span, it is recommended that several alternatives of the dome dimensions (radius of curvature, thickness, etc.) are compared. This is to determine the most optimum design with respect to structural integrity, safe yet efficient shape with a minimum amount of concrete.

No special code-related restriction is imposed on the shape of concrete dome; however the codes for the carbon steel liner that is attached inside the concrete dome should be also followed. For example, API 650 specifies that the radius of curvature of the liner should be in range from 0.8d to 1.2d. Some of the large in-ground LNG tanks where the roof dome is also exposed above the ground level have the radius of curvature close to 0.8d. Buckling safety of the carbon steel liner is also important and varies sensitively depending on the shape and placing method of the concrete dome.

Design of Ring Beam

The ring beam is located at the upper end of the wall. The primary role of the ring beam is to cope with the major portion of the thrust transmitted from the roof dome, thus reducing excessive deformation of the upper part of the wall as well as the roof dome. Therefore, dimensions of the ring beam and the amount of prestressing tendons inside the ring beam have a close relationship with the shape of the roof dome. From the geometrical consideration, a higher rise dome induces less thrust to the ring beam, and is therefore advantageous for the ring beam as well as the dome itself.

The construction sequence of roof dome placing and ring beam prestressing has a primary importance because the stress state of the ring beam is much affected, depending on the sequence. The procedure should be so planned as to avoid excessive compression or tension in the ring beam throughout all the phases.

Design of Outer Wall

The safety and serviceability of the outer wall should be checked in various modes of operation: (1) The construction stage, (2) Normal operation, (3) LNG leakage from the inner tank. It may be an important design issue to maintain a slender wall thickness and proportion when a tank size gets bigger with the wall height increased. To achieve this purpose, it is expected that the conventional design practice should be revisited and improvements be made as necessary. Some of the possible improvements are introduced here.

High strength concrete contributes to raising the section strength often resulting in a thinner section. As an example, a high strength of 60 MPa over conventional 40 MPa has been adopted for the wall of an LNG tank in Senboku, Japan. It is reported that the wall thickness was actually reduced by 30 cm by the strength increase.

Prestressing is a main design load in the construction stage, where the section is often subjected to a more critical stress state in the construction stage than the operation stage. During the prestressing sequence of horizontal hoop tendons, a large amount of section force (moment) and corresponding tensile stress is produced at the lower part of the wall from restraint of the wall deformation by the rigid bottom slab. When the wall becomes higher with the increased capacity, the hoop tendons become too congested in the lower part of the wall according to the conventional design practice of tendons (Fig. 2(a)). This would produce excessive or sometimes unallowable stresses at the joint of wall and bottom slab in the construction stage.

One straightforward countermeasure to reduce the stress is to increase the thickness in the lower part of the wall. Other than the thickness increase, one conceivable and powerful alternative is to revise the hoop tendon arrangement. According to the recent researches, it is effective to modify the conventional tendon-induced trapezoidal pressure distribution into other types of distribution, an example of which is shown in Fig. 2(b). This strategy can be achieved by adjusting the spacing of the hoop tendons in the lower part of the wall through some trial and error or optimization technique. The advanced pressure distribution produces moderate magnitude of the base moment during the construction stage as well as the normal operation. Of course, the required number of tendons can be reduced as an additional advantage.

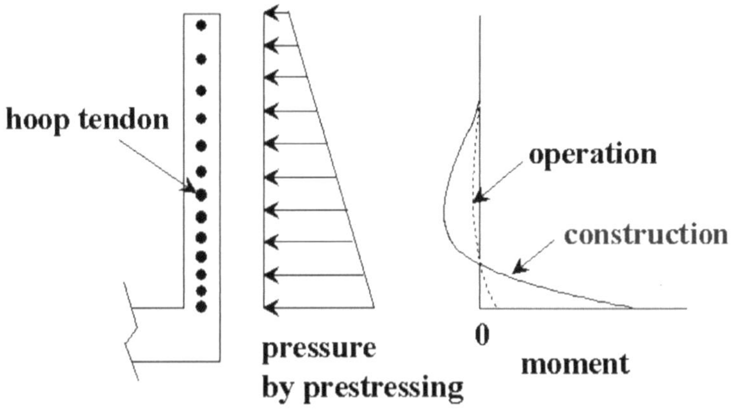

(a) Conventional Scheme

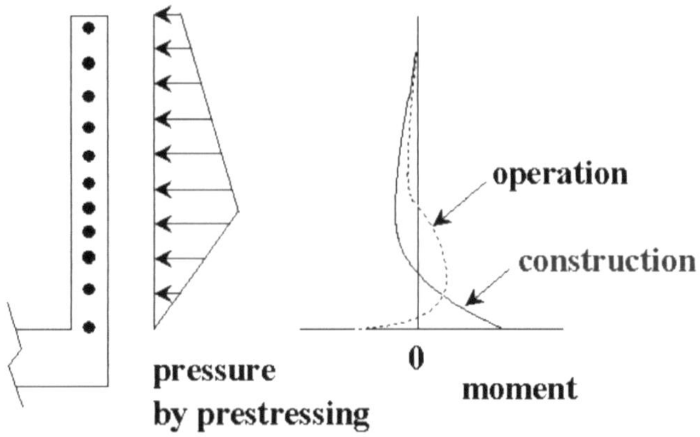

(b) Alternative Scheme

Fig. 2 - Conventional and Alternative Prestressing Schemes

Advanced Structural Analysis

During the past decades, computer-based structural analysis tools such as FEM (Finite Element Method) have achieved such great progress that the structural behavior can be estimated with high accuracy nowadays. It should be remembered that accurate and realistic structural analysis is always a key factor to design an LNG tank with optimized sections no matter what capacity it has. There are various structural analyses involved in the design of an LNG tank, some of which require more intensive and advanced techniques. Among them are heat transfer analysis of a structure subjected to the cryogenic temperature of LNG (Fig. 3(a)), internal or external fire analysis, and seismic analysis of the structure containing the fluid (Fig. 3(b)).

In the temperature-related analyses attention should be paid to establishing proper boundary conditions, since an LNG tank has complicated structural details inside including a suspended deck and various insulation layers. Especially, the favorable effect of the suspended deck which isolates the inner surface of the roof from direct exposure to the cryogenic atmosphere should not be overlooked.

In a detailed seismic analysis of an LNG tank, fluid-structure-soil-interaction should be considered. BS 7777 specifies that dynamic analysis should be carried out for the outer tank of double and full containment systems for areas with high seismicity. However, combined dynamic analysis considering the inner and outer tank as a whole could be required, e.g. at the client's request. In that case, the dynamic effect of the insulation layers between the inner and outer tank on the overall seismic behavior of the LNG tank makes the actual situation more complicated and needs more investigation.

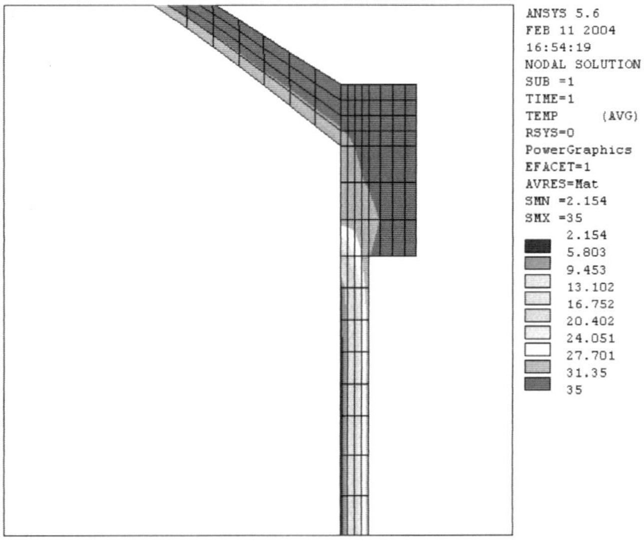

(a) Cryogenic Temperature Analysis

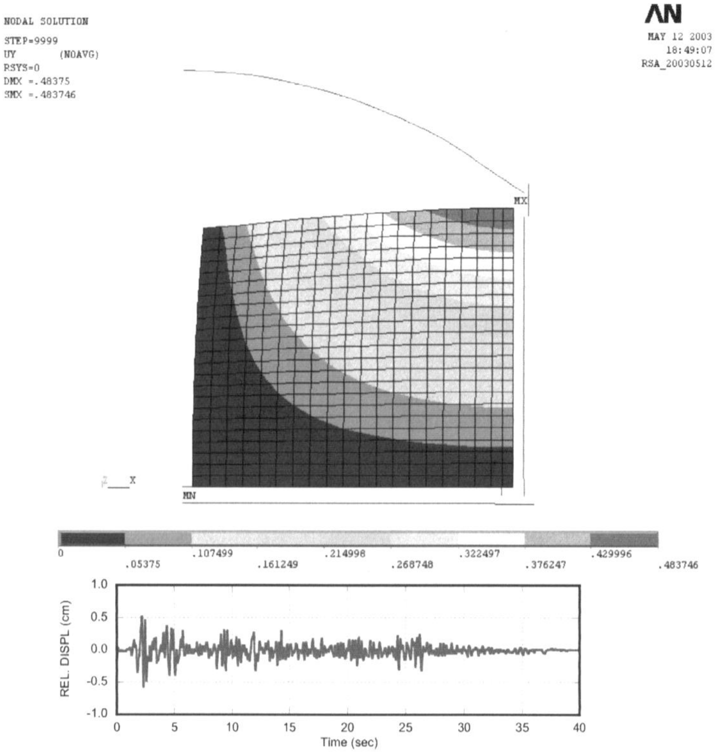

(b) Seismic Analysis

Fig. 3 - Examples of Structural Analysis of LNG Tank

CONCLUSION

Development efforts could secure engineering and analysis technology for a supersized full containment LNG tank (beyond 200,000m^3 storage capacity) of 9% Ni type based on its own advanced technology as a leading engineering and construction company in the LNG tank market.

The supersized LNG tank can achieve a decrease in the relative cost per unit of stored capacity and enable a more effective land use. A virtual project analysis result demonstrates that the cost effectiveness of the supersized LNG tank has been achieved through technology development in material and structural aspects and specific construction period of unit storage capacity.

REFERENCES

1. Se-Jin Jeon, Eui-Seung Park, "Toward a design of larger above-ground LNG tank", LNG Journal, March/April, 2004.
2. "BS 7777 : Flat-bottomed, vertical, cylindrical storage tanks for low temperature service", 1993, British Standard Institute.
3. "API Standard 620 : Design and construction of large, welded, low-pressure storage tanks", 2008, American Petroleum Institute.
4. "BS EN 14620 : Design and manufacture of site built, vertical, cylindrical, flat-bottomed steel tanks for the storage of refrigerated, liquefied gases with operating temperatures between 0°C and -165°C", 2006, British Standard Institute.

Biography of Speaker

Eui-Seung Park is a Executive Vice President and a Technical Fellow of Program Committee B of International Gas Union (IGU). He has worked for Daewoo E&C for 30 years plus and has been involved in many gas monetization and LNG projects. He graduated from Seoul National University with a BS degree in electrical engineering. He also completed Advanced Managerial Program (AMP). He is a head of Plant Division and currently in charge of managing LNG projects in the division.

E-mail : espark@dwconst.co.kr

Korean Regulations and Standards for LNG Storage Tanks in Comparison with Internationally Recognized Standards

31e

Seung Rim Lee, *Senior Researcher*
Institute of Gas Safety R&D
Korea Gas Safety Corporation, Seoul, Korea

Jang Sik Park, *Senior Researcher*
Institute of Gas Safety R&D
Korea Gas Safety Corporation, Seoul, Korea

Young-soon Lee, *Professor*
Dept. of Safety Engineering,
Seoul National University of Technology, Korea

Joseph Cho, *Director of Houston Tech. Center*
SK Engineering and Construction

2009 AIChE Spring National Meeting
9thTopical Conference on Natural Gas Utilization
Tampa, Louisiana, April 26-30, 2009

ABSTRACT

This paper presents the differences in the design criteria of Korean and internationally-recognized overseas standards for the design of LNG storage tanks. The overseas standards include API 620, NFPA 59A, EN 14620 series, EN 1473, RPIS and RPAS. This paper also addresses the need to modify the international standards and regulations.

INTRODUCTION

Since natural gas was first supplied to Korea in 1987, its annual demand increased by 15% on an average until 2002, when it was supplying to all major cities in Korea, and then the demand slowed down to 8.4%. Natural gas has been used as city gas and power plant fuel gas, and annual increase in city gas demand was 39.6% between 1987 and 2002, and then it drastically slowed down to 5.2%. On the contrast, on an average, the demand for power plant fuel gas has showed the tendency of annual increase of 13.3% due to increase in power demand caused by relative reduction of LNG price compared to heavy oil.

Table 1: Trend in Natural Gas Demand in Korea

(unit :1000 tons)

Classification	1987	1997	2002	2007	2008 (temp)	Annual increase rate (%)	
						'87~'02	'02~'07
City gas (Urban gas)	75	5,770	11,194	14,449	15,352	39.6	5.2
Power plant fuel gas	1,537	5,377	6,509	12,075	12,095	10.1	13.2
Total	1,612	11,147	17,703	26,524	27,447	17.3	8.4

It is expected that the natural gas demand will be increased at an annual increase rate of 1.5% from 26,524,000 tons in 2007 to 33,397,000 tons in 2002. City gas demand will be increased at the rate of 3.9% from 14,449,000 tons in 2007 to 25,734,000 tons in 2022, while demand for power plant fuel gas will be decreased at an annual average rate of 3.0% from 12,075,000 tons in 2007 to 7,663,000 tons in 2022. Regarding city gas, industrial consumption will be increased due to fuel conversion and distribution of LNG cargo trucks, but household consumption will slow down due to its saturated market. Especially, LNG consumption for peak-shaving power generation is expected to be drastically decreased after 2012 due to increase in other sources such as coal, unclear power and renewable energy.

Table 2: Outlook for Long-Term LNG Consumption in Korea

(unit: 1000 ton)

Year	City gas				Power plant fuel gas				Total
	Household	General	Industrial	Subtotal	Power plant	Co-generation	In-house	Subtotal	
2007(record)	7,153	2,813	4,483	14,449	9,234	2,219	623	12,075	26,524
2008(temp.)	7,480	2,930	4,942	15,352	9,012	2,529	554	12,095	27,447
2012	8,354	3,589	6,226	18,169	10,873	3,524	627	15,025	33,194
2015	8,917	4,093	7,233	20,243	7,251	3,549	627	11,428	31,671
2020	9,950	5,067	9,171	24,188	3,563	3,555	627	7,745	31,933
2022	10,319	5,461	9,953	25,734	3,487	3,549	627	7,663	33,397
Annual average increase rate	2.5%	4.5%	5.5%	3.9%	-6.3%	3.2%	0.05%	-3.0%	1.5%

LNG STORAGE TANKS IN KOREA

In recent years, full containment aboveground LNG storage tanks have formed the mainstream in the construction market. This is due to the economical benefits obtained through development of oversized storage tanks and enhanced land utilization on tank sites. A significant safety improvement by efficient leakage control of liquefied gas and vapor from inner tanks made aboveground storage tanks popular. In Korea, in recent years, there is a tendency to construct almost all storage tanks aboveground full containment type instead of aboveground membrane type which used to be in the earlier stages.

Table 3: Present Status of LNG Storage Tank Operation in Korea

Terminal	Number of tanks	Inner tank / outer tank	Dike	Position	Capacity (kℓ)	Design firms		Constructor	Remarks
						Mech.	Civil		
KOGAS Pyeongtaek (14)	10	Mem./ P·C	Yes	A.G	100,000	TGZ	TGZ	Hanyang	In operation
	4	9%Ni/ P·C	No	A.G	140,000	KGM	KGM	Daerim	In operation
KOGAS Tongyeong (10)	10	9%Ni/ P·C	No	A.G.	140,000	KHI TKK	Obayashi Diwidag	Daewoo + Daerim	In operation
KOGAS Incheon (18)	10 (#1 ~ 10)	9%Ni/ P·C	Yes	A.G	100,000	TKK	Diwidag	Daewoo + Daerim	In operation
	2 (#11,12)	Mem./ R·C	No	I.G	140,000	MHI	Kajima	Hyundai + Hanyang	In operation
	(#13 ~ 18)	Mem./ R·C	No	I.G	200,000	IHI KHI	Taisei Obayashi	Daewoo + Samsung + Hanyang	In operation
POSCO Gwangyang (2)	2	9%Ni/ P·C	No	A.G	100,000	KGM	KGM	Daerim	In operation
Total	44 tanks								

Note : A.G = aboveground, I.G = In-ground

According to the 9th long-term natural gas supply and demand program of Korean government (December, 2008), the stability of supply will be secured by increasing the storage-to-demand ratio from 9.2% in 2007 to 20% in 2022. At the same time the ratio of high-price spot purchase in winter season will be decreased by increasing storage capacity. The following table shows the long-term plan for construction of storage facilities.

Table 4: Long-Term Plan for Construction of Storage Facilities

(unit: 10,000 kℓ)

Classification		'07 (record)	\`08 ~ \`10	\`11 ~ \`12	\`13 ~ \`15	\`16 ~ \`20	\`21 ~ \`22
LNG tanks	Construction (Total)	2.8 (51.6)	19.6 (71.2)	17.4 (88.6)	28.0 (116.6)	- (116.6)	- (116.6)
East Sea Gas field	Storage scale (Total)	-	-	-	-	37.0 (37.0)	- (37.0)
Total	Construction (Total)	2.8 (51.6)	19.6 (71.2)	17.4 (88.6)	28.0 (116.6)	37 (153.6)	- (153.6)

* Supply quantity and supply facilities (200,000 kℓ) for in-house consumption of POSCO and K-Power are not included in the above table.

STANDARDS FOR LNG STORAGE TANK

International Standards

The present status of overseas codes and standards for LNG storage tanks and their characteristics is as follows:

Table 5: Present Status of Major Overseas Codes and Standards for LNG Storage Tanks and Their Characteristics

Code	Scope	Remarks
EN 1473 (2007)	o Installation and equipment for liquefied natural gas, design of onshore installations o Overall LNG plant system - i.e. general specification for LNG tanks, treatment facilities, pipelines, HAZOP, and other detection systems o Bases for risk assessment o Single containment, double containment and full containment type storage tanks are covered in addition to membrane type tanks and spherical storage tanks.	Performance based standard

NFPA 59A (2006)	○ Standards for production, storage and handling of liquefied natural gas (LNG) ○ Standards for liquefaction facilities, storage facilities, vaporizers, transmission and handling facilities of natural gas ○ Standards for design, layout, construction, maintenance and operation of LNG facilities ○ Extended application of double containment type and full containment type LNG storage tanks (revision) ○ Revision of seismic design in accordance with ASCE7 requirements	
EN14620-1~5 (2006)	○ Design and manufacture of site-built, vertical, cylindrical, flat-bottomed steel tanks for the storage of refrigerated liquefied gases at operating temperatures between 0□ and -165□ ○ Standard for design, construction, test, commissioning, operation, shutdown of storage tanks for refrigerated liquefied gases with their boiling points under atmospheric temperature ○ Standard for membrane type storage tanks (No bund wall required) ○ Supplementation of limits of BS 7777 (maximum shell thickness, hydrostatic test requirements, etc.) ○ Specified specification for corner protection ○ Gases to be stored are the same as covered in BS 7777.	
API 620 App.Q (2008)	○ Design and construction of large, welded, low-pressure storage tanks ○ Design and construction of large, low-pressure, carbon steel aboveground storage tanks (including flat-bottomed tanks) ○ Guideline for materials, design, and manufacture of storage tanks for liquefied ethane, ethylene, methane, etc. ○ Specification for single containment type tanks (single /double wall tanks)	Specification based standard
RPIS RPAS (2002)	○ Recommended practice for LNG in-ground storage ○ Recommended practice for LNG aboveground storage ○ Planning, design, construction and maintenance of in-ground and aboveground storage tanks ○ Guideline mainly for design ○ Specified detailed design specification (data) ○ Seismic design and emphasis on control	

Korean Regulations and Standards

Laws and Regulations

Korean statutory system for gas industry consists of 3 laws as shown in the following table. LNG storage tanks are covered by "Specified Facilities" in "High-Pressure Gas Safety Control Act."

Table 6: Korean Gas Acts

High Pressure Gas Safety Control Act (HPGSCA)	**High Pressure Gas**	• **Manufacture and use of Cylinders · Refrigerators · Specified Facilities** **Production · Filling ·Storage · Sale · Utilization Facilities**
Safety Control and Business of LPG Act	LP Gas (Propane,	• **Filling · Storage · Sale · Group Supply · Utilization Facilities** • **Manufacture and use of Gas Appliances**
Urban Gas Business Act	**City Gas (LNG, LPG+Air)**	• **Supply and Utilization Facilities for City Gas**

Specified facilities are manufactured in accordance with related codes and standards after being registered in accordance with the Enforcement Decree of High-Pressure Gas Safety Control Act, Article 5, Paragraph, and Sub-paragraph 3.

Korean standards and codes

Korean standards and codes are classified into Korean Industrial Standards which are voluntary standards while KGS Codes are mandatory. Codes for LNG storage tanks are applied and managed in accordance with the mandatory codes based on the High-Pressure Gas Safety Control Act, and their establishment status is as follows:

Table 7: Present Status of Korean Codes Covering LNG Storage Tanks

Code No.	Title	Year of Establishment
KGS AC115	Inspection Code for Manufacture of LNG Storage Tanks	2008
KGS PV015	Code for Full-containment LNG Storage Tanks	2002
KGS PV 000	Code for Above-ground Membrane Storage tank	2009(planned)

COMPARATIVE ANALYSIS

Among the 3 types (aboveground membrane type, aboveground 9% Ni steel type, and in-ground membrane type) of LNG storage tanks introduced into Korea, the aboveground 9% Ni steel type, became the mainstream design and enjoys international popularity. Korean, European and American major codes and standards for tanks of this type is compared in their major items in Table 8.

Table 8: Comparison of Major Design Parameters in LNG Tank

a) Tank Types and Material

구분	EN 14620	API 620	RPAS	KGS Code
Tank Types	Single Containment Double Containment Full containment Membrane Containment	Single-wall insulated tank or double-wall tank (Single containment)	Single-wall insulated tank or double-wall tank (Single containment)	Full containment
Stored material	Refrigerated Liquefied Gases	Refrigerated Liquefied Gases	LNG	LNG

b) Maximum Working Pressure and Minimum Working Temperature

Classification	EN 14620	API 620	RPAS	KGS Code
Maximum working pressure	500mbar(g)	15psi(g) (1,034mbar(g))	-	To be designed suitable for the maximum working pressure and minimum working temperature (To be specified by the purchaser)
Minimum working temperature	Below liquefaction temperature	-270□ (-167.8□)	-	

c) Design Allowable Stress of Inner Tank (9% Ni steel)

Classification	EN 14620	API 620	RPAS	KGS Code
In service	Min.(0.43 UTS, 0.67 YS)	Min.(1/3 UTS, 2/3YS)	Min.(0.25 UTS, 0.67 YS)	Min.(1/3 UTS, 2/3 YS)
Hydrostatic test	Min.(0.60 UTS , 0.85 YS, 340)	Min.(0.55 UTS, 0.85 YS)	Min.(0.5 UTS, 0.85 YS)	Min.(0.55 UTS, 0.85 YS)
Earthquake	OBE:1.33 σa SSE:1.0 YS	OBE : Min.(1.33 σa , 0.8 YS) SSE: Not specified	Level1: Min.(0.5fu, 0.85fy) Level2: Allowable ductility factor	OBE: 1.33 σa SSE: 1.0 YS

Note:
1) UTS : Ultimate tensile strength
2) YS : Yield strength

d) Thickness of 9% Ni Steel Shell Plate for Inner Tank

Classification	EN 14620	API 620	RPAS	KGS Code
Minimum nominal thickness	10mm (d〉60)	3/8 in(□ 9.5mm) (d≥220ft(□67m))	60〈d≤75: 10mm 75〈d: 12mm	10mm
Maximum thickness	50mm	2in(□ 50mm) (ASTM A553)	50mm JIS SL9N520/590	50mm

e) Minimum Nominal Thickness of 9% Ni Steel Bottom Plate for Inner Tank

Classification	EN 14620	API 620	RPAS	KGS Code
Minimum nominal thickness	5mm	3/16in(□4.8mm)	6mm	5mm

f) Minimum Nominal Thickness and Allowable Stress for Roof Plate

Classification	EN 14620	API 620	RPAS	KGS Code
Minimum nominal thickness	5mm	3/16 in (□ 4.8mm) (Diameter ≥ 220ft (□67m))	Aluminum alloy: 6mm Other materials: 4.5mm	4.8mm (D ≥67m)
Allowable stress	-	Min.(1/3 UTS, 2/3 YS)	Min.(1/4 UTS, 2/3 YS)	Min.(1/3 UTS, 2/3 YS)

g) Nondestructive Test of 9% Ni Steel Plate Welds

EN 14620	API 620	RPAS	KGS Code
100% RT	Partial RT	100% RT	100% RT

h) Hydrostatic & Pneumatic Test

Classification		BS 7777		API 620 App.Q		KGS Code
		Pneumatic test (1.25×DP [1)])	Hydrostatic test (F·H) [2)]	Pneumatic test	Hydrostatic test (F·H) [3)]	
Single containment type	Single wall	○	○	○ (1.25×DP)	○	<9% Ni steel type- aboveground > -Inner tank hydrostatic test: 1.25
	Double wall (outer tank)	○	×	○ (≥DP)	×	

Double containment		○	○	-	times water level equivalent to maximum operating liquid level pressure. - Outer tank pneumatic test pressure: 1.25 times design pressure
Full containment		○	○	-	
RPAS	Inner tank	- Aboveground type (metal double wall inner tank) ·Hydrostatic test: 1.25 times maximum working pressure + water level equivalent to LNG design weight or over ·Hydrostatic test: 1.5 times water level equivalent to design liquid level or over ·Pneumatic test: 1.1 times maximum working pressure (after draining test water)			
	Outer tank	- Pneumatic test: 1.1 times maximum working pressure - Vacuum test: on bottom welds (-400mmHg or under)			

Note:
1) DP: Design pressure
2) FH (Full height): Water level the same as design liquid level
3) Water level the same as design liquid level. However, hydrostatic test with water level equivalent to 1.25 design liquid level is allowed. (In case water cannot be filled up to the design liquid level due to the foundation or stress condition.)

i) Free Board Height

EN 14620	API620	RPAS	KGS Code
500 mm or more from shell top	Free board height ≥ sloshing height + 1ft	-	500 mm or more from shell top

j) Separation Distance between Tanks

EN 14620	API 620	RPAS	KGS Code
To be determined by Hazard Assessment (EN 1473)	1/4 sum of diameters of 2 adjacent storage tanks or over(NFPA 59A)	1/2 sum of diameters of 2 adjacent storage tanks or over	1/4 sum of diameters of 2 adjacent storage tanks or over

CONCLUSION

The demand for natural gas in Korea is on a continuously increasing though the trend has slow down. Korea is promoting its technical self-reliance through localization of design of LNG storage tanks and material production governed by Codes and Standards. In this connection, Korea is making a concerted efforts to internationalize the codes for LNG storage tanks. As the leader of Task Group 5 of ISO/TC 67/ WG 10, Korea is taking the initiative in internationally standardizing national codes and regional codes which are presently non-uniform and inconsistent.

REFERENCES

1. Ministry of Knowledge Economy, 9th long term plan for Natural gas supply and demand, 2008.
2. KGS, A study on Developing Code for Full-containment LNG Storage Tank, 2002.
3. KGS, A Study on Developing Code for Membrane Type LNG Storage Tanks", 2006.
4. Yoichi Kamata, "Japan's Proposal for Task 5(A Storage Tanks Guidance Document), ISO/TC 67/WG10 meeting, 2008.
5. Doug Miller, "Breifing to ISO TC67, WG 10, TG 5 Regarding American Petroleum Institute API Standard 625", CB&I, 2008.
6. BS EN 1473 "Installation and equipment for liquefied natural gas Design of onshore installations", 2007.
7. BS EN 14620-1 ~ 5 "Design and manufacture of site built, vertical, cylindrical, flat-bottomed steel tanks for the storage of refrigerated, liquefied gases with operating temperatures between 0□ and -165□", 2006.
8. API 620 "Design and Construction of large, Welded, Low-Pressure Storage Tanks", 2008.
9. NFPA 59A "Standard for the production, storage and handling of liquefied natural gas(LNG)", 2006.
10. JGA RPIS(Recommended Practice for LNG In-ground Storage), 2002.
11. JGA RPAS(Recommended Practice for LNG Above-ground Storage), 2002.

Author Biography

Name: Seung Rim Lee
Position: Senior Researcher
Company: Institute of Gas Safety R&D, Korea Gas Safety Corporation
332-1, Deaya-dong, Siheung-si, Gyeonggi-do, 429-712, Korea
Phone: +82.31.310.1532, Fax: +82.31.315.4363
E-mail: srlee@kgs.or.kr
Cellphone : +82.10.5290.6668

Seung-Rim Lee is a Senior Researcher in the Institute of Gas Safety R&D at Korea Gas Safety Corporation. He has been working for Korea Gas Safety Corporation for thirteen years. He has been involved research and development projects related with domestic standardization and risk assessment in LNG industry. Especially, he has been participating in development of Korean Standards and International Standards (ISO/TC67/WG10) related with LNG storage tank and LNG terminal from 1999. He obtained a Master of Science in field of Safety Engineering in 2001 and he is under the doctor´s course in field of Gas Engineering.

Completing the Liquefaction Train by Using Two-Phase LNG Expanders

Yoshitsugi Kikkawa
Senior Engineering Consultant
ykikkawa@ykh.chiyoda.co.jp

Masaaki Ohishi
Unit Manager, Mechanical Engineering Center
Division Technology & Engineering
mohishi@ykh.chiyoda.co.jp

Chiyoda Corporation
Yokohama, Japan

Hans E. Kimmel
Executive Director
hkimmel@ebaraintl.com

Ebara International Corporation
Sparks, Nevada, USA

2009 AIChE Spring National Meeting, Tampa, Florida, USA

Yoshitsugi Kikkawa has been working for Chiyoda Corporation for 44 years and is a well known expert in petrochemical processing. He has a Doctorial Degree in Engineering from Kyushu University, Japan

Masaaki Ohishi is the Unit Manager of Rotating Machinery Unit at Chiyoda Corporation in Yokohama, Japan. He graduated in 1984 from Waseda University Tokyo, Japan with M.S. in Mechanical Engineering.

Hans Kimmel is Executive Director Research and Development at Ebara International Corporation in Nevada, USA. He graduated in 1965 from the Technical University Munich, Germany in Mechanical and Process Engineering.

The modern natural gas
liquefaction process
is based upon the
Linde-Hampson Cycle
independently patented
in 1895 by
C. von Linde and W. Hampson

In three steps, the process
compresses
cools
expands
the gas to a lower temperature
This triple step process is repeated
until the gas condenses

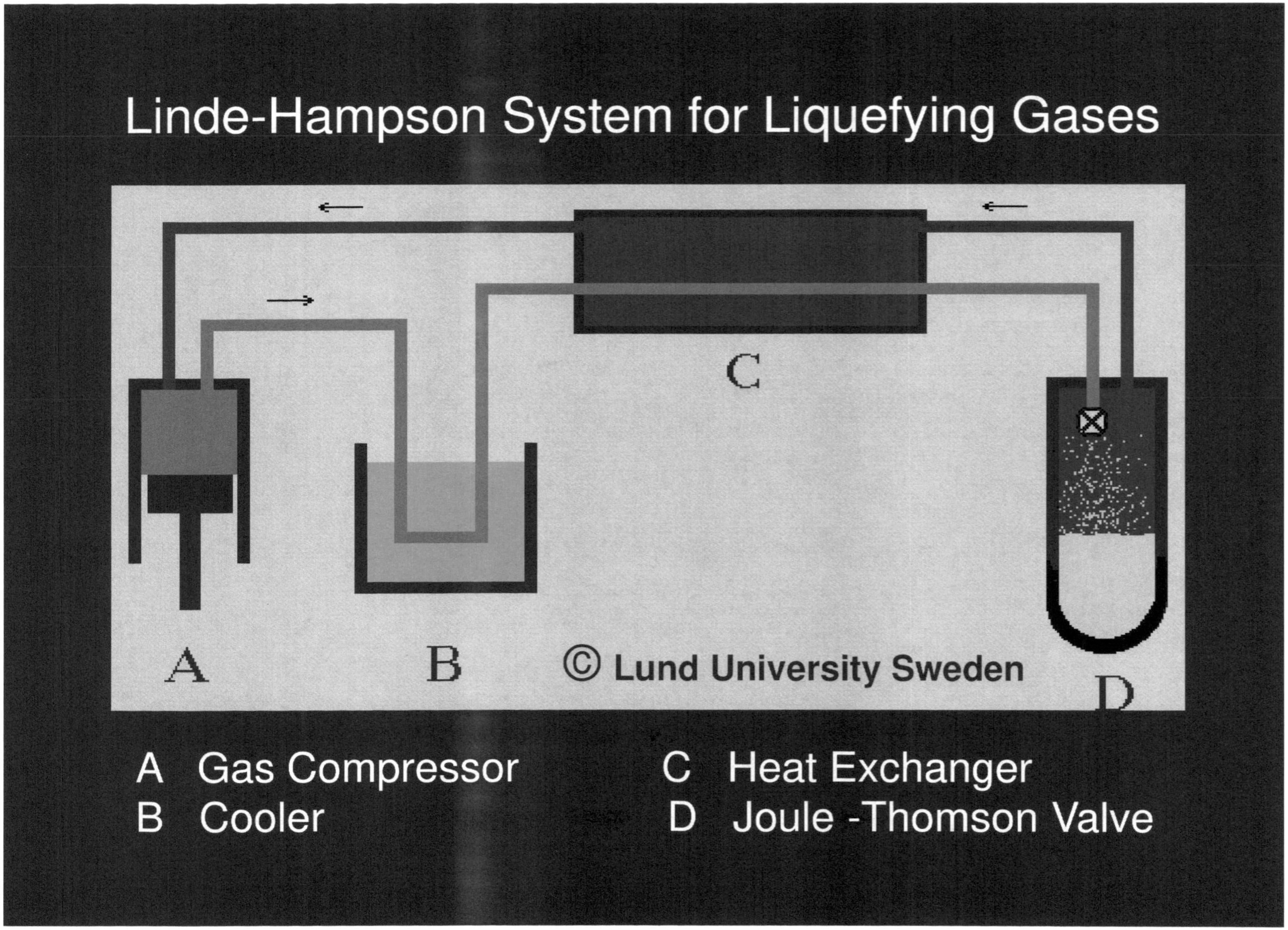
Linde-Hampson System for Liquefying Gases
C
A
B
© Lund University Sweden
D
A Gas Compressor
B Cooler
C Heat Exchanger
D Joule -Thomson Valve

Originally
Compression
was achieved

by piston compressor

Cooling

by heat exchanger

Expansion

by Joule-Thomson Valve

First air liquefaction plant from 1895
© German Museum, Munich

Oman LNG
© Shell
Contemporary LNG Liquefaction Plant

First improvements in
operation and efficiency:
Rotary Gas Compressors
and
Rotary Gas Expanders
replaced piston compressors
and gas expansion valves

Subsequent improvements in
operation and efficiency:
Since 1996
Liquefied Gas Expanders
replace
Joule-Thomson
Liquid Expansion Valves

A long overdue improvement:
Since 2003
Two-Phase LNG Expanders
replace
Two-Phase Expansion Valves

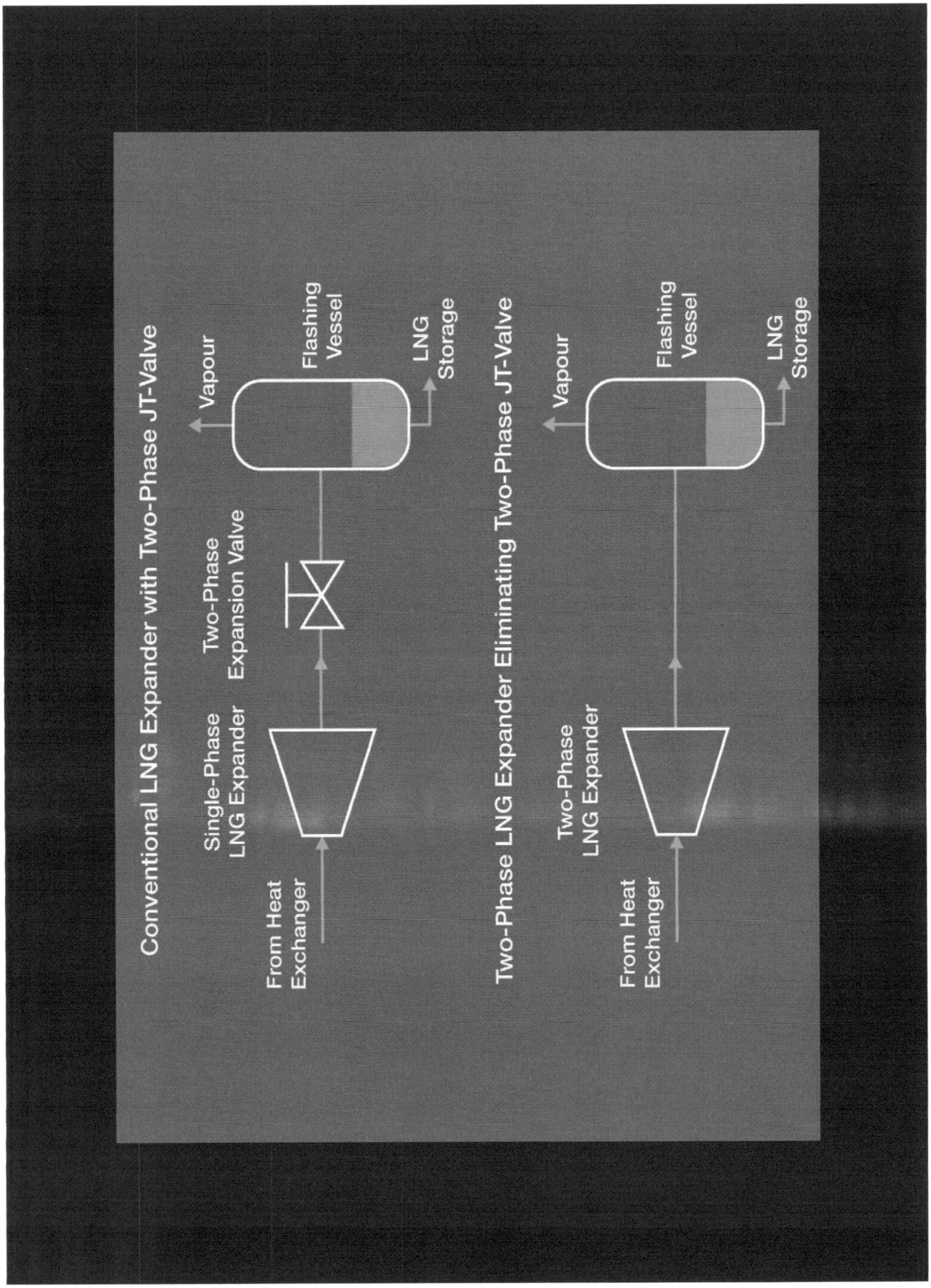
Conventional LNG Expander with Two-Phase JT-Valve
From Heat Exchanger
Single-Phase LNG Expander
Two-Phase Expansion Valve
Vapour
Flashing Vessel
LNG Storage
Two-Phase LNG Expander Eliminating Two-Phase JT-Valve
From Heat Exchanger
Two-Phase LNG Expander
Vapour
Flashing Vessel
LNG Storage

Completing the
Liquefaction Train
Two-Phase LNG Expanders
extract additional
energy from the stream
and increase the
amount of liquefied LNG

Current trends in LNG production are

Environmentally friendly

Large-scale

Small-scale

Floating

Two-Phase LNG expanders are important and necessary equipment for all existing and projected liquefaction plants of any size

© Statoil
Snøhvit, an environmentally friendly
LNG liquefaction plant in the Barents Sea

© Statoil
Dragados Offshore S.A.
A floating LNG process facility

Two-Phase Expanders
increase the plant efficiency,
reduce the overall equipment size,
reduce the pollution and
require minimal space

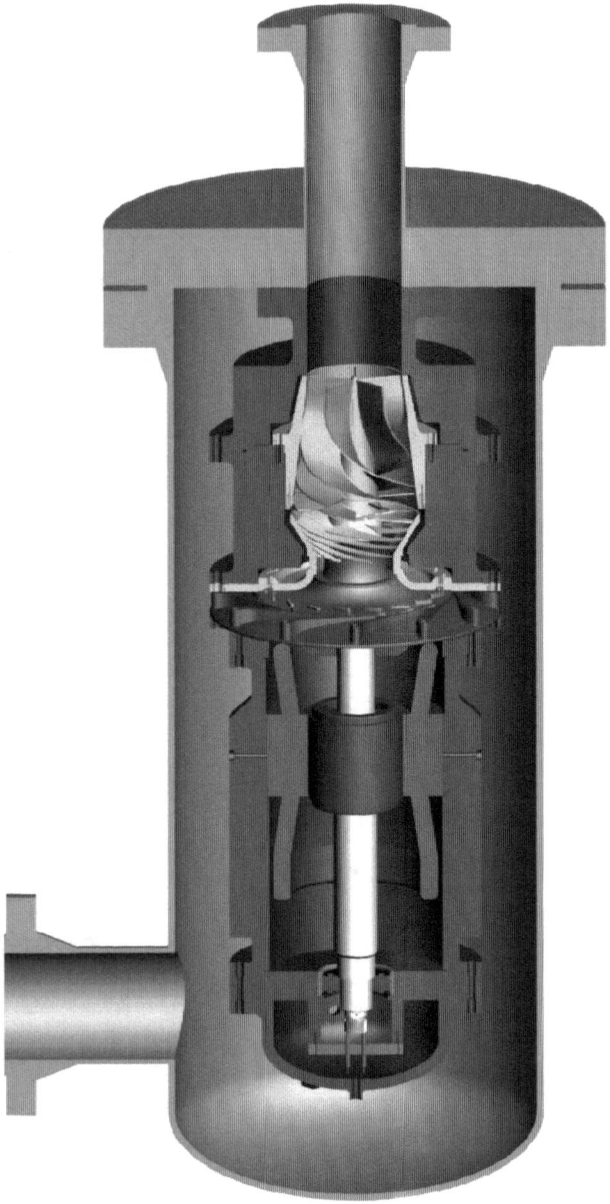

Cross section of Two-Phase LNG Expander inside pressurized containment vessel with lower inlet and upper outlet nozzle

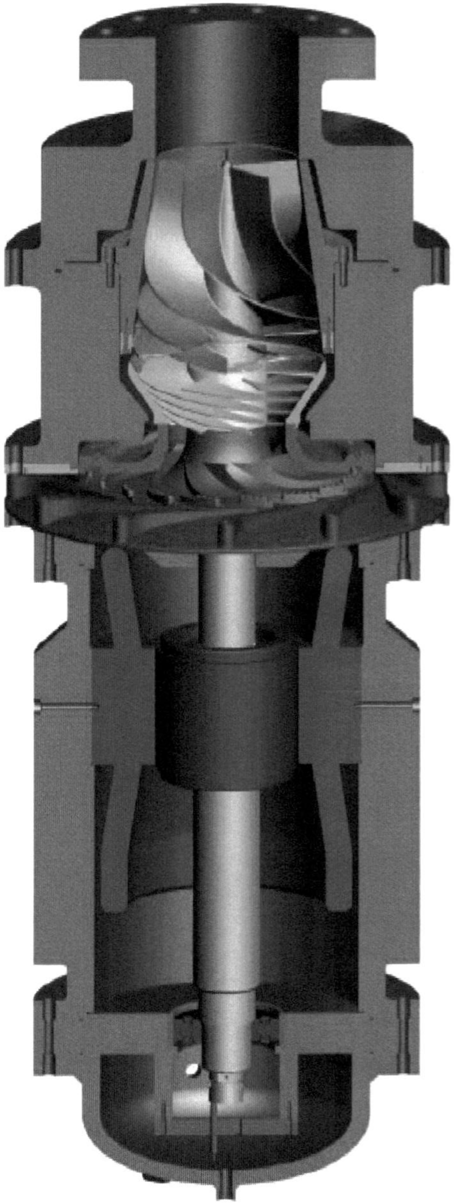

Cross section of Two-Phase LNG Expander with electric generator and two-phase hydraulic assembly mounted on one rotating shaft

Two-Phase Hydraulic Assembly
with
nozzle ring
(red),
turbine runner
(gold),
jet exducer
(yellow)
and two-phase
draft tube
(green)

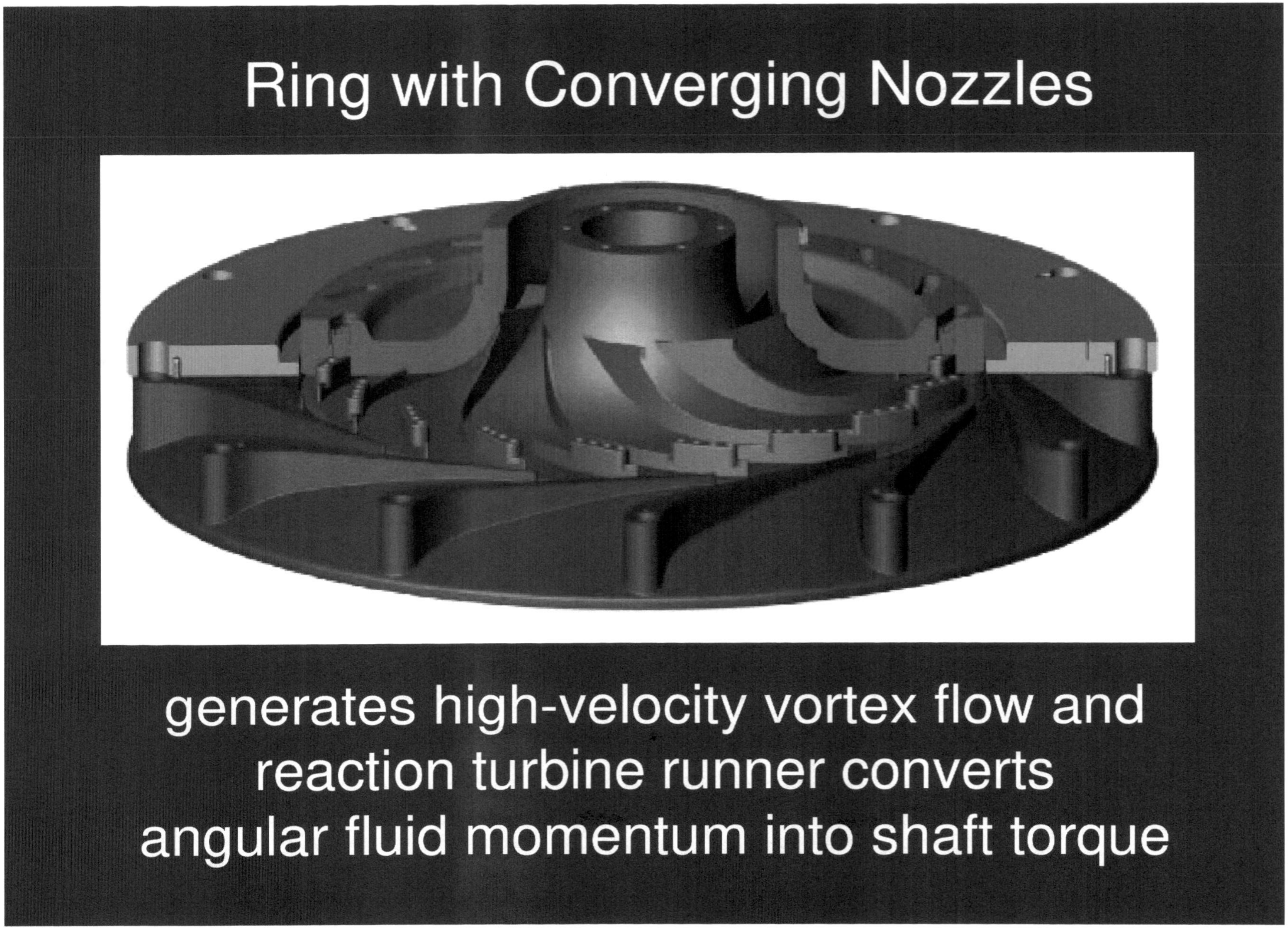
Ring with Converging Nozzles
generates high-velocity vortex flow and
reaction turbine runner converts
angular fluid momentum into shaft torque

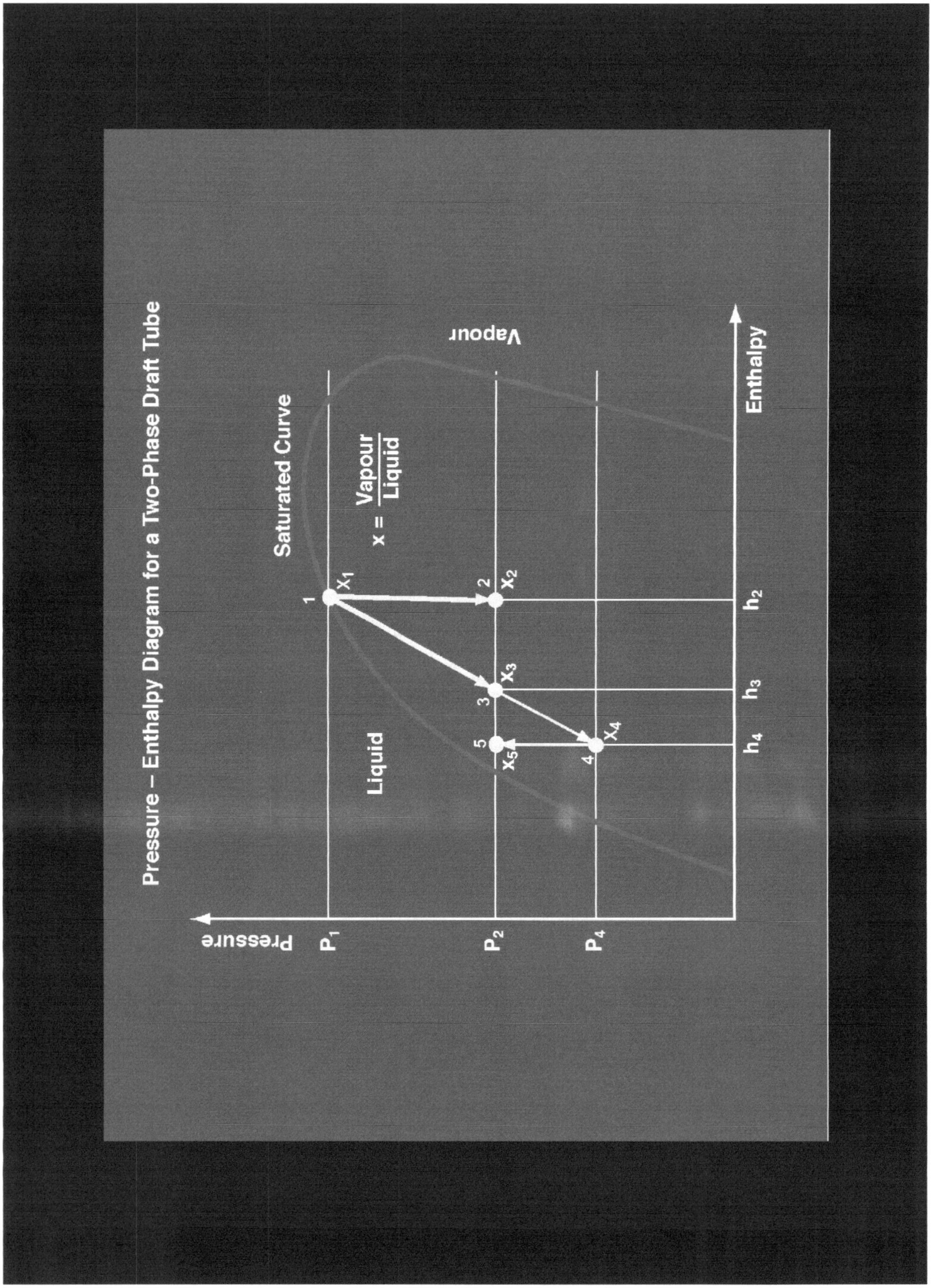
Pressure – Enthalpy Diagram for a Two-Phase Draft Tube
Pressure
Enthalpy
Saturated Curve
Liquid
Vapour
x = Vapour / Liquid
P1
P2
P4
h2
h3
h4
1
2
3
4
5
X1
X2
X3
X4
X5

Jet Exducer ™

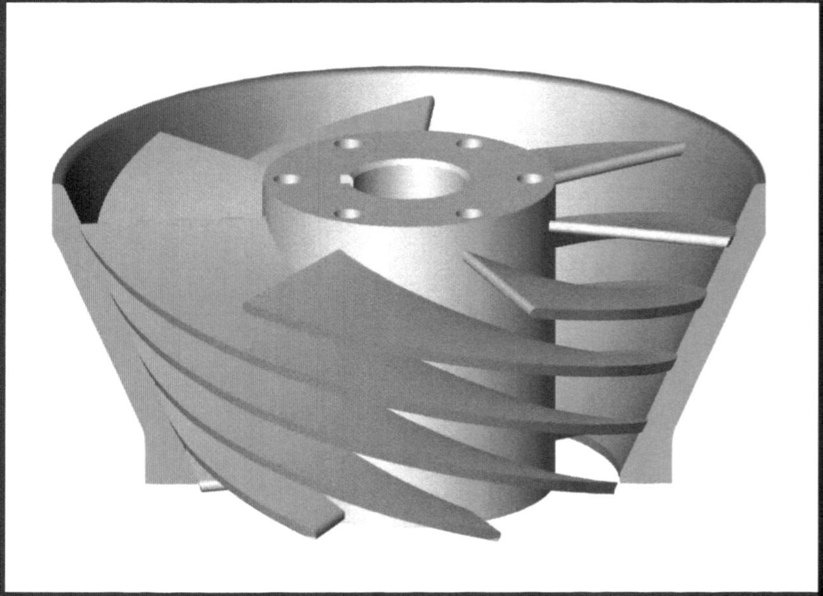

A radial outflow turbine for power generation by two-phase expansion from P_1 to P_4

Condensation Cone™

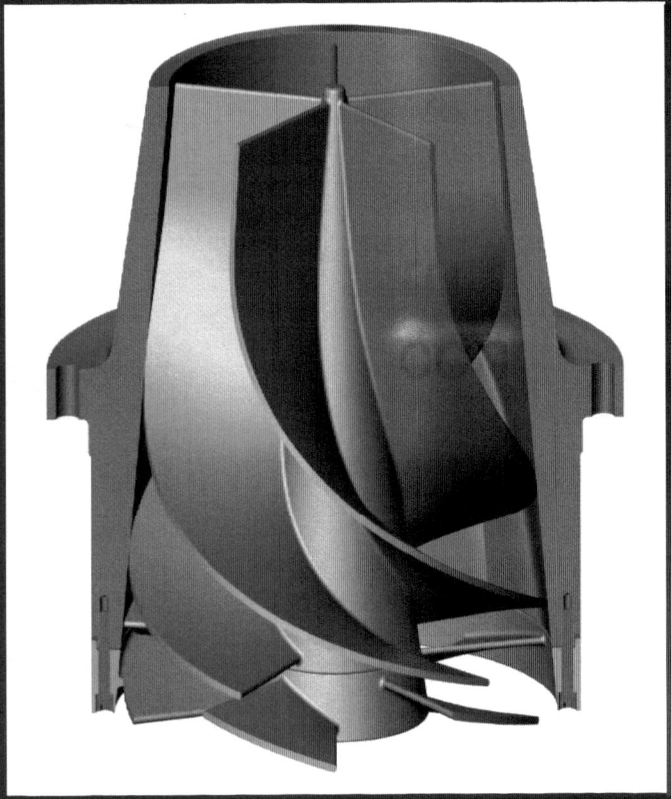

for pressure recovery with additional LNG vapor condensation from X4 to X5

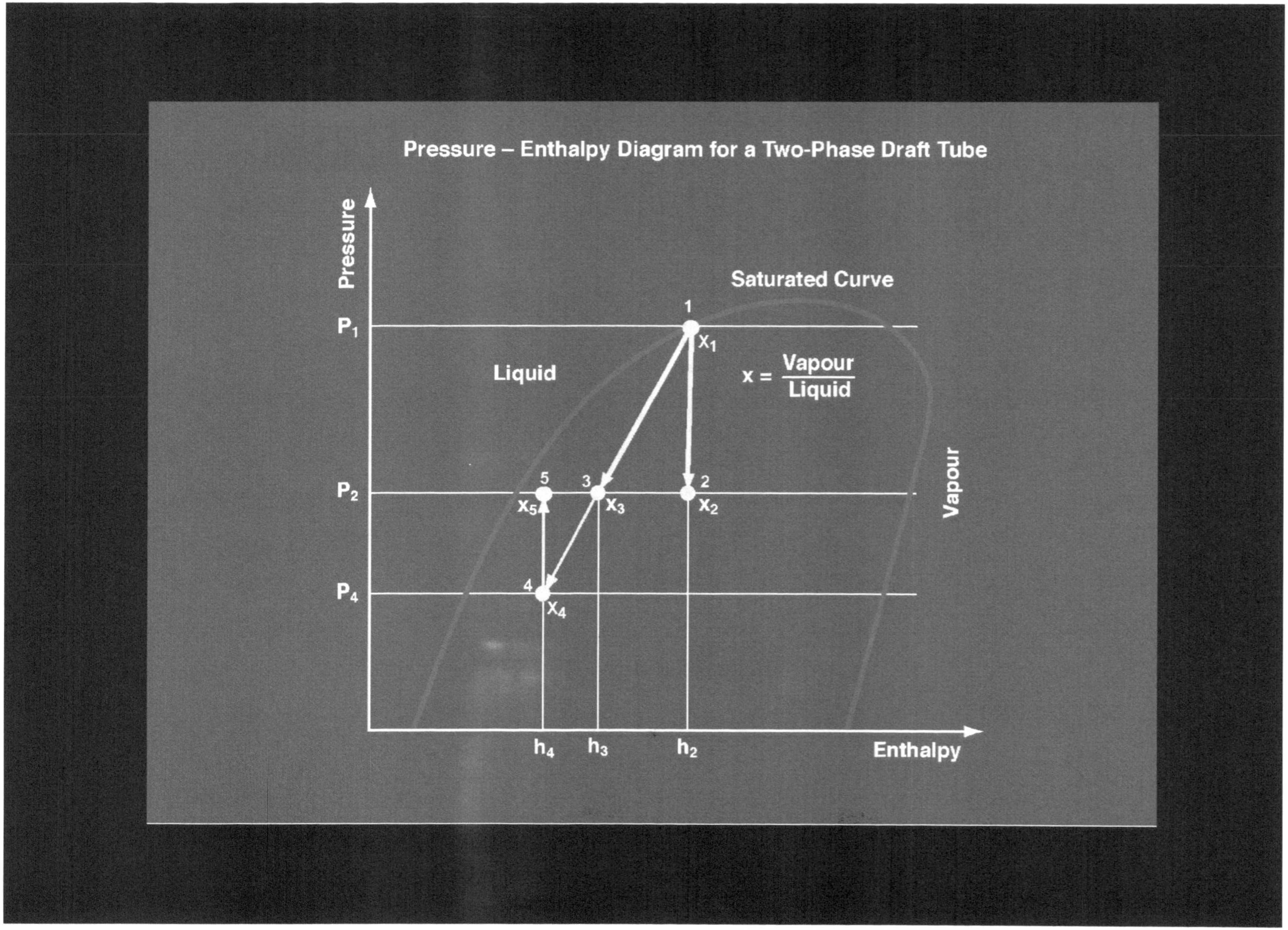
Pressure – Enthalpy Diagram for a Two-Phase Draft Tube
Pressure
P1
P2
P4
Saturated Curve
Liquid
Vapour
x = Vapour / Liquid
1
2
3
4
5
x1
x2
x3
x4
x5
h4
h3
h2
Enthalpy

Ebara Two-Phase LNG Expander

Convertible Design

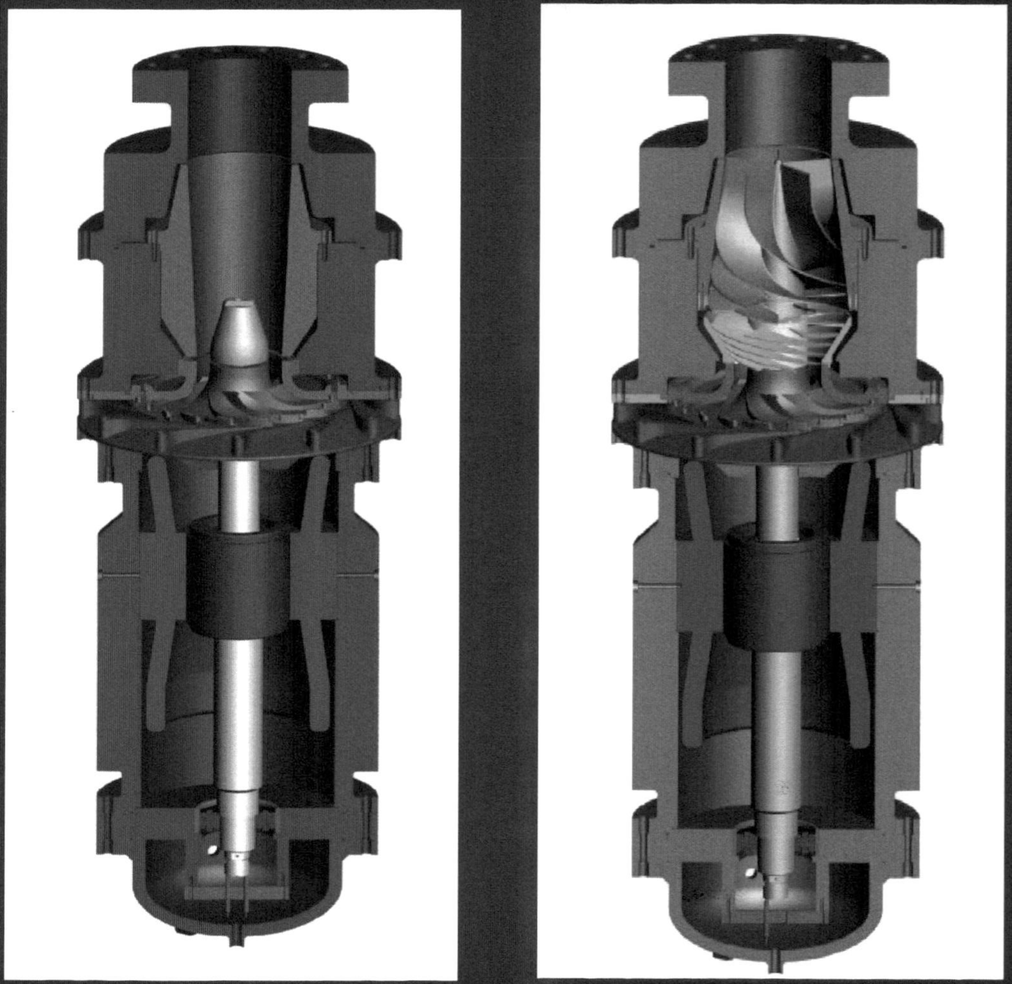

Convertible
Single Phase
and
Two-Phase
LNG Expander

Increase in LNG production using Liquid or Two-Phase Expanders

Liquid LNG Expanders: 5%

Two-Phase LNG Expanders: 8%

Epilogue

Two-Phase LNG Expanders complete the liquefaction process and are essential equipment in the LNG Liquefaction Train

Thank you

Yoshitsugi Kikkawa
Masaaki Ohishi
Hans E. Kimmel

Performance Prediction and Testing of Cryogenic Liquefied Gas Reaction Turbines

56c

Sarah Alison-Youel, Research and Development Engineer
Ebara International Corporation
Sparks, NV

AIChE Spring Meeting, April 2009
9th Topical Conference on Natural Gas Utilization
Tampa, Florida, April 26-30, 2009

Abstract

A new general performance equation has been previously presented for fixed geometry hydraulic reaction turbines and it has been shown to have a more accurate representation of liquefied hydrocarbon gas reaction turbine performance. This paper develops further equations relating the affinity laws and the conservation of energy and momentum laws to predict efficiency curves and the best efficiency point.

Due to the nature of the cryogenic environment in which liquefied gas reaction turbines operate, acquiring dependable test data presents a unique set of conditions. This paper also reviews the current testing techniques developed and used by Ebara International Corporation (EIC).

The presented equations are the most general, as well as, most accurate tools for turbine performance prediction. Actual test data from an EIC liquefied gas reaction turbine is used to investigate the presented equations.

Introduction

Lobanoff et al (1992) presented the affinity law relationships for pumps and suggested that they also apply to hydraulic reaction turbines. In 2008, Alison-Youel developed a partial model for liquefied gas reaction turbine performance. Also in 2008, Finley developed a method for predicting liquefied gas reaction turbine performance using the no-load and locked-rotor characteristics. This paper attempts to complete turbine performance modeling and prediction by investigating the available hydraulic power, developing a new general shaft power equation, and subsequent efficiency equation. These new developments are also investigated in order to predict the best efficiency point (BEP).

In order to achieve this, two very important assumptions are made. First, the turbine is of constant geometry; therefore, radius, cross-sectional area, and angles are fixed. Second, the working fluid is in-compressible; therefore, the fluid density is constant. Another key aspect to be addressed relates to the specific design of a submerged generator hydraulic liquefied gas reaction turbine and the testing of these types of machines.

Nomenclature

SG	Specific Gravity
H	Head
Q	Volume Flow Rate
N	Rotational Speed
α, β, γ, λ, δ	General Performance Equation and No-Load Constants
P_{hy}	Available Hydraulic Power
h	Unit Conversion Constant for Hydraulic Power Equation
M_{rot}	Rotational Momentum
v	Velocity
R, r	Outer and Inner Hydraulic Radius
P_{shaft}	Shaft Power
m, n	Summation Indices
a, b	Indices Representing Distinct Operating Points
k	Shaft Power Constant
ξ	BEP Constant

Testing Cryogenic Hydraulic Reaction Turbines

The inherent nature of a cryogenic environment coupled with the integrated turbine hydraulics and submerged generator design creates a special set of circumstances that drive the testing and data acquisition of liquefied gas reaction turbines.

Figure 1 is a simplified solid model of the most recent EIC single stage liquefied gas reaction turbine. Due to the submerged generator design it is not possible to separate the hydraulic performance from the generator performance without major design alteration. Therefore, the overall turbine efficiency is inclusive of the generator and the hydraulic efficiencies.

During testing, multiple pieces of data are measured and recorded. After testing is complete the raw data recorded during testing is then reduced into understandable and useable information. Table 1 tabulates the instrumentation used for the data collected and the subsequent units.

Fig. 1: Turbine Solid Model

Table 1: Recorded Test Data

Desired Data	Instrumentation	Raw Data Units	Reduced Data Units
Flow Rate	Venturi Flow Meter	inches of H_20	m^3/s or m^3/hr
Turbine Differential Pressure	Differential Pressure Transmitter	PSI	Bar
Vessel Inlet & Outlet Pressure	Pressure Transmitter (Bordon Tube)	PSIG	Barg
Speed	Eddy Probe and Target	Voltage pulse	RPS or RPM
Turbine Inlet & Outlet Temperature	1000 Ω RTD Probe	°C	°F
Vessel Inlet and Outlet Temperature	1000 Ω RTD Probe	°C	°F
Generator Output	Power Analyzer	kW, Hz, V, A	kW, Hz, V, A
Drive Output	Power Analyzer	kW, Hz, V, A	kW, Hz, V, A

Prior to turbine start-up the specific gravity (SG) of the fluid is measured and the relationship $SG = m \cdot T + c$ is calculated. Using this linear relationship, the temperature measurements are used to calculate the SG at each recorded test point. Following testing, the recorded data is used to calculate the flow, head, power output, and efficiency of the machine.

Data collected in this manner for an EIC turbine is utilized for demonstration and analysis in the following sections. The model used is LX14-07, with a nominal flow 1400 m3/hr, nominal head 700 m, and is designed to expand liquid natural gas (LNG). Test data was collected and reduced for turbine performance under the no-load condition and at the following speeds: 2400 RPM; 3110 RPM; 4000RPM.

General Performance Equation & No-Load Speed Curves

Using the above mentioned test data and the following General Performance Equation (Alison-Youel, 2008),

$$H = \alpha Q^2 + \beta N^2 + \gamma QN \tag{1a}$$

A least square error fit was applied to the full set of test data to find the constants, α, β, and γ, yielding the following surface fit equation,

$$H = 4174.85Q^2 + 0.115032N^2 - 9.42532QN \tag{1b}$$

This equation provides a three-dimensional surface fit for the performance of the turbine at multiple speeds and flow rates. Another important aspect to the performance of a turbine is when the machine is spinning, but the generator is not applying a torque. This is known as the no-load curve. Finley (2008) provides the following relationships between, flow, rotational speed, and head under the no-load condition,

$$Q = \lambda N \tag{2a}$$

$$H = \delta Q^2 \tag{3a}$$

Again, a least square error surface fit was performed, using only the no-load condition data to produce the following relationships,

$$Q = 0.002596N \tag{2b}$$

$$H = 17610.1Q^2 \tag{3b}$$

Figure 2 is a two-dimensional plot of the actual test data and the resulting surface fit curves for the no-load condition and three speeds.

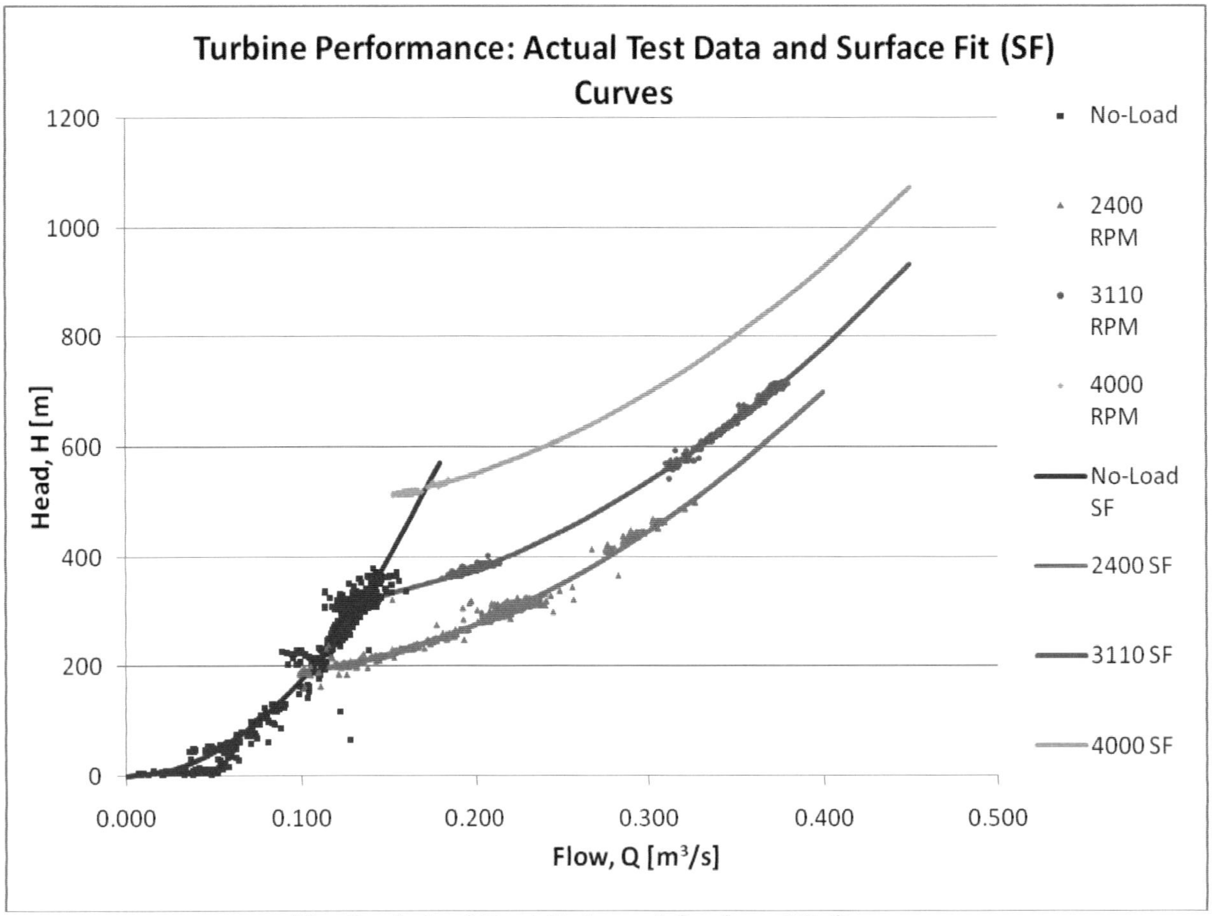

Fig. 2: Actual Test Data and Surface Fit Curves

Available Hydraulic Power

As Lobanoff et al (1992) states, the hydraulic power output of turbomachinery is proportional to the product of volume flow rate and head. Following the above calculations and surface fit equations, it follows that the available hydraulic power is,

$$P_{hy} = bQ(\alpha Q^2 + \beta N^2 + \gamma QN) \tag{4}$$

Where, the constant b is a unit conversion constant which includes the fluid density and gravity. As was previously mentioned, it is assumed that the operating fluid is incompressible; therefore, the density is a constant value.

Shaft Power

This section will develop the most general equation for shaft power that follows the conservation of energy and is based upon the conservation of momentum and the affinity law relationships.

Shaft power is a product of the rotational speed and torque experienced by the shaft. Test data provides rotational speed values leaving a relationship for torque based on test data to be found. The conservation of momentum principal applied to rotating machinery states that, torque is equal to the change in rotational momentum. The change in rotational momentum is a result of the change in mass flow, radius, and velocity. In the case of constant geometry, as is assumed in this paper, the radius, fluid flow passage area, and angels are fixed. Furthermore, due to conservation of mass, the mass flow rate is also constant. Hence, the change in rotational momentum is directly proportional to the change in velocity,

$$\Delta M_{Rot} \propto \Delta v \tag{5}$$

The change in velocity is a function of volume flow rate and rotational speed and can be developed into a general Taylor polynomial (Courant, 2000),

$$\Delta v = f(Q, N) = \sum_{m,n=0}^{\infty} CQ^m N^n \tag{6}$$

Applying this new polynomial for velocity (Eq. 6) to rotational momentum and substituting volume flow rate for mass flow rate, the following relationship is developed,

$$\Delta M_{rot} - Q\left[R \sum_{m,n=0}^{\infty} C_{1_{m,n}} Q^m N^n - r \sum_{m,n=0}^{\infty} C_{2_{m,n}} Q^m N^n\right] \tag{7}$$

Combining and simplifying, Eq. 7 becomes,

$$\Delta M_{rot} = Q \sum_{m,n=0}^{\infty} C_{mn} Q^m N^n \tag{8}$$

The shaft power is then,

$$P_{shaft} = NQ \sum_{m,n=0}^{\infty} C_{m,n} Q^m N^n \qquad (9)$$

The affinity law relationships for two distinct operating points (Lobanoff, 1992) are now utilized to solve Eq. 9 into a meaningful equation. The following are relationships from the affinity laws,

$$\frac{Q_a}{Q_b} = \frac{N_a}{N_b} = x \qquad (10)$$

$$\frac{H_a}{H_b} = \frac{Q_a^2}{Q_b^2} = \frac{N_a^2}{N_b^2} = y = x^2 \qquad (11)$$

As was mentioned above, hydraulic power is the product of head and flow. Applying this to the affinity law relationships,

$$\frac{P_a}{P_b} = \frac{H_a Q_a}{H_b Q_b} = xy = x^3 \qquad (12)$$

Applying Eq. 10, 11, and 12 to 9 for distinct operating point a,

$$P_{shaft_a} = N_a Q_a \sum_{m,n=0}^{\infty} C_{m,n} Q_a^m N_a^n = \sum_{m,n=0}^{\infty} C_{m,n} Q_a^{m+1} N_a^{n+1} \qquad (13)$$

Now applying Eq. 10, 11, and 12 to 13, a relationship for the power at each distinct operating point (a and b) is found,

$$x^3 P_{shaft_b} = \sum_{m,n=0}^{\infty} C_{m,n} X^{m+1} Q_a^{m+1} X^{n+1} N_a^{n+1} = \sum_{m,n=0}^{\infty} C_{m,n} X^{m+n+2} Q_a^{m+1} N_a^{n+1} \qquad (14)$$

Equation 14 is only a true statement if $x^3 = x^{m+n+2}$ is true; more simply if, $m + n = 1$. There are only two cases where this is possible (Table 2).

Table 2: Possible Cases for m and n

m	n
0	1
1	0

Applying the values for the indices in Table 2 to Eq. 14, a functional relationship for shaft power is found,

$$P_{shaft} - NQ(C_{1,0} + C_{0,1}N) \qquad (15)$$

However, at this point the constants $C_{1,0}$ and $C_{0,1}$ are not known, but the known relationships presented above for the no-load condition are applicable and reduce the number of unknown constants. Under the no-load condition torque is zero; therefore, the power is also zero. In addition, flow and rotational speed are related by the constant λ (Eq. 2a). Using this information, $C_{0,1}$ is solved for in terms of $C_{1,0}$, and λ. Putting it all together and simplifying,

$$P_{shaft} = kNQ(Q - \lambda N) \tag{16}$$

The Shaft Power Constant, k, at this point is still an unknown value. However, the value of k can be found using test data and will be further investigated in the following sections. Equation 16 is the most general equation that incorporates conservation of energy, conservation of momentum, and the affinity law relationships. More importantly, Eq. 16 was derived from a general polynomial to arrive at a specific relationship for shaft power, which supports the empirically presented equation for shaft power by Kimmel in 1997.

Power Curves & Efficiency

As mentioned above, due to the design and nature of testing cryogenic turbines the performance of the hydraulics and generator are fully integrated, and therefore, the measurable power output incorporates the efficiencies and losses in both the hydraulics and the generator. Therefore, the power output data measured during testing (in kW) is equal to the shaft power (Eq. 16). The full set of test data was again employed in a least square error fit to deduce the value of the shaft power constant, k. The resulting surface fit for the shaft power curves is shown in a two-dimensional plot (Fig. 3).

Following the deduction of the value for the Shaft Power Constant, the efficiency of the machine was analyzed. Turbine efficiency is the ratio of actual power output to available power input (Cengel, 2002). Applying the above developed equations (Eq. 4 and Eq.16) and applying them to this definition, the following relationship for efficiency, η, is found,

$$\eta = \frac{kN(Q - \lambda N)}{d(\alpha Q^2 + \beta N^2 + \gamma QN)} \tag{17}$$

This relationship was applied to the tested data and the results plotted (Fig. 4).

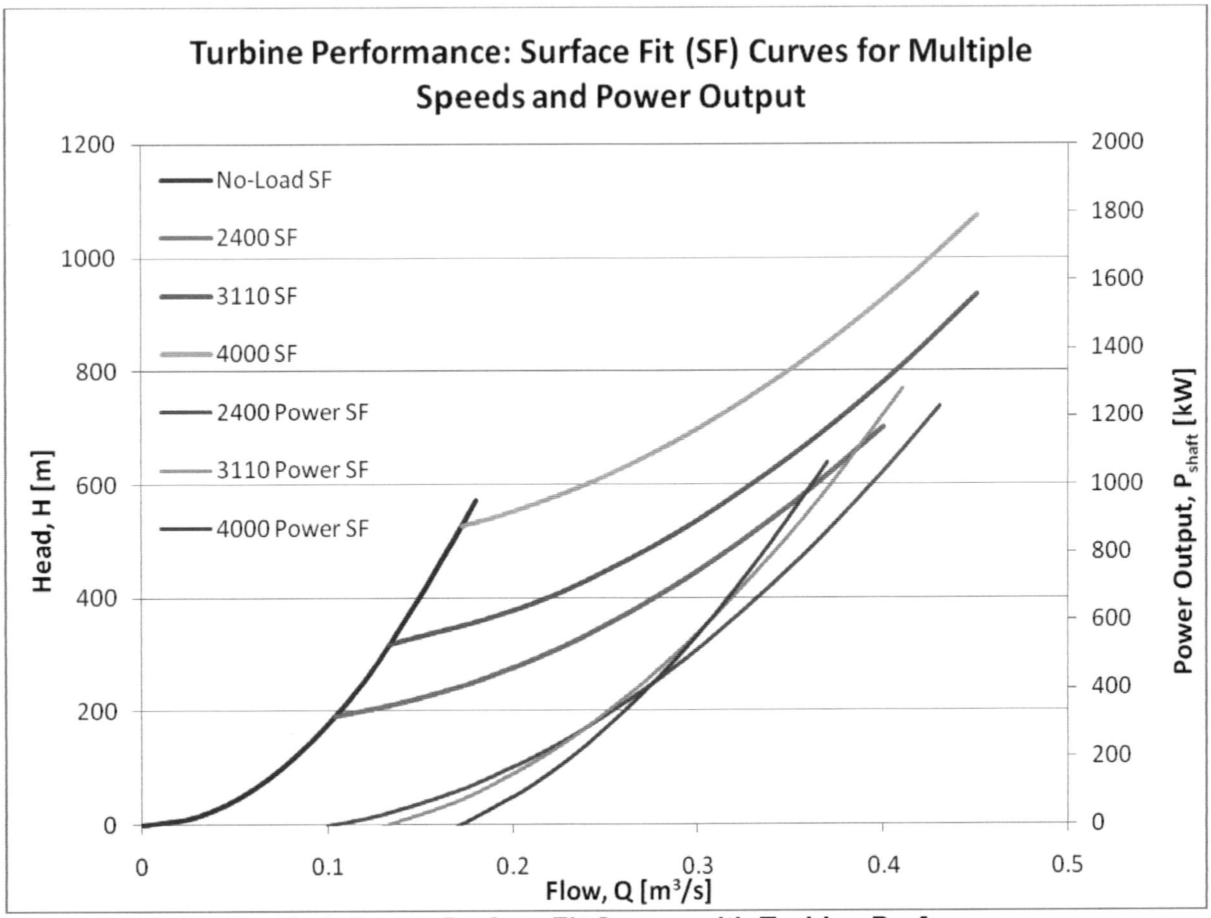

Fig. 3: Shaft Power Surface Fit Curves with Turbine Performance

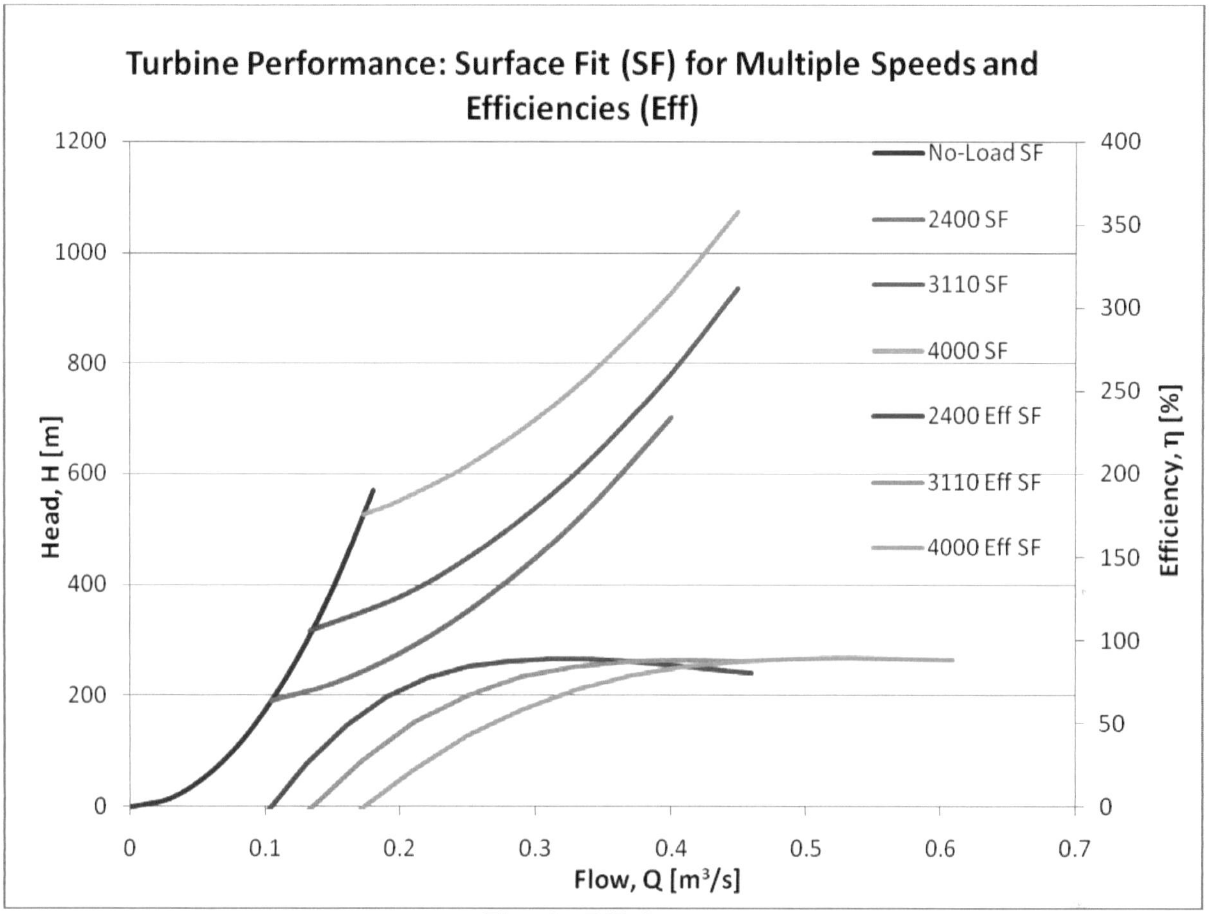

Fig. 4: Efficiency

Best Efficiency

The BEP corresponds to the flow and head at the maximum on the efficiency curve. To find the efficiency curve maximum, the first derivative of Eq. 17 with respect to Q is set equal to zero. It is found that the maximum efficiency flow rate, Q_{BEP}, is a function of all of the surface fit constants and the rotational speed,

$$Q_{BEP} = N_{BEP}\lambda\left[1 + \sqrt{1+\xi}\right] \tag{18}$$

Where, ξ is the BEP Constant and is defined as,

$$\xi = \frac{\beta + \gamma\lambda}{\alpha\lambda^{2}} \tag{19}$$

The BEP Constant signifies a very important relationship between the BEP and recirculation flow. If the value of ξ is greater than zero, γ is greater than the ratio of β/λ and the turbine functions as desired. However, if the value of ξ is less than zero, then γ

is less than the ratio β/λ and recirculation flow becomes a dominant factor in the turbine operation, resulting in undesired performance.

Figure 5 is a plot showing the results of the above generated efficiency and BEP equations and resulting values for 2400 RPM and 3110 RPM specifically.

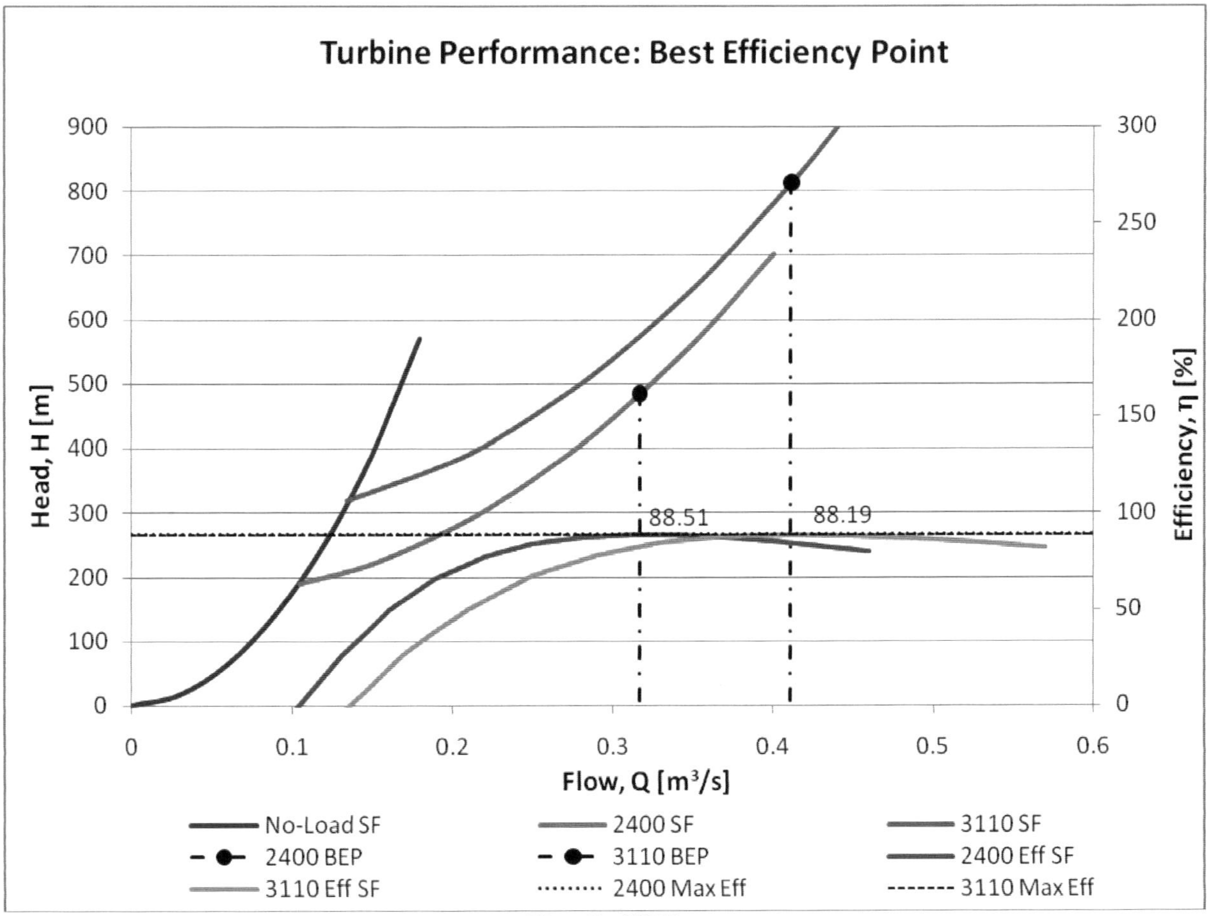

Fig. 5: BEP

Figure 5 shows that the tangent lines to the maximum efficiency curves are practically on top of each other and that the difference between the two maximum efficiencies is 0.32%. This confirms that at any flow rate, or head the rotational speed can be adjusted to maintain maximum efficiency.

Conclusions

The equations and results shown above provide complete liquefied gas reaction turbine performance in the most general form and based upon major principals including, conservation of energy, conservation of momentum, and the affinity law relationships.

Typically, older models used to define turbine efficiency have proven to be good a basis, but are generally accepted as somewhat inaccurate. The presented efficiency equation based on the developed hydraulic power and actual power equations represent, and more accurately model, actual turbine performance and efficiency. Thus, providing a better representation of performance overall.

The discovered BEP Constant provides important insight into turbine performance and the recirculation flow. Future turbine design will benefit from investigating and understanding this relationship.

It is also important to note that the speed curves are not evenly spaced and tend to curve up towards the next speed as flow increases (Fig. 2 – Fig. 5). This has been experienced in actual test data, but was not sufficiency accounted for by previous performance models. The General Performance Equation (Eq. 1a) models this observed result through the third term, γQN, which represents the recirculation that occurs within the hydraulics (Alison-Youel, 2008). It can be further concluded that the performance envelope created by the no-load and locked-rotor curves as developed by Finley (2008) will encompass the experienced, and now more accurately modeled speed curves.

The general hydraulic and shaft power equations coupled with the developed efficiency and BEP relationships presented in this paper, more accurately model actual liquefied gas reaction turbine performance. These and those presented by Finley (2008) represent a movement towards further turbine performance understanding and the capability to more accurately model performance in a simple and well supported manner.

References

Alison-Youel, S.D. "Observation and analysis of affinity law deviations through tested performance of liquefied gas reaction turbines." International Journal of Rotating Machinery (IJRM), Vol. 2008. Article ID 737285.

Cengel, Y.A., Boles, M. A. Thermodynamics: An Engineering Approach (4th Ed.). McGraw Hill, 2002.

Courant, R., Fritz, J. Introduction to Calculus and Analysis. Springer, 2000.

Finley, C.D. "Comparing predicted and tested performance of cryogenic reaction turbines under locked-rotor condition." 12th International Symposium on Transport Phenomena and Dynamics of Rotating Machinery (ISROMAC). Honolulu, Hawaii, February 17-22, 2008. ISROMAC12-2008-20252.

Kimmel, H. E. "Speed Controlled Turbines for power recovery in cryogenic and chemical processing." World Pumps, June 1997.

Lobanoff, V.S., Ross, R.R. Centrifugal Pumps, Design and Application. Gulf Publishing, 1992.

Biography of Speaker

Sarah Alison-Youel received a Bachelor of Science degree in Mechanical Engineering with a minor in Mathematics from the University of Nevada, Reno in 2006. Since joining Ebara in 2007 she has been working with prototype engineering, taking a project from concept through to final testing. Currently Sarah is the Project Engineer for the first two-phase Tandem Expander^{TM}, two of which will be installed in a Polish Oil and Gas LNG plant in Poland. Sarah has several research papers published in the American Institute of Chemical Engineers and the International Symposium on Transport Phenomena and Dynamics of Rotating Machinery conference proceedings. In addition, a research paper has been published in the International Journal of Rotating Machinery. salison-youel@ebaraintl.com

Techno-Economic Design and Optimization of NGL Plant: A Retrofit Study

56d

Aurora Hernandez Enriquez and Jin-Kuk Kim

School of Chemical Engineering and Analytical Science
The University of Manchester
PO Box 88, Manchester UK
email: Aurora.Hernandez@postgrad.manchester.ac.uk

AIChE Spring Meeting, April 2009
9th Topical Conference on Natural Gas Utilization
Tampa, Florida, April 26-30, 2009

Abstract

This paper proposes an approach that combines a process simulation model with Design of Experiments (DOE) and Surface Response Methodology (SRM) tools, for retrofit design. A case study considered in the study is a Natural Gas Liquid (NGL) Recovery Plant, and a steady-state model of the plant is set up for evaluating economic and environmental impacts arising from debottlenecking or process improvements made to the plant. Aspen Plus TM® is used for simulation, of which results agrees well with the measured data. The sensitivity analysis has been first performed and the results obtained have been taken into account to build the optimization problem in which distributed control system data and operators experience are also considered. Optimization has been performed by varying operating conditions (continuous variables) of the NGL plant, as well as by embedding different structural changes (discrete variables), leading to improvement of profitability. The optimal operating scenarios have been selected with promising structural changes. The proposed approach can provide deep understanding of system interactions, which allows substantial improvement of process performance.

Features of the gas processing industry operation

It is well known that the gas processing industry handles highly complex operations due to the following facts: it comprises energy intensive processes with several units (heat exchangers, distillation, and absorption columns, turbo-expanders, reactors, etc.) set in a very complex structure (heat exchanger networks, compression-expansion networks, pump-arounds, etc.); there are strong interactions between units and several products are handled (multi-component streams) resulting in many degrees of freedom for design or analysis; revamping is difficult (flexibility, controllability, hydraulic and pressure drop limitations), and due to extremely uncertain supply chain conditions, it is economically and environmentally challenging.

Natural Gas Liquids (NGL) recovering processes from Natural Gas are generally based on one of the following or a combination of them: use of external refrigeration, turbo-expansion, Joule-Thompson-expansion, or absorption. As previously referred, each specific NGL recovery problem is characterized by a large number of variables, some of those include inlet pressure, residual gas and the top of the demethanizer column, duty of external refrigeration, power capacity of the turbo-expansion system, temperature and pressure of the cold tank and of the bottom of the demethanizer column[1]. It has been stated that the refrigeration required to cool the inlet gas may give a significant cost in processes. There is a wide pool of coolants that can be used in the NGL processes ranging from single coolants to mixed refrigerants and cascade refrigeration cycles, even turbo-expansion or work expansion of the compressed natural gas feed with appropriate heat exchange[2].

Therefore, it becomes extremely difficult when looking at improving the economics and efficiency of that processing industry. Moreover, difficulties arise when it is intended to apply a robust optimization approach in the whole integrated production process. The present work aims to develop a practical systematic approach based on an integrated model, which identifies opportunities for improvement in an operating plant. This is basically known as operating plant retrofit design to maximise profit (energy vs. yield), minimise capital cost and consider environmental impacts.

Previous work

The developed retrofit design method is applied in Natural Gas Liquid recovery process (NGL). In such units different turbo-expansion processes[3,4] have been analyzed, based on capital analysis and operating limitations by using an ad-hoc simulator built and MINLP optimization technique. Parameter and structural changes studies have been performed using commercial simulator (e.g. HYSYS®) and a stochastic optimization (e.g. an advanced genetic algorithm program[5]), to determine optimal design. The main focus in revamping studies has been turbo-expander processes, and their flexibility to work in C_2 rejection mode. Mixed-integer nonlinear programming (MINLP) has been widely used for discrete–continuous optimization problems[6] showing the significant drawback of not giving a guarantee of convergence and finding the global optimum. Grossmann[7,8,9] proposed a global optimization method. In this method, applications have been presented in synthesis of integrated process water networks, complex distillation and crystallization systems, HENs, bioethanol plants and synthesis of Integrated Gasification Combined Cycle (IGCC) plants. It was observed that when applied to more complex streams, problems in finding the global solution appeared. Genetic algorithms[10], evolutionary algorithms[11], and simulated annealing[12], have been commonly applied to problems with multiple local optima.

"Design of Experiments" (DoE) is an important tool used not only to fit mathematical models based on experiments performed[13], but also as a powerful technique used mainly in the optimization of products and processes. DoE can be applied either to experimental or to simulation results as found in a survey [14] done on 77 studied cases of practical DoE engineering application, of which 77% used real data whereas 23% used simulated data. Most of the simulated data based on mechanical designs. The DOE methodology is a systematic approach that varies factor levels to assure that most of possibilities are taken into account to get a response. Selection of Designs used for experimental runs depends principally on aspects such as the purpose of the study, the number of factors and minimum levels, the restrictions on runs (experimental costs), the effects (single or interactions) to be studied (resolution), the order of the model to be fitted and the restrictions on the design shape (corner restrictions). After setting the proper design to be used, the experimental or simulation runs are performed and the results obtained are subjected to an "Analysis of Variance" which helps to determine if the means in a set of data differ when grouped by multiple factors, and if so, to determine which factors or combination of them are associated with the difference. The p-value is an important quantitative measure of this. It is defined as the probability of the test statistic being at least as extreme as the one observed given that the null hypothesis is true. A small p-value is an indication that the null hypothesis is false, which means that the factor is statistically significative. "Surface Response Methodology" (SRM) can then be applied to fit a reduced model based on the most important factors or combination of them and their effects on the studied response. Based on the final results obtained from SRM near optimal solutions can be found in practical periods of time[15].

A considerable amount of work found in the literature has been briefly summarized in the previous section. These studies aimed to find optimal solutions to the cases analyzed giving good results within most of the ranges explored. However, some serious weakness could be identified in particular for each methodology applied. In order to select the most suitable approach to be applied to this study, the criteria were as follows:

1) The principal advantages and drawbacks of each methodology:

a) Classical optimization (MINLP). Its principal advantage is that it is relatively faster than other methodologies, but it has two main drawbacks. The first is that if there are too many implicit functions in the model, difficulties to be solved can increase considerably. The second one that it generally finds local optima points.

b) Stochastic optimization (GA, SA, etc.) . Its advantage is that it generally finds global optima points having a broad cover of the system studied. However an important drawback is that it is highly time consuming when performed, the time increases exponentially with the complexity of the system.

c) Statistical optimization (Design Of Experiments and Surface Response Methodology). It can provide a reduced model with a high level of confidence (if the experiments or runs are well designed and properly performed), and based on this it finds near optimal points. An additional advantage is that it explains interrelations among factors involved and responses. The principal drawback of this methodology is that it can be highly time consuming when performing the runs, the time increases exponentially with the number of variables implied.

2) The features requested by the final users and an analysis of whether they were met:

a) The approach must achieve the interests of the management team: all the methodologies can match this request.

b) It must provide useful and reliable information: it was discussed and concluded that the classical optimization cannot match this request due to the fact that does not give global solutions. Moreover, it is highly probably that there will be difficulties in finding a solution with the large number of intrinsic functions in the model (thermodynamic properties).

c) It must give a reasonable response time: at first glance, this request could only be matched by the classical optimization. However the drawbacks of having difficulties to be solved and the restrictions of finding local solutions mean this methodology cannot be applied in the context of this study. Stochastic optimization strictly cannot match this request due to the highly computational load needed to find a global solution. Statistical optimization cannot match this request either at the fist step in finding near optimal solutions because of the large number of simulation runs needed at the beginning point. However, in most cases this is just at the beginning of the approach, and as the most important variables are known during the application of the methodology, this number is reduced resulting in time reductions.

d) It must work in a user friendly environment. This feature can only be achieved with DOE-SRM, as it can done based on commercial software (simulator and statistical tools) and does not require an expert computer programming to do the study.

As a consequence of these facts, it was decided that the best methodology to adopt for this investigation was the Statistical Optimization. It covers the broadest spectrum of points required for this analysis with a similar number of drawbacks to the Stochastic Optimization. The present work aims to improve the economics and efficiency of the process for the recovery of NGL by using a combined strategy Simulation-DOE-SRM based.

A Proposed Approach

In order to provide insight into understanding the steps carried on the strategy applied, Figure 1 summarizes the information flow among all the steps involved in the proposed approach, and a brief description of each one is presented as follows:

Figure 1. Steps involved in proposed approach

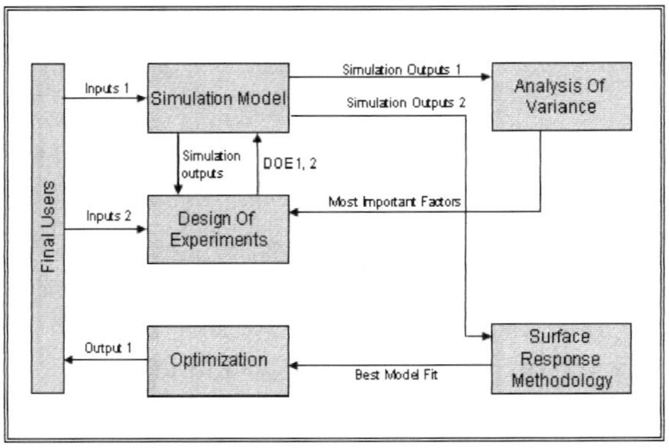

1) First at all, the steady state simulation model is built under the users needs and it is used to evaluate the response(s) they are interested in (objective function). At this point, most of the factors (variables) to be studied together with their type (continuous or discrete) are identified.
2) A Sensitivity Analysis is carried out in order to determine the behavior of the response(s) (objective function) to factor variations (perturbations). At this stage the factors that most affect the response(s) and their ranges can be defined.
3) DOE is applied in order to select a suitable set of experiments (DOE1) or simulations to be performed by taking the results obtained in the Sensitivity Analysis and the knowledge of the model (Plant DCS data and operators experience).
4) Following this, the set of simulations are performed and their results are subjected to an Analysis of Variance (ANOVA) where the Most Important Factors (MIF) that affect the response(s) are statistically identified and ranked. Those of which can principally be main factors, or interactions between 2 factors (ie higher level interactions have a very low probability to be important).
5) Once the MIF for the studied response(s) have been obtained, the best settings for each of the k MIF are found. Followed to this, a new DOE (DOE2) is required in order to find proper sets of simulations that let the MIF be fitted to reduced models (i.e. linear or non-linear equations).

6) These new sets of simulations are performed and their results are fitted into most representative reduced equations (SRM) by using a least squares fitting algorithm (regression).
7) The reduced models are then optimized by applying a NLP or LP solver(s) depending on the order of the reduced equations.

The final optimum solutions are then issued to the end users for review and comments on those. A new cycle must be initiated as requested starting from the most appropriate step.

Case Study

The Natural Gas Liquid recovery (NGL) process is addressed in this work, in a current operating plant. Figure 2 gives an schematic illustration of the general NGL process.

Figure 2. General flowsheet of NGL recovery process studied

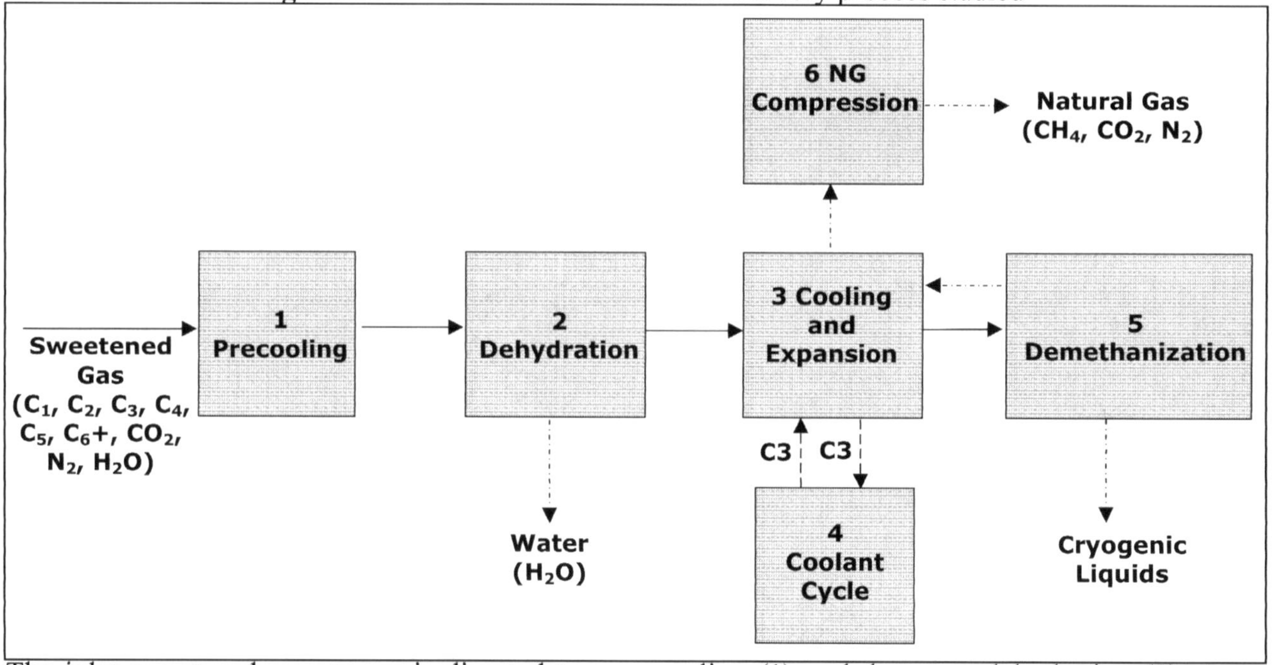

The inlet sweetened gas stream is directed to a precooling (1) and then to a dehydrating (2) stages. Following to this, the inlet gas is directed to a series of cooling gas/gas and gas/Propane heat exchangers (3) which are needed for chilling the feed gas in the unit, those are coupled with two turbo-expanders and separator tanks between each series, here a propane refrigeration closed cycle (4) helps to further lower the temperature. The outlet stream from this section then flows to the demethanizer column (5). The overhead vapors produced in the demethanizer, mainly Methane, flow back to the series of gas/gas heat exchangers (3), where they provide heat exchange to cool down the inlet gas stream, then it is compressed (6) to the desired pipeline pressure set by the final customers. As the final pressure is high, its associated temperature is hot enough to be heat integrated with the demethanizer column bottoms, it provides an amount of the total duty requested by reboiler to perform the complete separation, reducing the total low pressure steam consumed. The interest in this plant is focused on the recovery of C_3+ hydrocarbons. The bottom demethanizer product (cryogenic liquid) is pumped to the desired pressure, and then flows to a tank where is mixed with other products of the production centre and then pumped to final customers. A large amount of energy consumed in the plant is related to the high pressure steam used in the NG compression stage (6), the efficient energy use in the propane medium pressure compressors, and low pressure steam reboiler of demethanizer principally.

Simulation Model

The simulation model was built under ASPEN PLUS platform 2006.5 SM (steady-state). PRSV equation of state was set for the calculation of thermodynamic properties[16]. A number or assumptions were made based on the normal operating strategy of the plant, and the results show a good level of agreement with real operating data in average.

Sensitivity Analysis

In order to determine the behaviour of the objective function (PROFIT response) a sensitivity analysis was performed to the model by taking into consideration the controllable operating conditions (e.g. through DCS), the operating experiences, some suggestions from literature, and setting the operational limits of each controllable variable (based on industrial code). The most important independent variables and their best settings were found to be:

- Pressure in stage 1 of demethanizer, -10%, psi (Factor A)
- Power generation capacity of second turboexpander, +5%, hp (Factor B)
- Power generation capacity of first turboexpander, +5%, hp (Factor C)

Figure 3a, 3b, 3c and 3d present the behaviour of the Profit response as a deviation from the normal operating Profit (Effect on profit = 100* Profit response / Normal operating Profit) versus demethanizer pressure (PDA101), relative volatility C_2/C_3 versus demethanizer pressure (PDA101), power generation capacity of second turbo-expander (GC102), and power generation capacity of first turbo-expander (GC101) respectively.

Figure 3a. Effect of demethanizer pressure on the Profit response

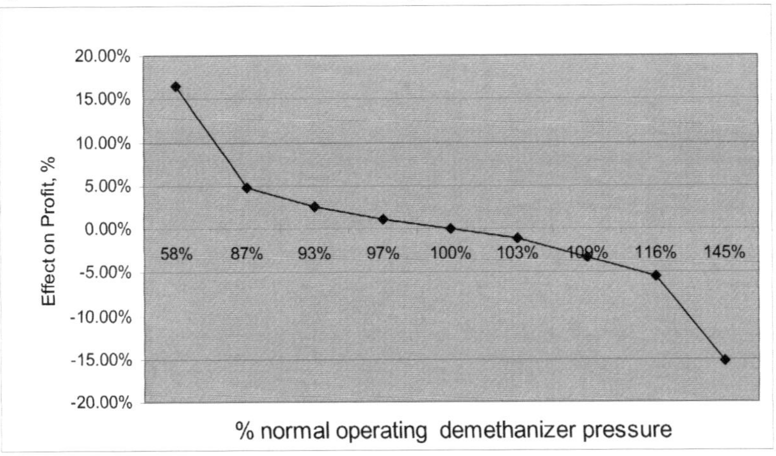

Figure 3b. Effect of demethanizer pressure on the relative volatility C_2/C_3 (α)

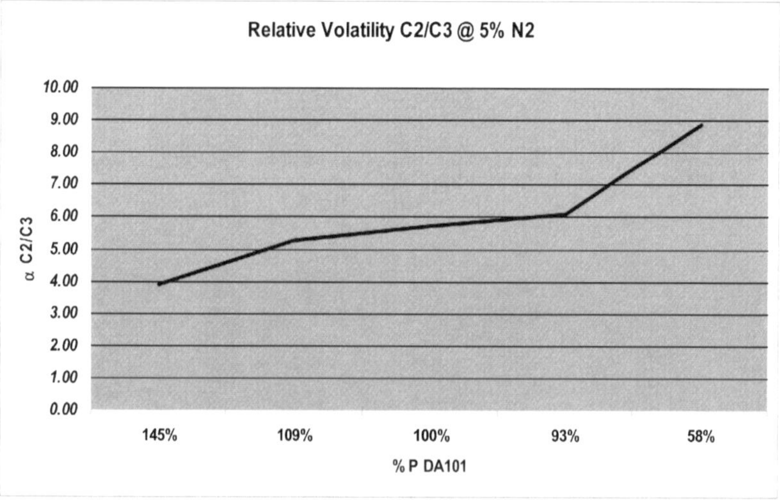

As can be seen in Fig. 3a, when the column pressure is decreased, the Profit response value increases considerably. This occurs because lower column pressure increases relative volatility C_2/C_3 as can be observed in Figure 3b making fractionation easier; the reboiler duty requested in the column to carry on the separation decreases, bringing an increase in Profit response. Profit response is decreased when pressure is increased from normal operating point, this fact is basically due to a reduced relative volatility C_2/C_3 in the column; higher duty is needed by demethanizer reboiler to do the separation. It can be concluded from the above figures that the influence of demethanizer pressure on the hydrocarbon liquid recovering is improved with lower pressures. However, because of strict operating conditions, the lowest level of pressure in demethanizer was set at the minimum operating condition.

Figure 3c. Effect of and power generation capacity of 2nd turbo-expander (GC102) on the Profit response

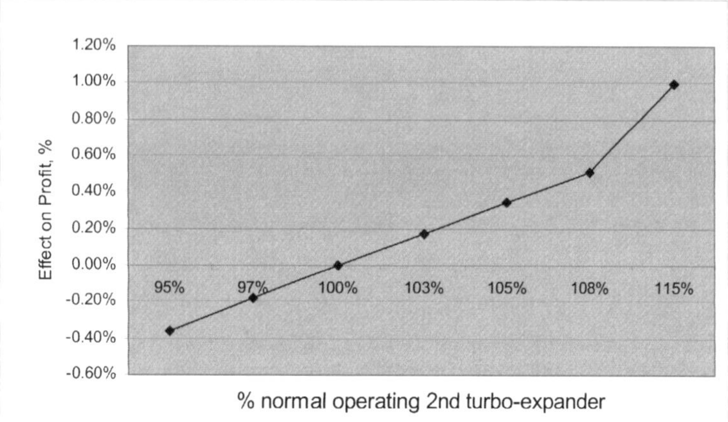

As shown in Fig. 3c, the objective function increases against the increasing power generation capacity of second turbo-expander GC102 . This is because of two basic effects: 1) decreasing the outlet pressure-temperature in outlet expander, which increases the liquid recovered in the cold tank and sent to demethanizer, increasing the total liquid recovery in the system, and 2) extra compression added to the NG stream which reduces the high pressure steam consumed in NG compressors. As a result of both of them effects, the profit value will be increased. Nevertheless, because of turbo-expander operating capacity, the highest level of power due to expansion in GC102 was fixed at that value.

Figure 3d. Effect of and power generation capacity of first turbo-expander (GC101) on the Profit response

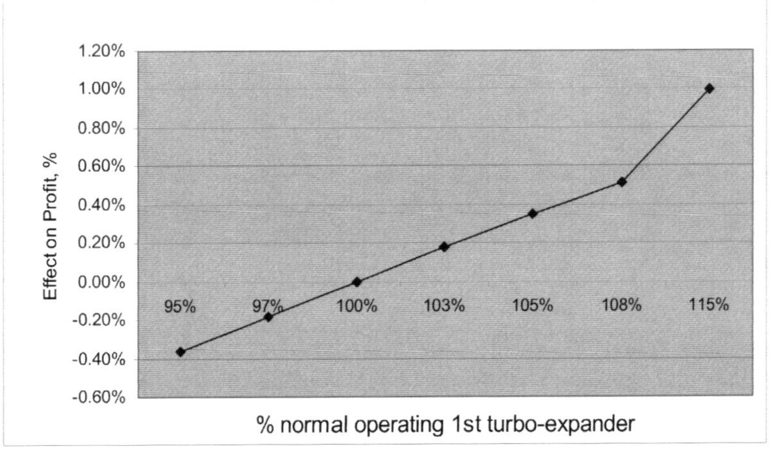

Fig. 3d illustrates the relationship between the objective function and the power generation capacity of first turbo-expander GC101. It shows that the Profit response value rises at the beginning, and reaches a maximum. This behaviour is, similar to the previously explained second turbo-expander, due to two facts: 1) decreasing the outlet pressure-temperature in outlet expander, which increases the liquid recovered in the following separation tank and sent to demethanizer, increasing the total liquid recovery in the system, and 2) extra compression added to the NG stream which reduces the high pressure steam consumed in NG compressors. Because of turbo-expander operating capacity, the highest level of power due to expansion in GC101 was fixed at that value.

Optimization Results

Case A

Once the ranges of values of the independent variables had been defined, a first optimization was performed with built-in AspenPlus optimizer, where Profit value was maximised subject to the variation of these 3 important variables found in the sensitivity analysis. The maximum value of fitness (profit) showed an increase of 4% from the normal operating point. The corresponding values of the independent variables could be achieved by this single optimization, setting their values in their best settings found in the sensitivity analysis (PDA101 -10%, GC102 +5% and GC101+5%).

The profit improvement was not attractive enough; it was necessary to go for more profitable improvements based on changes to it. Changing the model typically falls into one of three categories: changing the model's input parameters, changing the model's initial conditions, and, changing the system design. It was decided to go for optimizing parameters and structural changes together as the initial conditions are set by default and cannot be moved by the users.

Before proceeding with the application of the proposed approach, it was decided to test it by taking the first previous explained optimization as a reference. Following this, there were set the three variables found in the sensitivity analysis which showed to have more impact in the Profit (response) of the unit, and there were ranged from -10% to +10% of their normal operating point. DOE was then applied with a screening purpose. In order to do that a Full Factorial Design (FFD) at 2 levels (2^3) was selected, resulting a total of 8 simulations to perform. Once obtained, the simulations results of all the 8 points, an analysis of variance was performed in Matlab 7.0.1®. The results obtained for the response Profit were: Single factors A, B and C are highly significant, with A as the most important factor, then C and last one B. The rest of the interactions between two factors and among three factors did not show importance. Surface Response Methodology was applied subsequently to obtain the best fitted equation for the response PROFIT based on simulation data. In order to do this, a Central Composite Design was applied based on the previous FFD 2^3. There were randomly performed five central simulation runs to verify the central calibration of the model while proceeding (i.e. no settings were moved away from the initial point). The best fit model was a pure quadratic equation which is shown as follows:

$$\%IMPROVEMENT = 90.98 - (2.74X_A) - (0.33X_B) + (0.04X_C) - (0.19X_A^2) - (0.03X_B^2) - (0.04X_C^2) ----(1)$$

The Relative Root Mean Square Error (RRMSE) of 0.08% indicates a good level of reproduction for the Profit response. After obtaining this equation the Profit was maximized based on varying the 3 most important factors in the ranges found. Based on a NLP optimization a total increase of 4 % in the Profit could be obtained. The results of this second optimization carried out can be positively compared with those of the first done under Aspen Plus Optimizer. It can be concluded that there was not a significant difference between the two optimization results obtained in both total Profit and parameters. This means that both methodologies obtained the same solution. The aim to test the proposed approach with this second optimization was achieved, verifying that it was in good agreement with the results obtained in the first one.

As stated previously, in the previous work section of this text, to increase the profitability of the plant, a pool of structural changes from where to start to determine whether they strongly affect the Profit response is needed. At first glance we looked at 2 principal sources in order to obtain the pool: a) What has been found in studies with this kind of plant applicable to the current one, and b) the experience of people involved in the operation of the plant. However, as the number of structural changes was very large it was necessary to select a group that contained the feasible changes that could be made in this case. Temperature and pressure profiles in the model were taken into consideration to address this, together with space availability in the plant. Figure 4 provides the final structural changes which were applied based on the previous considerations mentioned. The system factors that represent those structural changes are discrete variables.

Figure 4. Structural changes to optimize

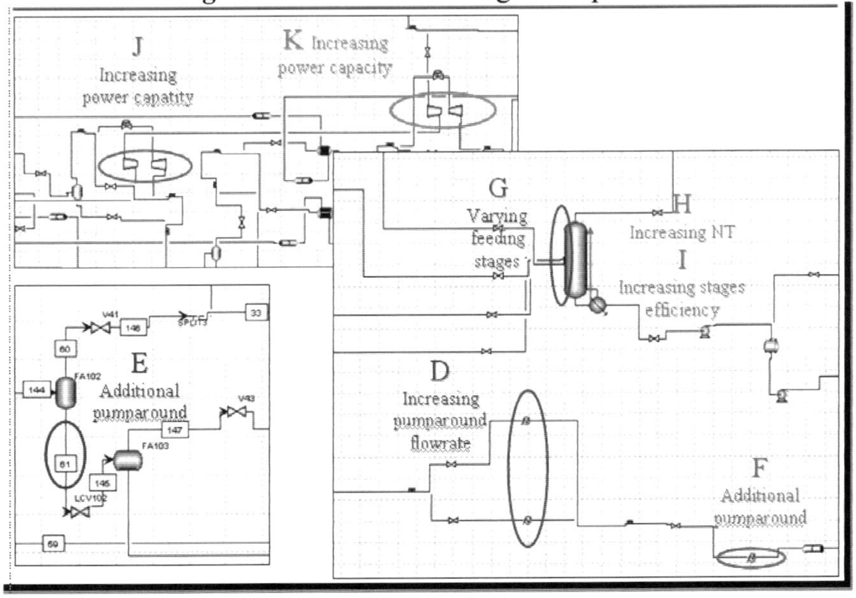

The total number of factors was 11, of those, 4 were continuous variables and 7 were discrete variables. For the purpose of screening, a two-level Fractional Factorial Design was applied to those factors. There were chosen the generators: ‘a’ ‘b’ ‘c’ ‘d’ ‘e’ ‘abcde’ ‘cde’ ‘bde’ ‘ade’ ‘bce’ ‘ace’, with a level of resolution IV to minimise confounding. This resulted in a minimum of 32 simulations. It was assumed that interactions among three or more factors had a lower effect in the Profit response so they were not taken into consideration. Table 1 describes the most important factors that resulted from the analysis of variance applied in order of importance according to their p-value. All of them were main factors, and interactions of second and third order were not statistically significant.

Table 1. Results of the analysis of variance

Most Important Factor	P value < 0.05	Description	Scenario
F	0.0006	Additional pumparound from demethanizer to heat exchanger (P-3)	1
E	0.0059	Additional pumparound from demethanizer to outlet liquid of second separator tank (P-2)	2
A	0.0117	Pressure in stage 1 of demethanizer (PDA101)	0
J	0.0130	Additional turboexpander in parallel with first turboexpander (GC101)	3

Once the set of scenarios were obtained, Surface Response Methodology was applied to each one of those, giving the following results:

Scenario 1

The best fit model was the pure quadratic:

$$\%IMPROVEMENT = 98.94 - (3.60X_A) + (0.44X_B) + (0.06X_C) + (1.33X_A^2) + (1.53X_B^2) + (1.55X_C^2) \text{-------------} (2)$$

The RRMSE of 0.09% indicated a good level of reproduction for the Profit response. The Profit was then maximized based on varying the 3 most important continuous factors identified in the first step in the ranges let. There was achieved a total increase of 19 % in the Profit. This increase was principally due to: a) a 56.3% decrease in demethanizer reboiler duty derived from the heat provided by the new pumparound, this is translated into exceeding low pressure steam that can be exported (sold) as a co-product positively affecting the Profit response, and b) a 0.9% increase in C_3+ recovery due to the low operating pressure of the demethanizer which enhances the volatility coefficients. Nevertheless it was presented a 1.7% increase in high pressure steam consumed by the NG compression to final customers coming from the low power provided by second turboexpandor set at the overall optimum point.

Scenario 2

The best fit model was:

$$\%IMPROVEMENT = 98.94 - (3.60X_A) + (0.44X_B) + (0.06X_C) + (1.33X_A^2) + (1.53X_B^2) + (1.55X_C^2) \text{---------} (3)$$

The RRMSE of 0.13% indicated a good level of reproduction for the Profit response. The Profit was maximized then based on varying the 3 most important continuous factors identified in the first step. There was obtained a total increase of 8% in the Profit. This increase was principally due to: a) a 24% decrease in demethanizer reboiler duty derived from the heat provided by the new pumparound, this exceeding low pressure steam could be exported (sold) as a co-product positively affecting the Profit response, b) a 1.1% increase in C_3+ recovery due to the low operating pressure of the demethanizer which enhances the volatility coefficients, and c) a 4.8% decrease in high pressure steam consumed by the NG compression to final customers, coming from the increasing power provided by both turbo-expanders at maximum capacity.

Scenario 3

The best fit model was:

$$\%IMPROVEMENT = 98.94 - (3.60X_A) + (0.44X_B) + (0.05X_C) + (0.14X_A^2) + (0.34X_B^2) + (0.36X_C^2) \text{----------} (4)$$

The RRMSE of 0.10% indicated a good level of reproduction for the Profit response. After obtaining this equation the Profit was maximized based on varying the 3 most important continuous factors identified in the first step in the ranges let. There was obtained a total increase of 8% in the Profit. This increase was principally due to: a) a 33.2% decrease in demethanizer reboiler duty derived from the heat provided by the new pumparound, this exceeding low pressure steam could be exported (sold) as a co-product helping to increase the Profit response, b) 0.8% increase in C_3+ recovery due to the low operating pressure of the demethanizer which enhances the volatility coefficients, and c) a 14% decrease in high pressure steam consumed in the NG compression to final customers, coming from the increasing power provided by the additional turboexpander.

From all the 3 scenarios plus the first optimization performed, we can see that the most profitable scenario to be applied is the number 1, which increases in 19% the total profit response.

Case B

In this case, as a result of the increasing Nitrogen composition present in the inlet gas ocurring at this time, it was necessary to estimate the Nitrogen effect on the results found so far. At this point, it was thought two possible options: the first was, if Nitrogen composition is an important factor, all the scenarios found in the previous section needed to be re-evaluated so as to identify the best scenarios with Nitrogen factor taken into account. The second was, if Nitrogen composition was not a main factor, all the scenarios found were still valid.

To assess the importance of the inlet Nitrogen, it was added as a factor to the total number of factors examined in case A, resulting in 12 factors, of which 5 were continuous variables and 7 were discrete variables. It was applied a two-level Fractional Factorial Design to do a screening, and there were chosen the generators: 'a' 'b' 'c' 'd' 'e' 'f' 'def' 'abc' 'bcde' 'bcdf' 'abef' 'acef' to minimise confounding, the level of resolution was IV, and this resulted in a minimum of 64 simulations requested. It was assumed that interactions among three or more factors had a lower effect on the Profit response so those interactions were not taken into consideration. Table 2 presents the most important factors with its p-value identified when the simulations and the analysis of variance had been done. As there were three p-value set in 0 by Matlab, and to avoid the difficulty of setting the order of importance based on this value, an effect estimation column was added. Once this had been done, it was possible to rank the order of importance of the MIF.

Table 2. Results of the analysis of variance

Factor	Description	P value $\leq$ 0.05	Factor's Effect
D	Inlet Nitrogen	0	-8,099,624
A	Pressure of demethanizer	0	-231,346
G	Additional pumparound from demethanizer to heat exchanger	0	229,859
K	Additional turboexpander in paralell with first turboexpander	0.0001	140,678
L	Additional turboexpander in paralell with second turboexpander	0.0117	84,300

The ANOVA showed, as expected, that Nitrogen composition was statistically significant (MIF) for the Profit response when compared with all the factors. This can be explained by comparing the Nitrogen with the Methane physical properties. Nitrogen is a heavy component (MW 28) compared with Methane (MW 16), besides this it has a lower Specific Heat (Cp=0.249 Btu/(lbF)) than Methane (Cp=

8.001 Btu/(lbF)), and lower Latent Heat of Vaporization (LHV= 85.6btu/lb) than Methane (LHV= 219.3 btu/lb) what makes a stream rich in Nitrogen a "colder stream" than one poor in Nitrogen and rich in Methane (lower CP-MW). All of these properties make Nitrogen presence in the system (from 5 to 30% N_2) to have effects that affect the Profit response:

Effects that negatively affect the Profit response:

1) As the Nitrogen composition is increased in the system, the relative volatility for Ethane/Propane components is decreased, making fractionation more difficult, and giving a reduction in total liquid recovery (C_3+).
2) As inlet Nitrogen is increased and the liquid recovery is reduced, the total gas produced is heavier and volume increased requiring additional power to compress this, therefore, consuming more high pressure steam.

Effects that positively affect Profit response:

1) The outlet pressure-temperature in the cold tank (located at the second outlet expander) is increased, increasing the vapour fraction inside this and reducing the liquid recovered that is sent to the demethanizer, this results in decreasing the total duty required by the reboiler, which means less low pressure steam consumed.

2) The significant reduced amount of liquid sent to the demethanizer and total reboiler duty make to reduce the temperatures profiles in the demethanizer; this increases the driving force temperature differences, and makes the heat integration in the system easier.

All the above effects affect positively and negatively the Profit response resulting in Nitrogen as an MIF in the range studied. Moreover, there have occurred interesting changes with the additional pumparound from demethanizer to outlet liquid of second separator tank (P-2), it has been moved out of the MIF group and instead of this, the additional turbo-expander in parallel with the second turbo-expander (GC-102) has been included in the group. Based on the previous comparison of Nitrogen with Methane properties, it considered a stream rich in Nitrogen a "colder stream" than one poor in Nitrogen and rich in Methane (lower CP-MW). This specific fact makes the driving force differences in temperatures (ΔT) from the outlet liquid of second separator tank and the pumparound of the demethanizer (P-2) to decrease at a point that is not enough to make it statistically significant for the Profit response. On the other hand, the additional turbo-expander in parallel with second turbo-expander (GC-102) has been turned statistically significant due to the fact that with Nitrogen in the system, the liquid recovered in the cold tank is increased and sent to demethanizer, this brings three effects on the Profit response in the demethanizer column: a) increasing the total duty requested by the reboiler because of increased total flows inside the column, which decreases Profit response, b) improving the liquid recovery in the system that increases Profit response, and c) giving and additional power with the coupled compressor set in the additional turbo-expander (GC-102) that reduces the required high pressure steam used to compress the natural gas produced. As a final balance among those effects, the final result is that the additional turbo-expander in parallel to second turbo-expander (GC-102) is a statistically significant effect (MIF).

Once the set of MIF were identified, it was applied Surface Response Methodology to obtain an optimum scenario. In order to do this, a Central Composite Design was built based on the previous Fractional Factorial Design. There were set complementary simulations at α values of +/- (number of factors)$^{1/4}$ and at the central point to verify the central calibration of the model while proceeding (i.e. no settings were moved away from the initial point). Because of the strong effect of Nitrogen in the Profit response, the SRM was applied at 4 fixed levels of inlet Nitrogen in the system (5, 10, 15 and 20% mole). In order to obtain the highest Profit based on the pressure of demethanizer column, it was set at its lowest possible operative value. Therefore, the SRM applied considered the remaining most important factors G, K, and L. Table 3 presents the best fit models for each one of the Nitrogen levels, and their Relative Root Mean Square Error (RRME). The response is shown as an improvement of the normal operating Profit, and the factors G, K and L are coded (ranged from 0 to 1.316).

Table 3. Best fit models at Nitrogen levels

% mole inlet Nitrogen	Best Fit Model	% RRME
5	$\%IP = 118.1 + 0.1\,K + 0.2L - 0.1KL + 0.1\,L^2$	0.13
10	$\%\,IP = 111.7 + 0.05\,K - 0.25G^2 - 0.25K^2 - 0.11\,L^2$	0.01
15	$\%IP = 183.8 + 0.6K + 1.1L + 0.1\,L^2$	0.01
20	$\%IP = 1.4 - 0.6\,K - 1.4L + 0.1\,L^2$	-0.24

The Relative Root Mean Square Errors (RRMSE) indicated a good level of reproduction for the Profit response. It was possible to maximize each fit model based on a NLP optimization. Table 4 visualizes the total increase of Profit achieved for each one with respect to each base case, and the percentages of variation that caused that Profit increase.

Table 4. Improvements in Profit and variations for each fit model

Inlet N2	% Variation			
% mole	ΔProfit	ΔDemethanizer reboiler duty	ΔC3+ recovery	ΔNG compression power
5	+8.4%	-12%	+3.9%	-40%
10	+9.0%	-13%	+5.1%	-39%
15	+9.5%	-19%	+5.5%	-38%
20	+9.8%	-22%	+6.2%	-36%

These increases were principally due to: a) a decrease in demethanizer reboiler duty derived from the heat provided by the new pumparound, this translates into exceeding low pressure steam that is exported (sold) as a co-product affecting positively the Profit response. b) an increase in C_3+ recovery due to the low operating pressure of the demethanizer which enhances the volatility coefficients, and to the increased liquid entering the demethanizer coming from the outlet liquid once the expansion power provided by first and second turbo-expanders has been increased with the addition of two more in parallel.

From all the 4 scenarios plus the first optimization performed, it can seen that the most profitable scenario to be applied strongly depends on the inlet Nitrogen. Nowadays, the current conditions for the inlet Nitrogen are nearly 9 %mole, therefore the closest scenario to apply would give an increase of nearly 9% of the total profit response. However, as Nitrogen content in the inlet gas is highly probably to be increased, there must be carefully selected the scenario to be applied.

Conclusions and further work

The purpose of the current study was to propose an approach for the determination of profitable retrofit solutions considering DoE and simulation. The approach applied is based in a steady state model (Aspen Plus), which agrees well with the measured data. DoE has been used to estimate the benefits and losses from structural changes made to the unit (retrofit design). Most important factors in affecting the profit response have been identified by analysis of variance and used for developing optimization strategies based on simplified models obtained from SRM and then applying non linear optimization. The approach was applied to a NGL recovery unit identifying the group of models that increased the total profit depending on the inlet Nitrogen to the system, and need to be complemented by further capital investment estimation.

Future work must be done in order to improve the efficiency and efficacy of the approach. This work should include the following:

a) To develop complementary links and algorithms to reduce the time of solution and,
b) To perform a complementary analysis to achieve global optimal solutions.
c) To replicate the study in a different production process in order to ensure robustness and systematicness.
d) To perform multi-objective optimization by taking into account environmental issues.

References

1. John D. Wilkinson, Hank M. Hudson, Kyle T. Cuellar, Richard N. Pitman (1998). "Next Generation Processes for NGL / LPG Recovery." Ortloff Engineers, Ltd. Midland, Texas USA.
2. Smith, R. (2005). "Chemical process design and integration." Chichester: Wiley. ISBN 0471486817.
3. S. Diaz, A. Serrani, A. Bandoni and E. A. Brignole(1996), "A study on the capital and operating alternatives in an Ethane extraction plant." Computers chem. Engng Vol. 20, Suppl., pp. S1499-SI504.
4. M. S. Diaz, A. Serrani, J. A. Bandoni,† and E. A. Brignole (1997), "Automatic Design and Optimization of Natural Gas Plants." Ind. Eng. Chem. Res., 36, 2715-2724.
5. Mehdi Mehrpooya,Farhad Gharagheizi, Ali Vatani1 (2006), "An Optimization of Capital and Operating Alternatives in a NGL Recovery Unit." Chem. Eng. Technol. 2006, 29, No. 12, 1469–1480.
6. Grossmann, I. E., & Kravanja, Z. (1997), "Mixed-integer nonlinear programming: a survey of algorithms and applications." In L. T. Biegler, et al. (Eds.), Large-scale optimization with applications, part II: optimal design and control (pp. 73-100). Berlin: Springer.
7. Sangbum Lee, Ignacio E. Grossmann (2003), "Global optimization of nonlinear generalized disjunctive programming with bilinear equality constraints: applications to process networks." Computers and Chemical Engineering 27: 1557-1575.

8. Karuppiah, R. and I.E. Grossmann (2008), "Global Optimization of Multiscenario Mixed Integer Nonlinear Programming Models arising in the Synthesis of Integrated Water Networks under Uncertainty." Computers and Chemical Engineering, 32, 145-160.
9. Ponce-Ortega, J.M., A. Jiménez-Gutiérrez, and I.E. Grossmann "Optimal synthesis of heat exchanger networks involving isothermal process streams." Computers & Chemical Engineering, 32, 1918-1942 (2008).
10. J., L. and Acevedo (2004), "Processes Synthesis and Design of Distillation Sequences Using Modular Simulators: A Genetic Algorithm Framework." Computers and Chemical Engineering 28: 1223-1236.
11. B., Gross and Roosen P. (1998), "Total Process Optimization in Chemical Engineering with Evolutionary Algorithms." Computers and Chemical Engineering 22: s229-s236.
12. Li P. et al. (2000), "Integration of Simulated Annealing to a Simulation Tool for Dynamic Optimization of Chemical Processes." Chemical Engineering and Processing. 39: 357-363.
13. Gaia Franceschini, Sandro Macchietto(2008), "Model-based design of experiments for parameter precision: State of the art." Chemical Engineering Science 63: 4846-4872.
14. Laura Ilzarbe, Marıa Jesus Alvarez, Elisabeth Viles and Martin Tanco(2008), "Practical Applications of Design of Experiments in the Field of Engineering: A Bibliographical Review." Qual. Reliab. Engng. Int.; 24:417–428.
15. Montgomery, Douglas C.(1997), "Design and analysis of experiments." New York ; Chichester: Wiley. ISBN 0471157465.
16. Aspen Plus v21.0 User Guide, Aspen Technology Inc., 2007, (www.aspentech.com).

Design LNG Facilities to Minimize Risks from Cryogenic Exposure

56e

Michael Livingston, P. E. and Richard Gustafson, P. E., C. S. P.
WS Atkins, Houston, TX 77079

Phillip Guy, P. E. and Luis J. Padilla, P. E.
AMG Engineering, Inc., Houston, TX 77084

Craig Bloom, Kamal Shah, P. E., and Victor H. Edwards, Ph. D., P. E.
Aker Solutions US Inc.
3600 Briarpark Drive
Houston, TX 77042-5206

Correspondence: vic.edwards@akersolutions.com

AIChE Spring National Meeting 2009
9th Topical Conference on Natural Gas Utilization
Tampa, FL April 26-30, 2009

The cryogenic nature of LNG facilities poses the risk of potentially injurious low temperature exposure of personnel, structural steel, equipment, and instrumentation, control and power cabling. The probability of cryogenic exposure due to loss of containment of LNG is inherently greater than the probability of fire exposure due to loss of containment because of the many precautions taken to eliminate ignition sources in LNG facilities.

This paper contrasts the hazards of cryogenic and fire exposure to personnel and facilities using examples of consequence modeling of pool and jet releases. Practical measures to eliminate or mitigate risk from cryogenic and fire exposure will be presented. Strategies for both the onshore and offshore LNG facilities will be discussed.

Introduction

Liquefied Natural Gas (LNG) is a safe and practical way to transport natural gas by sea from remote locations to user distribution systems. LNG is also an effective means for storing natural gas at peak-shaving plants during periods of low demand. Aker Solutions has designed and built state-of-the art onshore and gravity-based offshore LNG receiving, storage, and regasification terminals.

As with any hydrocarbon processing facility, fire prevention and protection are important considerations in LNG facilities. Because of its cryogenic nature (atmospheric boiling point approximately – 260°F), LNG also poses hazards of personnel injury and damage of structures and equipment from cryogenic exposure.

The design and operation of LNG terminals minimizes ignition sources, resulting in cryogenic exposure being more likely than fire exposure. This is particularly true in the high pressure processing areas where the inventory of fluid is lower but where the higher pressure creates greater potential for cryogenic exposure to personnel and the surrounding area.

Cryogenic exposure of personnel causes freeze burns; cryogenic exposure of carbon steel causes embrittlement, possibly resulting in structural failure. Consequently, protection from cryogenic exposure, as well as from fire exposure, is needed.

Protective measures should be chosen that are effective for both fire exposure and cryogenic exposure. Protective measures add cost; thus, they should only be applied to those parts of facilities where the possibility of harm exists. Consequence modeling can be used to predict the extent of potential fire and cryogenic exposure so that protection can be applied where necessary.

For the most part, onshore LNG facilities have generous spacing of equipment, so significant cost savings in fire and cryogenic protection can be achieved without compromising safety. In addition, relocation of personnel to a safe area is usually not an issue and the decision to provide facility thermal protection becomes an asset protection/capital investment question.

Offshore LNG facilities have comparatively close spacing because of the high cost of building offshore, so fire and cryogenic protection must be applied to a much higher proportion of equipment and structural steel. Egress and relocation to safe refuge areas are also significant factors in this evaluation. If the structure of the offshore platform is compromised, it would have to be abandoned using egress chutes, davit boats, freefall boats, life rafts, etc.

Two philosophies can be applied to fire and cryogenic protection. One philosophy is to protect all structural steel and equipment supports that could be exposed to fire and/or cryogenic temperatures. A second philosophy is to protect structural steel and equipment supports only where failure could lead to escalation of the incident.

This paper presents an overview of the integration of fire and cryogenic protection for onshore and offshore LNG receiving and regasification facilities. The principles illustrated here can also be applied to liquefaction and peak-shaving facilities.

Hazards of Cryogenic Exposure

Hazards to Personnel

Exposure of personnel to LNG and cold gas can cause severe cryogenic burns, resulting in tissue damage similar to frost bite or thermal burns. Contact with non-insulated and even insulated parts of equipment or vessels containing cryogenic fluids can also result in frost bite. Unprotected skin may stick to low-temperature surfaces and flesh may be torn upon removal. These hazards should be controlled by separation, guarding, insulation, and personal protective equipment such as gloves, safety glasses, and face shields.

Inhalation of cold vapor can damage the lungs and may trigger an asthma attack in susceptible individuals. Asphyxiation is a serious hazard because vaporized LNG is usually odorless. Air contains 21 percent oxygen. If the oxygen content falls below 18 percent, adverse effects such as loss of mental alertness and performance may result. At six to ten percent oxygen or less, exertion is impossible; collapse and unconsciousness occurs. At six percent oxygen or below death would occur in six to eight minutes.

Personnel in the vicinity of an LNG release can quickly be enveloped by cold hydrocarbon vapors resulting in oxygen deficient zones. The expansion ratio of LNG is approximately 600:1. Therefore the release of 1 m^3 of LNG will produce 600 m^3 of 100 percent natural gas.

Hazards to Structures and Equipment

Carbon steel, which is widely used in process plant structures and in the hulls of LNG carriers, loses its ductility and becomes brittle when exposed to LNG or cold natural gas. Figure 1 shows that AISI 4130 steel loses half of its impact resistance at 60°F below zero. Some other carbon steels become brittle at temperatures of 20°F below zero. LNG has a boiling point of 260°F below zero or 200°R

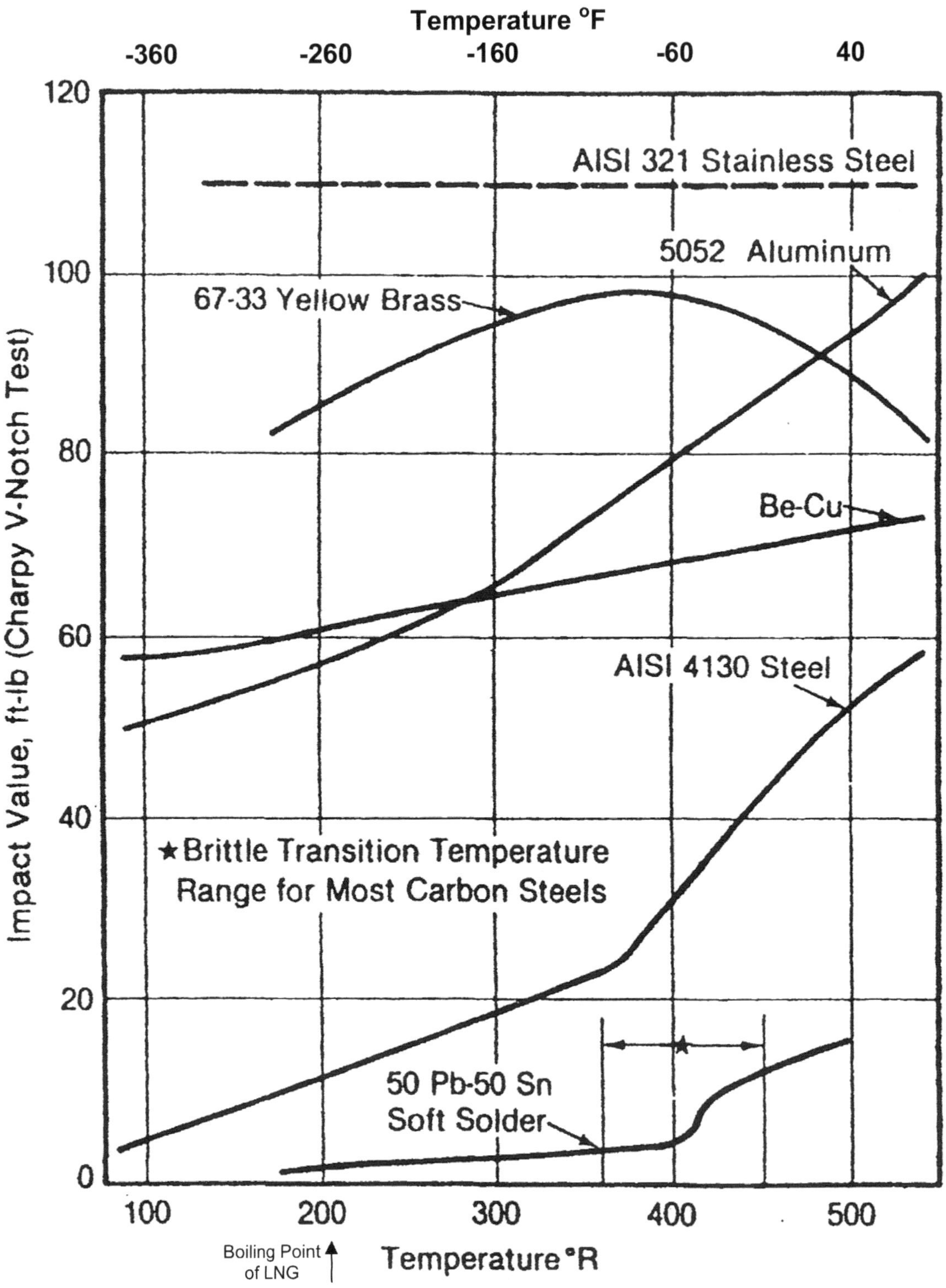

Figure 1 – Low Temperature Impact Strength of Metals (Flynn, 2005).

Since beginning LNG tanker trade in 1969, there have been eight marine incidents resulting in spillage of LNG with some hull damage due to cold fracture. However, to date there have been no cargo fires. Figure 2 shows a 2 m crack in the deck of an LNG carrier exposed to a 30 liter LNG spill.

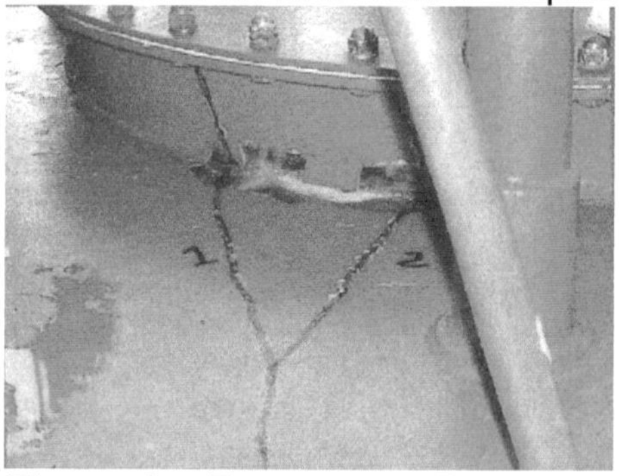

Figure 2 – Damage to Deck of LNG Carrier by LNG Spill.

Direct contact of LNG with structural steel can rapidly cool the steel to below embrittlement temperature. Experiments have demonstrated that immersion of 1/2 in and 1 in pieces of painted steel in LNG can completely cool the steel to LNG temperatures in less than two minutes. When combined with suggested failure criteria for structural steel sections due to embrittlement, these high heat transfer fluxes predict steel section failure in as little as one to five seconds. Vapor heat transfer due to contact with cold natural gas velocities is predicted to be much slower.

The cooling rate of structural steel depends on the amount of LNG available for cooling the steel per surface area, i.e. the LNG liquid flux in the jet. The LNG liquid flux is controlled by the flow rate and the location of the steel relative to the origin of the LNG release.

Because cooling rates are so rapid, early leak detection and system isolation and shutdown have little effect on managing cryogenic LNG hazards in the immediate area of the release. By the time the detection and shutdown system has activated, the cryogenic damage is complete within the LNG exposure hazard envelope. Thus cryogenic protection requires changing position, changing material of construction, or adding protection such as cryogenic insulation or shielding. It should be noted that rapid detection and process isolation will serve to limit the total volume of LNG released, lowering the potential for the LNG to spread over an even greater area, thereby reducing the exposure of even more equipment and structures to cryogenic conditions.

Polymeric materials, such as plastics and elastomers, are also subject to rapid brittle fracture on exposure to LNG, compromising some equipment components and electrical insulation.

In the United States, NFPA 59A "Standard for the Production, Storage, and Handling of Liquefied Natural Gas (LNG)" is one of the key design documents for the design of LNG facilities. In Europe, EN 1473 "Installation and Equipment for Liquefied Natural Gas – Design of Onshore Installations" is normally used. Both NFPA 59A and EN 1473 require that equipment, controls, and structures whose failure would result in escalation of the incident must be protected from cryogenic embrittlement.

Hazards of Fire Exposure

In contrast to cryogenic hazards, fire hazards associated with vaporization of LNG releases can be substantially reduced by rapid detection of releases, followed by shutdown and isolation of equipment. Experience has shown that fire impinging upon structural steel takes a few minutes of exposure to threaten the steel's integrity. Figure 3 illustrates the rate of temperature rise of steel plates exposed to a gasoline pool fire (API 521, 2007). Heating rate would be more rapid for direct impingement of jet fires. The heat flux associated with large pool fires would be approximately 120 kW/m^2 for fires larger than the object exposed and approximately 85 kW/m^2 for pool fires comparable in size to the exposed object. The heat flux associated with jet fires would be approximately 250 kW/m^2 maximum (DiNenno, 2002).

Due to the low pressures during loading, unloading and storage of LNG, and due to the rapid detection and shutdown system, large jet fires are limited to the high pressure pumps, vaporizers, and export gas pipeline sections of an LNG receiving, storage, and regasification terminal. Low pressure LNG releases from isolated low pressure sections of the terminal are expected to give rise to local pool fire hazards if ignited. Pool fires can be controlled with sloping, curbing, and trenching.

LNG releases that discharge at a pressure less than 4 barg are assumed to form liquid pools rather than jets.

Guidelines for fire protection include the following references:
API RP 14G (2007), API 2218 (1999), API 2510A (1996), NFPA 59A (2009), EN 1473 (2007), API 2FB (2006).

Modeling the Consequences of Loss of Containment of LNG

In order to design necessary thermal protection measures for an LNG facility, it is necessary to predict the consequences of leaks of LNG and natural gas from the facility. This can be done using either software correlation models such as PHAST® and CANARY® or with computational fluid dynamics (CFD) models. The work illustrated here used PHAST® (Process Hazards Analysis Software Tool, licensed from DNV Technica), which models leak rates for various

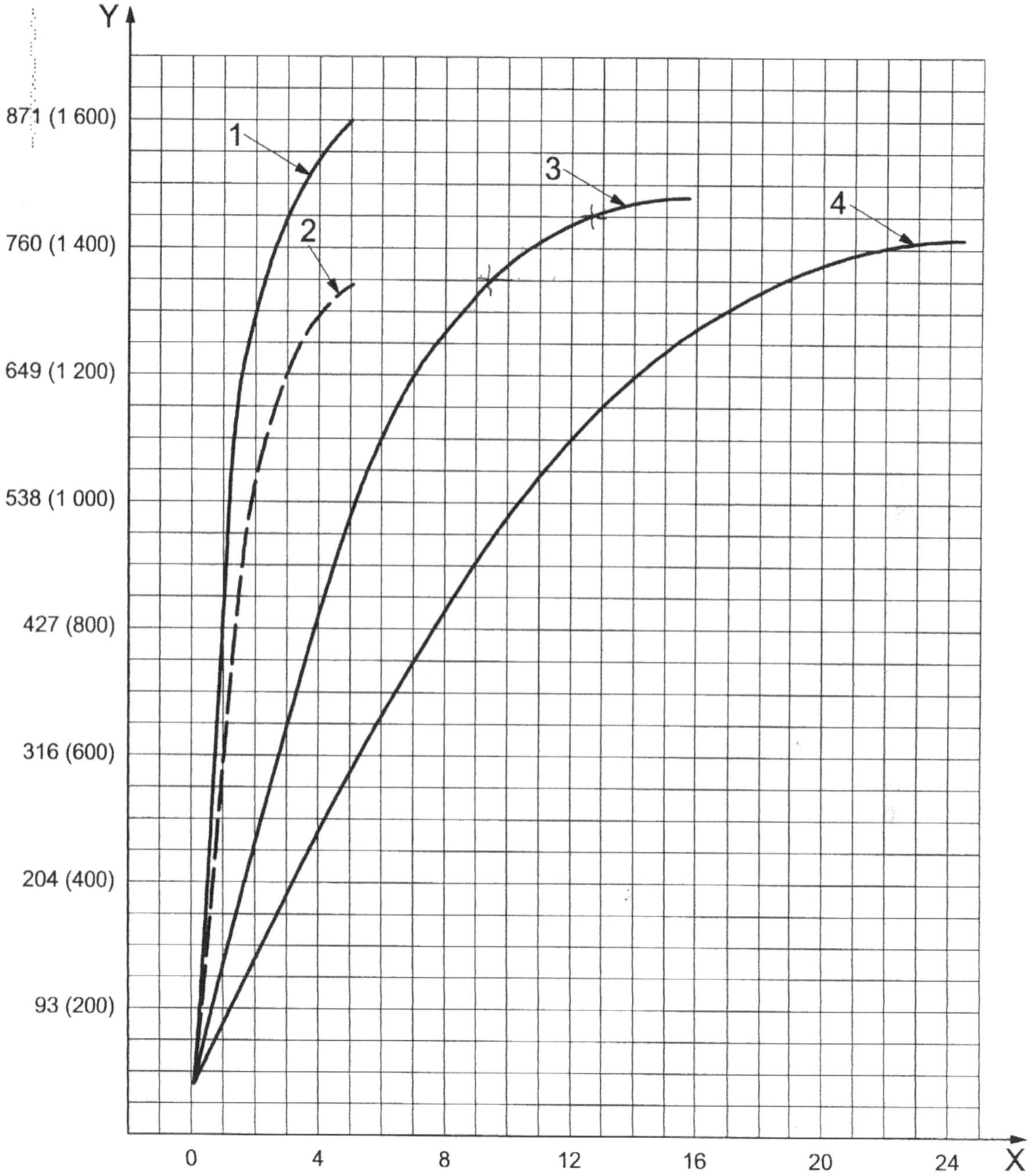

Key

X time after start of the fire, expressed in minutes

Y plate temperature, averaged over 2,3 m^3 (24 ft^2), expressed in degrees Celsius (degrees Fahrenheit)

1 plate 3,2 mm (1/8 in) thick, as computed

2 plate 3,2 mm (1/8 in) thick, observed

3 plate 12,7 mm (1/2 in) thick, as computed

4 plate 25,4 mm (1 in) thick, as computed

Average rate of heating steel plates exposed to open gasoline fire on one side

Figure 3 Temperature-time profiles from API Standard 521 (2007).

scenarios, atmospheric dispersion and rainout of releases, and the potential effects of resulting fires and/or explosions. This consequence modeling allows the design team to establish which portions of the LNG facility require thermal protection from cryogenic and external fire exposure.

Identification of Credible Scenarios

The first step in consequence modeling is the definition of credible leak scenarios to be modeled. Each LNG facility has its own isolation criterion. In the onshore LNG facility illustrated here, two minutes should be enough time to detect, isolate, and shutdown the facility in the event of an unplanned release or leak. Each isolated segment of the facility then contains an inventory of LNG, natural gas, or a mixture of the two phases. The segment that has the leak will then empty at a rate that is dependent on variables such as the initial inventory, pressure and temperature of the segment and the size, location, and orientation of the hole.

Failure rates and hole size distributions for LNG service are not readily available. The Gas Research Institute prepared a report (GRI, 1990) that collected failure rates and types of failure for LNG service equipment, but the report did not provide specifics on the size or distribution of holes. The report indicated that most of the major leaks were a result of either vaporizer tube ruptures or pump failures and that most major fires involved vaporizers.

In the current work, modeled hole sizes for onshore facilities include 0.12 in, 0.5 in, 0.75 in, 2 in, and 4 in. The 0.12 in release was selected to represent small leaks such as a leaking seal. The 0.5 in and 0.75 in leaks were modeled to represent flange leaks and small bore fitting ruptures such as instrument connections and drains. The 2 in and 4 in holes were modeled to represent fatigue, dropped objects, severe localized corrosion and large bore holes. However, for the onshore facility illustrated here, scenarios leading to 2 in and 4 in holes and larger are not considered credible for provision of thermal protection because of (1) the cleanliness of LNG service, (2) piping containing LNG is often welded rather than flanged, (3) most of the instrument connections are ½" and ¾" taps, (4) all piping and flanged connections to equipment containing LNG will be protected with insulation, jacketing and stainless steel straps, (5) a guillotine-type full diameter rupture of welded piping is not considered a credible scenario, and (6) a complete loss of a gasket in a flange for a large diameter LNG pipe is not considered credible.

Isolatable Segments and Jet Fire Potential

Table 1 summarizes typical isolatable segments for one onshore LNG receiving, storage, and regasification facility. Note that not all of these segments will produce a jet fire of duration long enough (in this work assumed to be five minutes) to cause failure of uninsulated heavy structural steel, as shown by Table 2. The expected failure time could be less, depending on the dimensions of

Table 1 – Typical Isolatable Segments for an Onshore LNG Receiving, Storage, and Regasification Facility

Section	Description	Assumed Initial State	Pressure (psig)	Temp (°F)	Vol (ft^3)
1	Unloading Arm	Subcooled LNG with a bubble point of 2.9 psig	68	-256.9	572
2	Unloading Arm Header to Shore Station ESDV	Subcooled LNG with a bubble point of 2.9 psig	63	-256.9	10798
3	In-Tank Pumps Discharge Line to HP Pumps ESDV	Subcooled LNG with a bubble point of 2.9 psig	140	-254.6	1043
4	LNG Inlet Piping from ESDV in Tank Inlet Manifold to Tank	Subcooled LNG with a bubble point of 2.9 psig	44	-256.9	4700
5	Upstream Vaporizer Piping (LNG)	Saturated LNG at 1500 psig	1480	-246.5	238
6	Downstream Vaporizer Piping (Gas)	Methane gas at the specified temperature and pressure	1400	42.4	7177
7, 9	LNG Surge Drum, BOG Condenser & Recirculation Line	Subcooled LNG with a bubble point of 2.9 psig	100	-254.3	5847
8	ESDV Upstream of HP Pumps to ESDV Upstream of Vaporizers	Subcooled LNG with a bubble point of 2.9 psig	1480	-246.5	2280
10	LNG Tanks Unloading Line Header	Subcooled LNG with a bubble point of 2.9 psig	63	-256.9	3488

Note: LNG modeled as methane

the steel member, its loading, and its design safety factor. Even including two minutes duration before isolation and shutdown is achieved, the duration of the leaks is less than five minutes for many of the credible scenarios. The duration of a leak is the two minute shutdown and isolation time plus the time required for the pressure to fall to atmospheric pressure in the isolated and leaking segment. Leakage rates will drop fairly soon after isolation because LNG is relatively incompressible, it has a vapor pressure of only 2.9 psig in most segments and the segments will be insulated. Note that autorefrigeration will occur during the initial leakage, partial vaporization, and depressurization of the isolated segment. These will rapidly bringing the LNG saturation pressure to atmospheric pressure. Once depressurization is complete, the leak rate will be due primarily to drainage and very slow vaporization caused by ambient heating of the insulated segment.

For this work, ignited releases were assumed to be a credible jet fire hazard if they existed for at least five minutes at a source pressure greater than 4 barg (58 psig). As explained earlier more conservative numbers may be warranted. Jet fire hazards are summarized in Table 3 (below).

Table 2 – Duration of Leak Required for Contents to Reach Atmospheric Pressure in a Typical Onshore LNG Receiving, Storage, and Regasification Facility

Section	Description	Hole size in (mm)	Duration (min)
1	Unloading Arm	0.12 (3)	4.2
		0.5 (12.7)	2.1
		0.75 (19.05)	2.1
2	Unloading Arm Header to Shore Station ESDV	0.12 (3)	42.
		0.5 (12.7)	4.2
		0.75 (19.05)	3.0
3	In-Tank Pumps Discharge Line to HP Pumps ESDV	0.12 (3)	8.5
		0.5 (12.7)	2.4
		0.75 (19.05)	2.2
4	LNG Inlet Piping from ESDV in Tank Inlet Manifold to Tank	0.12 (3)	16.
		0.5 (12.7)	2.8
		0.75 (19.05)	2.3
5	Upstream Vaporizer Piping (LNG)	0.12 (3)	120.
		0.5 (12.7)	10.2
		0.75 (19.05)	4.9
6	Downstream Vaporizer Piping (Gas)	0.12 (3)	>7900
		0.5 (12.7)	>430
		0.75 (19.05)	>200
7,9	LNG Surge Drum, BOG Condenser & Recirculation Line	0.12 (3)	32.
		0.5 (12.7)	3.7
		0.75 (19.05)	2.8
8	ESDV Upstream of HP Pumps to ESDV Upstream of Vaporizers	0.12 (3)	53.
		0.5 (12.7)	4.9
		0.75 (19.05)	3.3
10	LNG Tanks Unloading Line Header	0.12 (3)	15.
		0.5 (12.7)	2.7
		0.75 (19.05)	2.3

Note: These durations do not account for elevation differences within a segment.

Releases of LNG that discharge at a pressure of less than 4 barg (58 psig) are assumed to form liquid pools. As the pressure continues to drop, wind and rainout become factors and active fire fighting measures can effectively control the magnitude and exposure of the event. Process hazards analysis software such as PHAST® can predict when pool formation is expected and suitable precautions can be taken to protect structural elements and equipment.

Prediction of Cryogenic Exposure

As discussed previously, ordinary structural steel can fail rapidly when exposed to LNG. In this work, PHAST® was used to predict the range and liquid content of unignited jets of vaporizing LNG. These predictions provide a basis for identifying those structural members requiring thermal protection (Table 4) within the hazard envelopes.

Whereas in Table 3, thermal protection from jet fires was not warranted in segments 1, 2, 3, 4, and 10, Table 4 shows that thermal protection from cryogenic exposure may be warranted in all segments. This is because cryogenic exposure can lead to failure before the time needed to detect and stop a leak. Protective measures will be described in the next two sections.

Table 3 – Jet Fire Hazard Summary @ Five Minutes for a Typical Onshore LNG Receiving, Storage, and Regasification Terminal

Section	Hole Diameter in (mm)	Pressure psi @ 5min	Flame Length ft @ 5min	Comment
1	0.12 (3)	2.9	N/A	Pressure < 58 psig
2	0.12 (3)	56.	N/A	Pressure < 58 psig
3	0.12 (3)	51.	N/A	Pressure < 58 psig
4	0.12 (3)	31.	N/A	Pressure < 58 psig
5	0.12 (3)	1480.	42.	
	0.5 (12.7)	1480.	99.	
	0.75 (19.05)	1480.	140.	
	2 (50)	2.9	N/A	Duration < 5 min
6	0.12 (3)	696.	15.4	
	0.5 (12.7)	696.	54.	
	0.75 (19.05)	672.	76.	
	2 (50)	435.	146.	
7 / 9	0.12 (3)	84.	28.	
8	0.12 (3)	1317.	42.	
10	0.12 (3)	42.	N/A	Pressure < 58 psig

Risk-based Protection of Onshore Facilities

There are dozens of onshore LNG facilities around the world. Figure 4 is a photograph of a new LNG terminal that Aker Solutions, in partnership with IHI, is building for Sempra LNG at Cameron, LA.

Most onshore LNG facilities are comparatively open, equipment is not congested, and risk to life is low. In those cases, risk management becomes primarily a matter of asset protection. One risk-based philosophy that minimizes initial capital cost is to protect all assets that could be exposed to cryogenic fluids or fire, whose failure could lead to escalation of the incident. Because of wide spacing, and because many assets can fail without causing escalation, only

Table 4 – Cryogenic Hazard Summary for a Typical LNG Receiving, Storage, and Regasification Terminal

Section	Description	Hole Diameter in (mm)	Jet Hazard Length ft
1	Unloading Arm	0.12 (3)	4.2
		0.5 (12.7)	18.8
		0.75 (19.05)	27.
2	Unloading Arm Header to Shore Station ESDV	0.12 (3)	4.9
		0.5 (12.7)	18.8
		0.75 (19.05)	27.
3	In-Tank Pumps Discharge Line to HP Pumps ESDV	0.12 (3)	5.3
		0.5 (12.7)	19.8
		0.75 (19.05)	27.7
4	LNG Inlet Piping from ESDV in Tank Inlet Manifold to Tank	0.12 (3)	5.7
		0.5 (12.7)	18.8
		0.75 (19.05)	25.6
5	Upstream Vaporizer Piping (LNG)	0.12 (3)	5.2
		0.5 (12.7)	22.
		0.75 (19.05)	35.
6	Downstream Vaporizer Piping (Gas)	0.12 (3)	0.79
		0.5 (12.7)	3.7
		0.75 (19.05)	5.8
7/9	LNG Surge Drum, BOG Condenser & Recirculation Line	0.12 (3)	6.2
		0.5 (12.7)	20.6
		0.75 (19.05)	27.3
8	ESDV Upstream of HP Pumps to ESDV Upstream of Vaporizers	0.12 (3)	5.7
		0.5 (12.7)	23.1
		0.75 (19.05)	38.
10	LNG Tanks Unloading Line Header	0.12 (3)	4.8
		0.5 (12.7)	18.8
		0.75 (19.05)	27.2

some of the assets will require protection. Of course, loss of even a part of a facility can cause extended loss of production with a major impact on overall cost over the life of the facility; the choice is a business decision.

Onshore concrete and cementitious fireproofing materials have traditionally been the preferred choices due to cost, but other options are available. Cementitious fireproofing products such as Pyrocrete 241 provide economical protection of structural steel against short-term cryogenic exposure. Unfortunately at this time, not all potential fireproofing products have been tested for their ability to withstand cryogenic exposure, fire exposure or a combination of these hazards.

Industry testing has been conducted on intumescent and subliming fireproofing coatings such as Chartek, Thermolag, and Firetex. These materials used in conjunction with a cryogenic insulating coating can provide good protection from both cryogenic exposure and jet fire exposure but are more expensive to apply than most cementitious systems. With any of these systems corrosion under the insulation/fireproofing material needs to be considered and managed in the design.

There are no current industry standard tests for the performance of the coating materials to resist the effects of LNG cryogenic exposure and then a subsequent fire. The current testing data available is generally the result of manufacturers' independent research, and ongoing construction project testing

Figure 4 – LNG Receiving, Storage, and Regasification Facility Under Construction at Cameron, LA

One of the primary difficulties in designing for LNG release scenarios is that depending on the specific scenario, there could be a release resulting in a cryogenic exposure, a fire exposure (jet, pool, or spray), or a combination of events.

If pooled LNG does not ignite, then the bases of columns and equipment supports could be exposed and then fail due to cryogenic exposure. A spill containment system consisting of curbing, sloped paving, and troughs should be provided under all LNG lines and equipment in the plant. This type of containment system limits the area that can be affected by an LNG spill and the exposure duration. Limiting exposure duration keeps insulation requirements from becoming too thick and impractical. The containment area layout should consider the potential exposure areas that could result from a pressurized release of LNG.

When designing the LNG spill containment system, consider the Leiden frost effect which leads to higher liquid velocities due to the creation of a vapor film between the solid spill containment system and the boiling LNG. These higher liquid velocities when compared to flowing water could cause splashing around obstructions and overshoot a sloped trough at turns and changes in elevation. Where structural steel and/or critical equipment supports are within the curbing and/or drainage paths, one option is to support them on a suitable concrete base that prevents exposure of the steel to the pooling, splashing, or draining liquid. Stainless steel pedestals or support members have also been used where preferable.

Protection of instrument and electrical cabling is normally not done because these systems are designed to be fail-safe and/or redundant. However, specific review of the potential exposure to the shutdown and blowdown system controls should be conducted. Direct exposure from cryogenic spray to shutdown/blowdown valves or actuators could result in the failure to isolate or deinventory the process. In general, the probability of the cryogenic spray impinging on the specific equipment should be considered for the overall probability of the equipment's failure on demand in order to evaluate the need for additional protective measures.

Protective Measures - Offshore Facilities

Although less common and more expensive than onshore LNG facilities, LNG facilities may be located offshore when there are no suitable sites onshore. Aker Solutions has designed and constructed the first gravity-based offshore LNG receiving, storage, and regasification facility for installation in the Adriatic Sea to serve the Italian natural gas network (Figure 5).

Because of the close spacing offshore, protection of all assets that could be damaged by exposure to cryogenic fluids or fire is recommended. Because of weight restrictions cementitious based fireproofing is not practical. An example of a suitable system, developed by International Protective Coating, is the "Chartek Duplex System", a combination of Chartek 1709 (intumescent fireproofing) and Intertherm 7050 (thermal insulation). This is a lightweight ablative layering system that provides both fire and cryogenic protection. Insulation suitable for offshore structural steel, decking, and equipment must be resistant to salt water as well as to cryogenic and fire exposure. This type of epoxy-based system can

Figure 5 – Tow Out of the Adriatic LNG Facility from Spain in September 2008.

provide both cryogenic and jet fire protection; it can also serve as a coating to inhibit corrosion effects.

When compared with onshore LNG facilities, the significantly smaller areas associated with an offshore facility increase the potential for LNG release incidents to impair occupant evacuation and also to escalate damage to the facilities. One method to reduce the potential exposures (both cryogenic and pressurized fire) is to provide flange guards on specific flange connections. Such flange guards serve to reduce the potential spray area, and to prevent well-formed jets from occurring.

As noted above, cryogenic spray exposure on shutdown and blowdown valves has the potential to cause failure of the actuators prior to the valves moving to the safe position. This is an acknowledged low probability event since it would take direct spray onto an actuator to induce the failure. However, the probability and consequence of the event should be reviewed for critical valves, and protection provided if necessary. In general, fire exposure to such valves is less likely to prevent the valve from moving to the safe position because the embrittlement effects of the cryogenic LNG spray can induce a more rapid failure.

When LNG vaporizes it creates a condensation cloud in the air around the natural gas cloud. This cloud is often mistaken for the natural gas itself, but is merely condensed water vapor resulting from the cryogenic release. The creation of potentially large fog clouds during LNG releases can impair the occupants' ability to see. Occupants may not be able to see pooled LNG on the deck if it is obscured by a condensation fog cloud. Based on this, all portions of the process unit should be reviewed to assure that the occupants have access to more than one evacuation route back to a Temporary Refuge no matter where an incident may occur. In addition, the emergency response team members should be specifically trained on LNG release characteristics so that they are properly prepared to respond to LNG incidents.

Conclusions

LNG provides a safe and economical method of transporting natural gas from remote areas to natural gas markets. Many materials of construction, such as structural steel and electrical insulation are embrittled by exposure to cryogenic fluids like LNG and very cold natural gas. Protection from cryogenic exposure, as well as fire exposure is essential in LNG facilities. This paper presents a rational approach to engineering the fire and cryogenic protection systems for LNG facilities.

References

1. 33 CFR Part 127-Waterfront Facilities Handling Liquefied Natural Gas and Liquefied Hazardous Gas, U. S. Code of Federal Regulations, Washington, DC (2009).
2. 49 CFR Part 193-Liquefied Natural Gas Facilities: Federal Safety Standards, U. S. Code of Federal Regulations, Washington, DC (2009).
3. Center for Chemical Process Safety, "Guidelines for Fire Protection in Chemical, Petrochemical, and Hydrocarbon Processing Facilities", American Institute of Chemical Engineers, New York (2003).
4. Philip J. DiNenno, Editor-in-Chief, "The SFPE Handbook of Fire Protection Engineering", 3rd Edition, National Fire Protection Association, Quincy, MA (2002).
5. EN 1473, "Installation and Equipment for Liquefied Natural Gas – Design of Onshore Installations", European Committee for Standardization, Brussels, Belgium (2007).
6. "Fireproofing Practices in Petroleum and Petrochemical Processing Plants", API Publication 2218, Second Edition, American Petroleum Institute, Washington, DC (August, 1999).
7. "Fire-Protection Considerations for the Design and Operation of Liquefied Petroleum Gas (LPG) Storage Facilities", API Publication 2510A, Second Edition, American Petroleum Institute, Washington, DC (December, 1996).
8. Thomas M. Flynn, "Cryogenic Engineering", 2nd Edition, CRC Press, Taylor and Francis Group, Boca Raton, FL (2005).

9. "Pressure-relieving and Depressuring Systems", ANSI/API Standard 521, Fifth Edition, American Petroleum Institute, Washington, DC (January, 2007).
10. "Recommended Practice for the Design of Offshore Facilities Against Fire and Blast Loading", API Recommended Practice 2FB, American Petroleum Institute, Washington, DC (2006).
11. "Reduction of LNG Operator Error and Equipment Failure Rates", GRI 90/0008, Gas Research Institute, Chicago, IL (April 20, 1990).
12. "Standard for the Production, Storage, and Handling of Liquefied Natural Gas (LNG), NFPA 59A, 2009 Edition, National Fire Protection Association, Quincy, MA.

Dynamic Simulation for High Pressure Gas Pipeline Network System

56f

Hiroki Hayashi, Tokyo Gas Corp
Kenichi Matsuoka, Yamatake Corp
David Hill, Chemstations, Inc.

AIChE Spring Meeting, April 2009
9th Topical Conference on Natural Gas Utilization
Tampa, Florida, April 26-30, 2009

1. Introduction

Pressure, flow controllability and stability in high pressure gas pipeline networks need to be analyzed when new pipelines with pressure control valves are connected to an existing pipeline network system. Herein, the process interaction between an existing pipeline network system and a new pipeline is studied before it is connected to the existing pipeline system, and the optimum control parameters are re-calculated to give a stable operation.

Previously, the existing pipeline system had a problem in terms of controllability when a new pipeline with a pressure control valve was connected to it.

At this time, the pressure control valve in the pipeline at the Kitano valve station and the pressure control valve in another pipeline at the Sodegaura valve station interacted with each other.

The controllability of pressure and flow on the pipeline network system will become more complicated when the new pipeline is connected to the existing pipeline system at the end of 2008.

Therefore adequate PID control parameters of pressure are needed to operate the high pressure gas pipeline with stability. We studied the controllability of pressure and flow of the entire high pressure pipeline system by using a dynamic simulation model. Herein, the dynamic process model was made by using the process simulator "CHEMCAD", and PID parameters were calculated by using the internal model control (IMC) algorithm method.

The following figure shows the flow chart of process analysis on this system.

<Fig.-1.1: Flow chart of analysis by dynamic simulation>

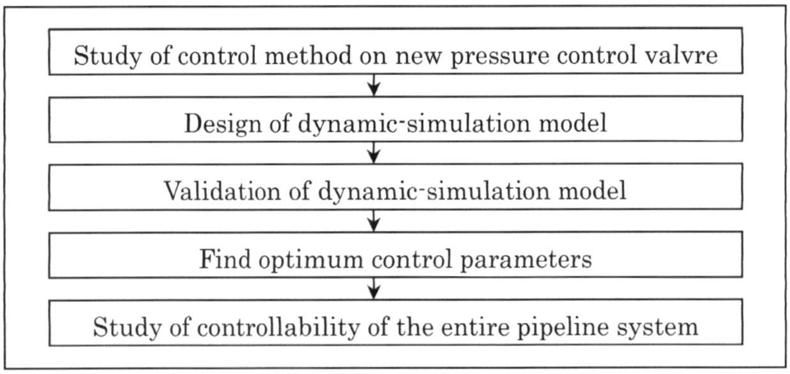

2. Contents of study

2.1 Scope of process

Figure 2.1.1 shows the area of high pressure gas pipeline we are concerned with, for this dynamic simulation model. The distance between the Sodegaura LNG terminal (Gasification plant) and the end of the high pressure pipeline is approx.300km, and the pipeline network connections are many and complex.

<Fig.-2.1.1: Entire high pressure pipeline networks>

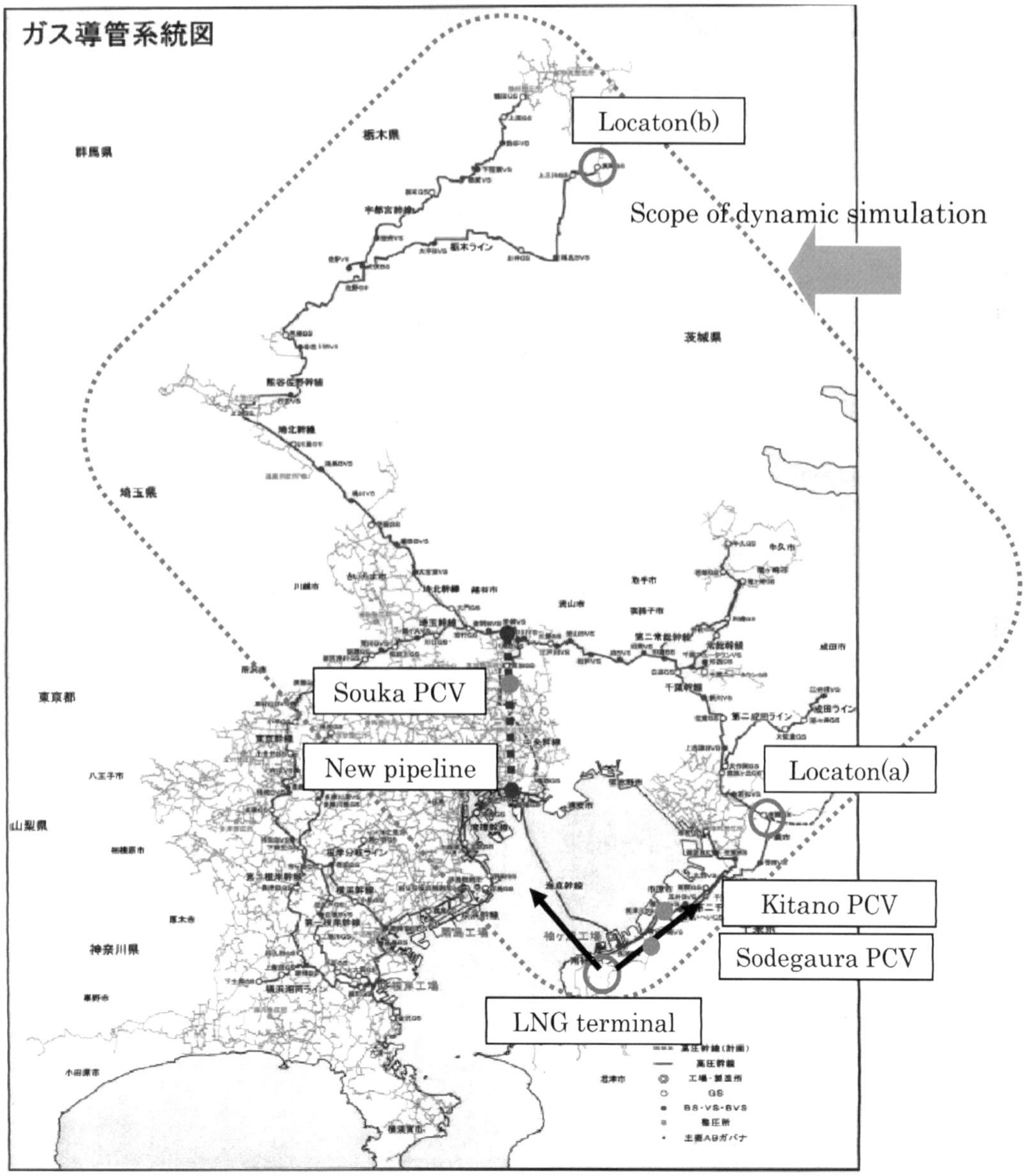

Location(a); approx.60km distance from LNG terminal(Gasification plant)

Location(b); approx.300km distance from LNG terminal(Gasification plant)

<Problem of controllability during start-up of plant for existing pipelines>:

The Sodegaura PCV and the Kitano PCV interacted with each other during the plant start-up operation. In order to prevent this phenomenon, the PID control parameters were tuned on site to minimize the interaction.

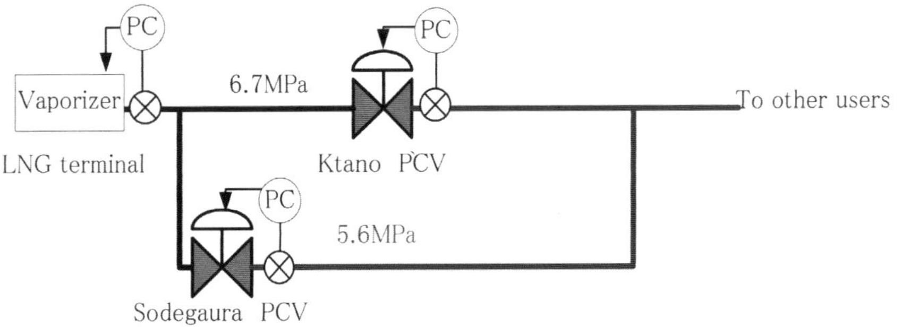

<Problem of controllability during start-up of plant for new pipelines>:

The controllability of pressure and flow in the pipeline network system will become complicated when the new pipeline is connected to the existing pipeline system. Therefore, the controllability on the entire system should be studied before the plant start-up.

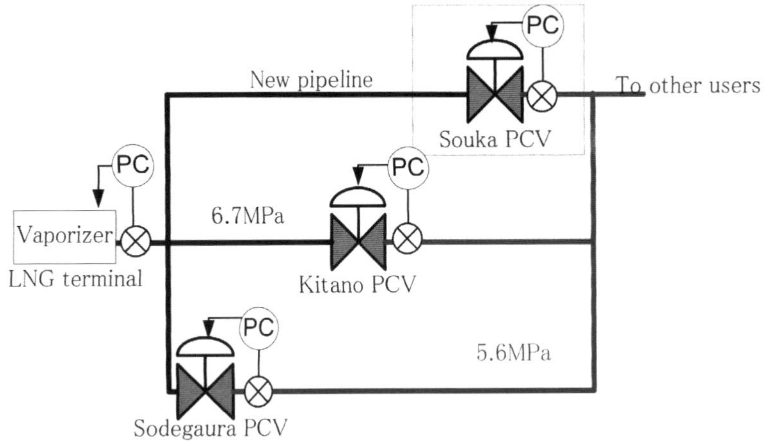

2.2 Case study

The dynamic behavior of the pressure control valve at the Souka valve station, is located in the new pipeline, and the controllability of the entire high pressure gas pipeline system are studied based on the following procedure.

(1) Validation of the existing process simulation model:

Simulation based on actual operational data

By the analysis of dynamic pressure, flow and valve controllability in the current high pressure pipeline system (excluding the new pipeline currently under construction), construct an adequate dynamic process model in which the simulation results match the actual data (pressure, flow rate and valve opening).

(2) Controllability of the entire high pressure pipeline including the new pipelines:

The following studies are based on an extension of the dynamic process model above.

1. Optimum PID parameters
2. Operational control in normal situation
3. Dynamic behavior when emergency valve is closed

3. Process analysis

3.1 Simulation based on actual operational data

We have been able to simulate the dynamic behavior of the process on the existing high pressure gas pipelines over their entire length with some accuracy. The parallels between the actual data and the simulation results are shown below.

The following graph shows the comparison of pressures at location (a), approx.60km downstream from the LNG terminal.

<Fig.-3.1.1:Pressure at location (a) approx.60km downstream from the LNG terminal>

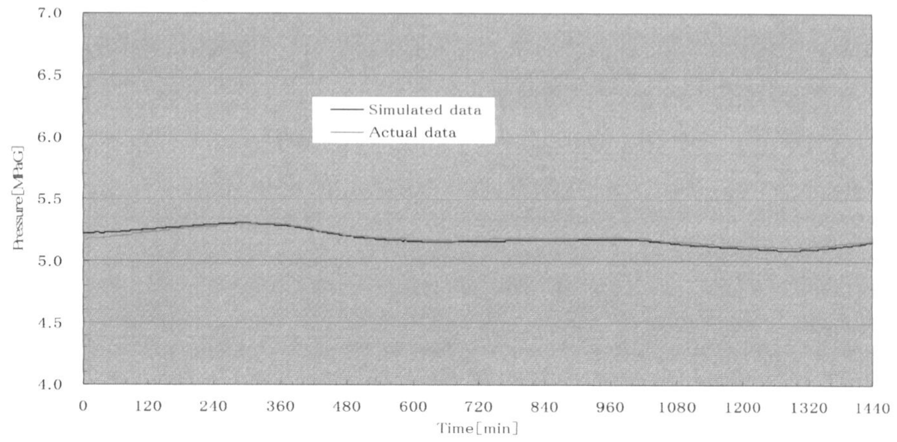

The following graph also shows the comparison of pressure at location (b), approx.300km downstream from the LNG terminal.

<Fig.-3.1.2: Pressure at location (b) approx.300km downstream from the LNG terminal>

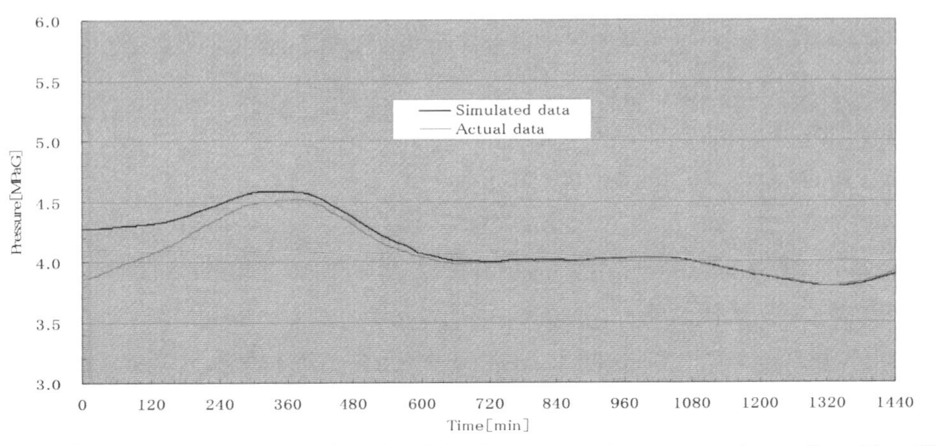

Consideration of dynamic behavior at initial stage:

The above figure-3.1.1 and 3.1.2 shows some differences between simulation results and actual data from initial stage (00:00) to several hours (maximum approx. five). This is caused by the difference at the initial stage between assumptions in the simulation and the actual process. In reality, the actual process has constant dynamic behavior and the dynamic simulation results will converge with the actual process behavior after several hours due to the momentum and material balance over the entire pipeline system. Therefore, differences in process behavior will disappear if the simulation is run for several days.

3.2 Extent of interaction between new pipelines and existing pipeline system

The process interaction between existing pressure control valves and the new pressure control valve was studied, assuming completion of the new pipeline.

Herein, the dynamic behavior of the opening of existing pressure control valves and the downstream pressure at the above valves were calculated when the new pressure control valve was opened manually as follows.

It was assumed that the gas demand was constant during the simulation time and the existing pressure control valves were controlled automatically.

<Fig.3.2.1-The change of valve opening of Souka Pressure control valve>

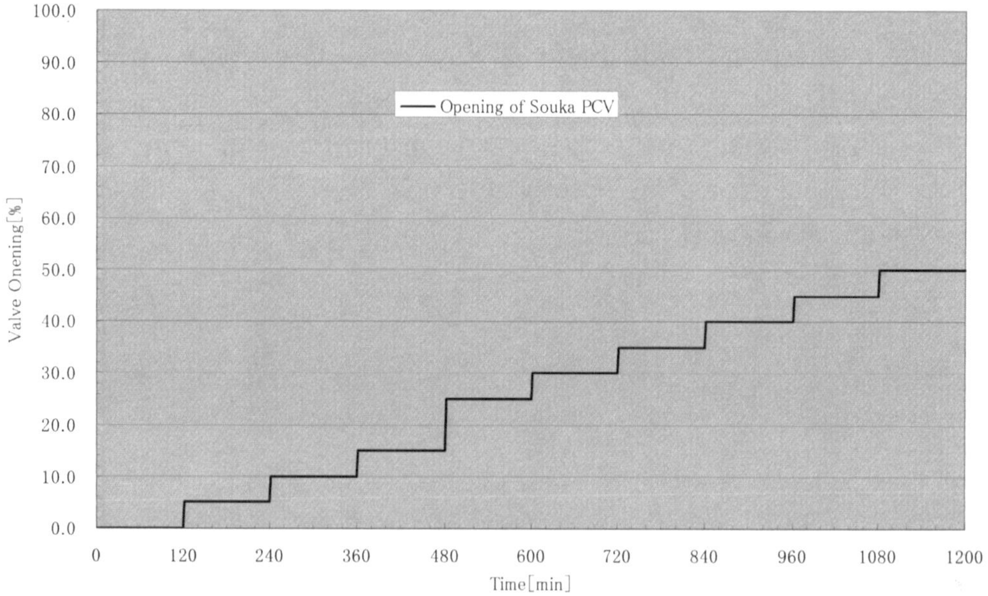

The following figure shows the change of flow-rate for each pressure control valve based on the above process disturbance.

The flow-rate of the new pressure control valve,"the Souka valve", increases when the valve is opened manually. Conversely, the flow-rate of the existing pressure control valve, "the Kitano valve", decreased. As shown in the following figures, the Sodegaura valve was not effected by the opening of "the Souka valve".

<Fig.-3.2.2: The change of flow-rate caused by the opening of the Souka valve>

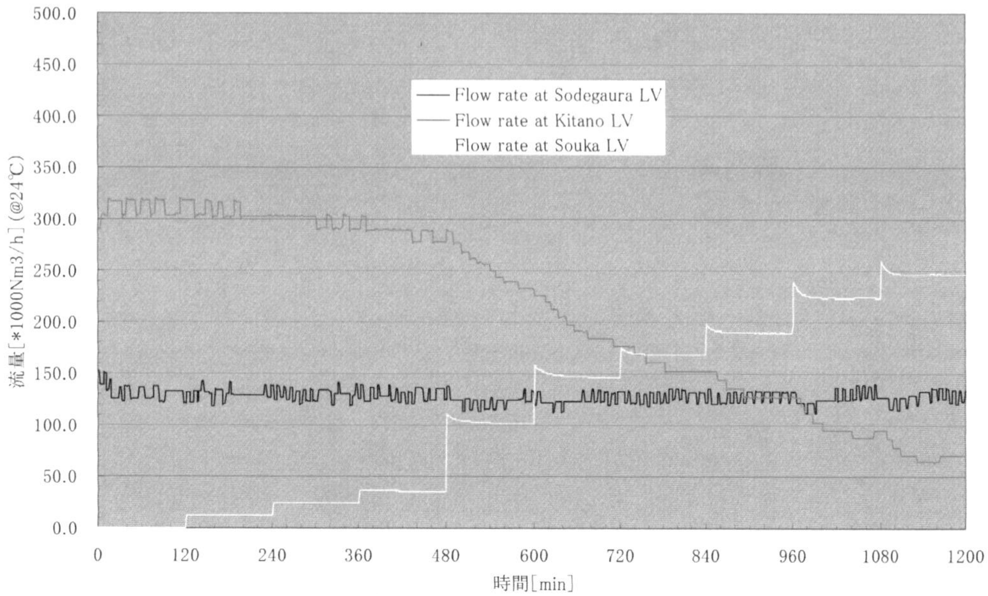

3.3 Initial calculation of optimum PID control parameters

The optimum PID control parameters were calculated, using the above dynamic simulation model (Chapt. 3.2) where the new pressure control valve was opened in increments of 5%. .

The following equations are derived from IMC(Internal Model Control) and used to calculate the initial value of the optimum PID parameters.

-Process transfer function Gp(s):

$$\therefore Gp(s) = \frac{\Delta P}{\Delta cv} = \frac{K_p}{s} \tag{3.3.1}$$

where, ΔP ; Change of downstream pressure [%/min]

Δcv ; Change of stroke [%]

K_p ; Process gain [-]

The initial value of optimum control parameters are shown as follows:

$$\therefore K_c \left(\equiv \frac{1}{PB} \right) = \frac{1}{\lambda K} \tag{3.3.2}$$

where, K_c ; Controller gain [%/min]

PB ; Proportional Band [min/%]

λ ; IMC filter [min]

For this dynamic simulation model, the parameter values shown below were assumed.

$\therefore PB = 13.0$ [%]

$\therefore TI = 2.0$ [min]

3.4 Dynamic simulation of completed pipeline system

Assuming completion of the new pipeline (Kasai station Souka station), the interaction between the pressure control valves in the high pressure pipeline system is studied when the control valve in the new pipeline is used in the pressure control mode. The following were taken into consideration in this simulation.

1. Stability of flow rate and pressure at the LNG terminal
2. Optimum PID Control parameters on the new pipeline without affecting the entire pipeline system.
3. Pressure changes at location (b) in Figure 2.1.1, at the termination of the high pressure pipeline.

<Fig.-3.2.1:Pressure at outlet of pressure control valve in the new pipeline>

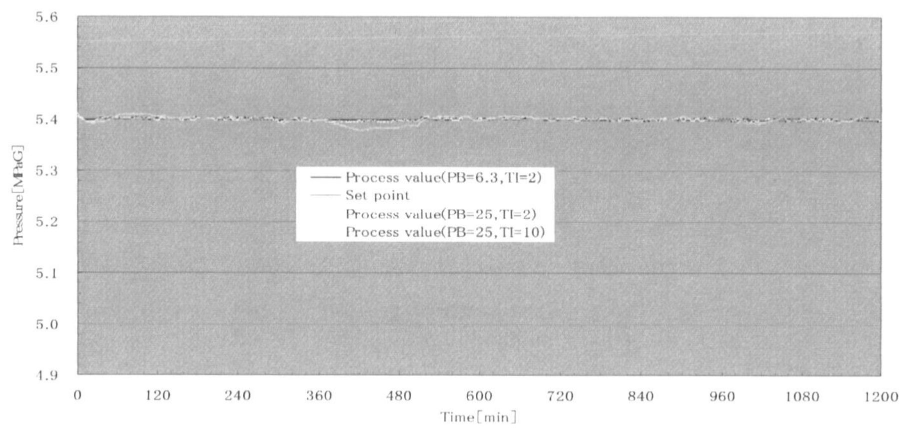

The above graph shows the simulated pressure at the pressure control valve in the new pipeline. The following graph shows the simulated pressure at location (b) approx.300km downstream from the LNG terminal.

<Fig.-3.2.2:Pressure at location (b) approx.300km downstream from the LNG terminal>

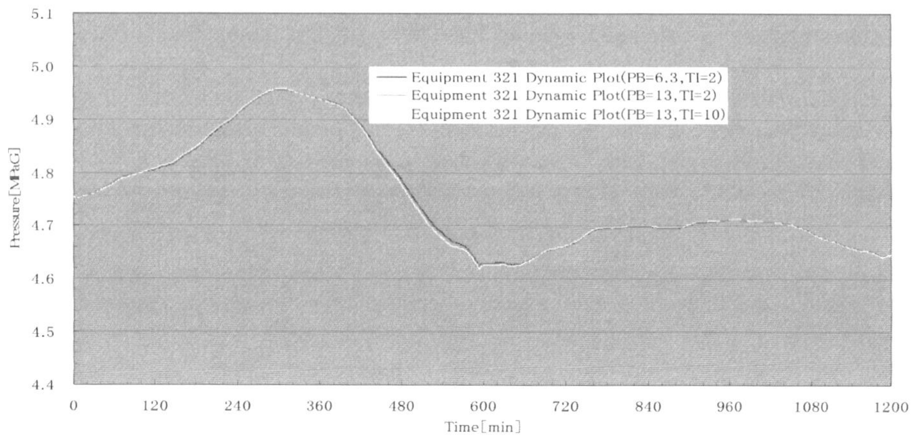

3.5 Dynamic simulation of closed emergency valve scenario

The dynamic behavior at the pressure control valve in the new pipeline, and how the other control valves in the existing pipelines back the entire system up, were studied when the pressure control valve at the Sodegaura valve station (near the LNG terminal in figure.2.2.1) was closed in an emergency.

The following graph shows the behavior of pressure at the outlets of the pressure control valves.

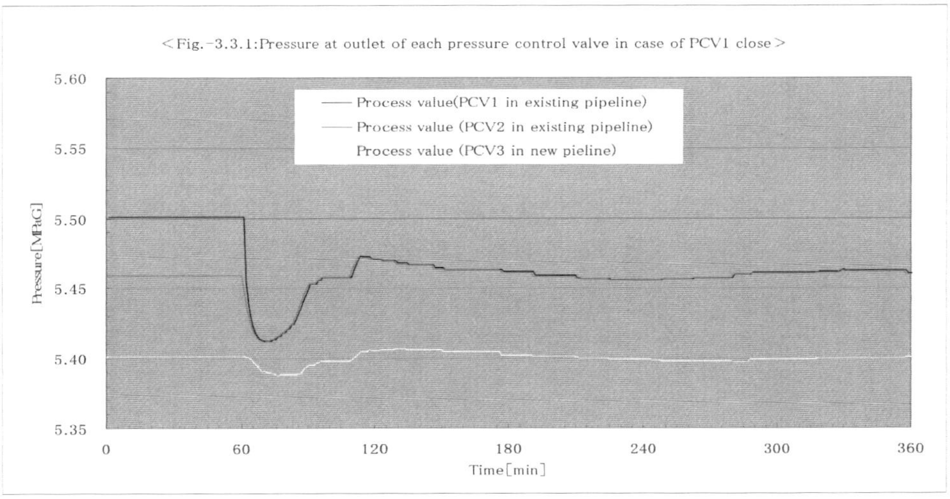

The following graph shows the change of flow-rate at the pressure control valves, when there is an emergency at one of the valves.

It is shown that the other valves compensate for the sudden drop in the flow-rate.

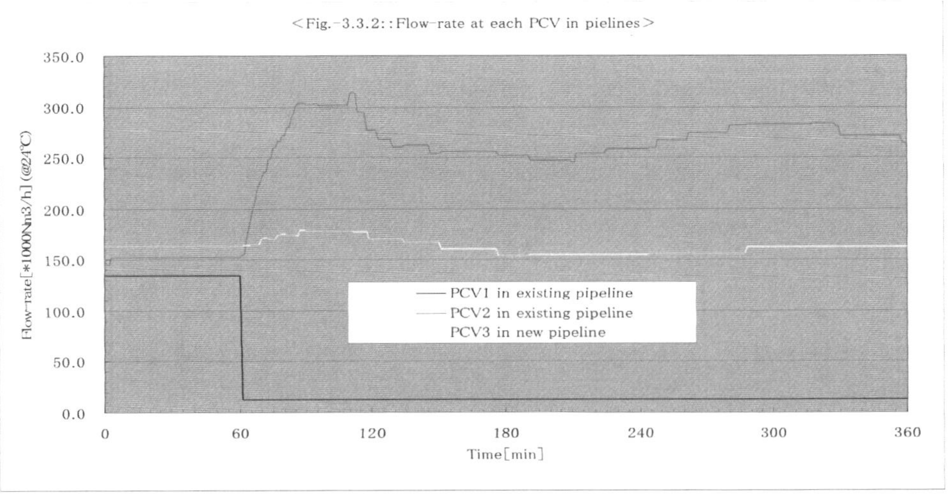

4. Conclusion

(1) Validation of the dynamic process model:

As shown in Fig.-3.1.1 and 3.1.2, we have been able to simulate the dynamic behavior of the process in the existing high pressure gas pipelines over their entire length with some accuracy.

(2) Controllability of the entire pipeline system after completion of the new pipeline

The interaction between the pressure control valves in the pipeline system is minimal. The controllability of the entire pipeline system remains stable, using PID parameters based on the dynamic process model and the internal model control algorithm method.

It is anticipated that the period of time for the tune-up of plant control during start-up on site can be reduced by studying the dynamic behavior of the process, using the above optimum control parameters as initial values.

literature Cited

Luyben, Process modeling, simulation and control for Chemical engineer, second edition, McGraw Hill, USA

Solving New Demands in LNG Technology using Field-Proven LNG Expanders and Pumps

72a

Joel V. Madison, C.E.O.
Ebara International Corporation
Cryodynamics Division
Sparks, NV

AIChE Spring Meeting, April 2009
9th Topical Conference on Natural Gas Utilization
Tampa, Florida, April 26-30, 2009

Abstract

With the increasing utilization of expander and pump technology in all aspects of the LNG industry, new demands in expander and pump technology naturally arise. There is a need for expanders and pumps that are designed for larger flow and head conditions. In response to this demand, several design configurations have been developed using already field-proven machines. These configurations not only solve the requirement of larger flows and heads, they provide numerous additional benefits including greater flexibility, increased efficiency, and compact configurations to minimize the footprint.

Compact Pump – Expander Configuration

Two examples of the compact pump-expander configuration are shown in Fig. 1 and Fig. 2. The design consists of a pump and expander separated by a seal and mounted on a common shaft with a common induction motor/generator. Due to the separation between the expander and pump the machines may operate in different fluids, the pump for the LNG stream and the expander for the MR stream. The motor/generator may be submerged in the pump fluid flow (Fig. 1), or in the expander fluid (not shown), or separated by seals from both the pump and expander fluid (Fig. 2). In the third case the motor/generator may be cooled using another fluid readily available in a liquefaction plant, such as liquid nitrogen or liquid propane. The induction motor/generator operates as a motor if the power generation from the expander is not enough to drive the pump, or it may operate as a generator when the power generated by the expander is larger than the input required to operate the pump. Thrust is balanced using the trademark TEM design, common to all Ebara International Corporation, Cryodynamics Division (EIC) pumps and expanders. Depending upon the process and design, the TEM device may be located in the pump side, the expander side, or within the separated generator cavity and balanced using the liquid nitrogen or liquid propane.

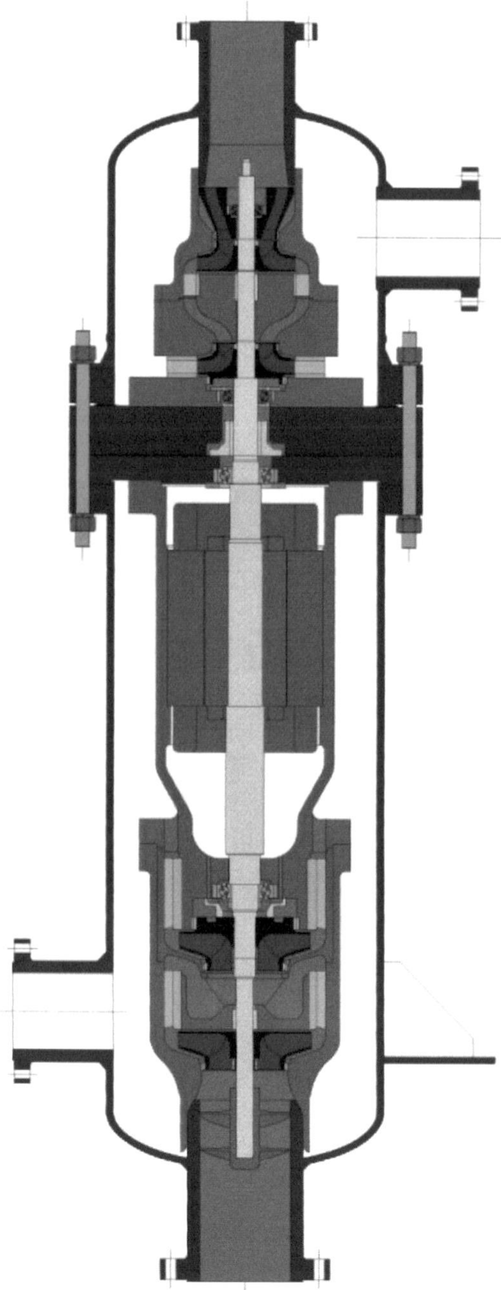

Fig. 1: Compact Pump-Expander Configuration with Submerged Motor/Generator

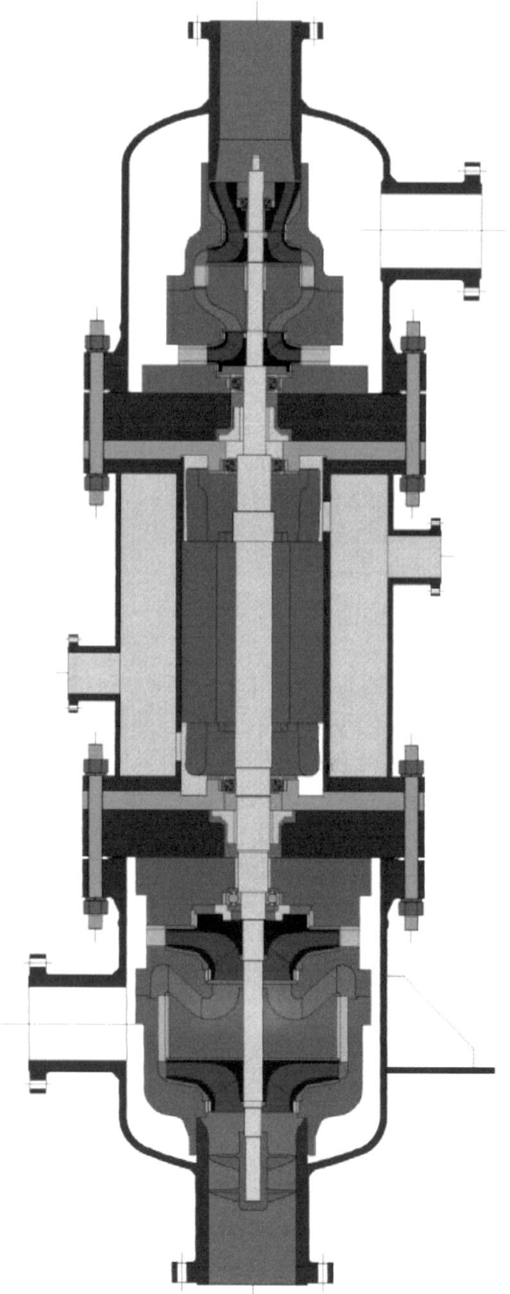

Fig. 2: Compact Pump-Expander Configuration with Separated Motor/Generator

Compact Expander-Expander Configuration

The compact expander-expander design incorporates two expanders mounted on a common shaft with a common generator and separated by a seal (Fig. 3). One expander operates in the LNG stream and the other in the MR stream. The generator may be submerged in the MR fluid (Fig. 3), or in the LNG fluid (not shown), or be separated and cooled in another fluid similar to the pump-expander design (Fig. 2). Also, like the pump-expander design the expanders in the expander-expander configuration are balanced with the TEM. The TEM may be located in the MR, LNG, or liquid nitrogen or liquid propane fluid. The expanders may be multiple stage, single phase, or two-phase depending upon the process requirements.

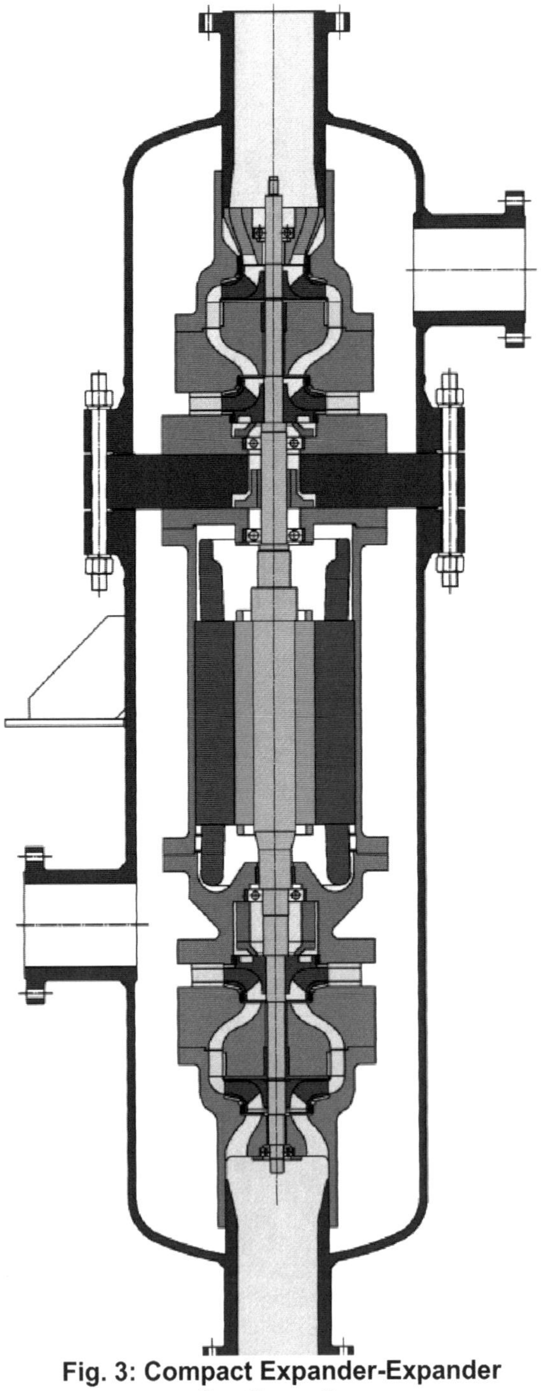

Fig. 3: Compact Expander-Expander Configuration

Advantages of Compact Configurations

The compact pump-expander and compact expander-expander designs have a multitude of advantages. Due to mounting the machines on a common shaft, but with thrust in opposite directions, overall thrust is minimized; therefore, less fluid is required to balance which increases hydraulic efficiency. By separating the motor/generator from the process fluids, no heat is transferred to the process fluids, thus increasing the process efficiency. If the motor/generator is submerged in a process fluid, it is more desirable to put it in the MR stream, rather than the LNG. When separating the motor/generator from both fluid streams, it is possible to also balance the thrust using the third cooling fluid, which maximizes the hydraulic efficiency. The compact design allows for a smaller footprint that is more readily integrated into an existing plant.

Specifically, the compact pump-expander design increases electrical efficiency and reduces, or eliminates the electrical power required for the pump because of the power generation of the expander. It is also possible for a positive power output, if the expander generates more electrical power than is required for the pump, which will increase plant efficiency even more.

In both the pump-expander and the expander-expander configurations, the compact nature of the design reduces cost and the increased process efficiency reduces the payback period and increases production. These designs also serve two needed plant functions in one small and compact design that minimizes the overall footprint required, and the piping required, while increasing process efficiencies and the overall plant efficiency.

Tandem Expander

The tandem expander design consists of two expanders located within one vessel and operating in series. Each expander is equipped with a submerged generator and a variable speed drive. Figure 4 is a simplified view of a possible configuration. The first expander in the fluid flow (A) is a single phase expander, which may be multi-stage. Expander A expands approximately half of the overall desired head. The second expander in the fluid flow (B) is a two-phase expander and may also be multi-stage. Due to mass conservation, each machine sees the same flow rate. Expander B may also be a single phase machine, depending upon the process conditions and requirements.

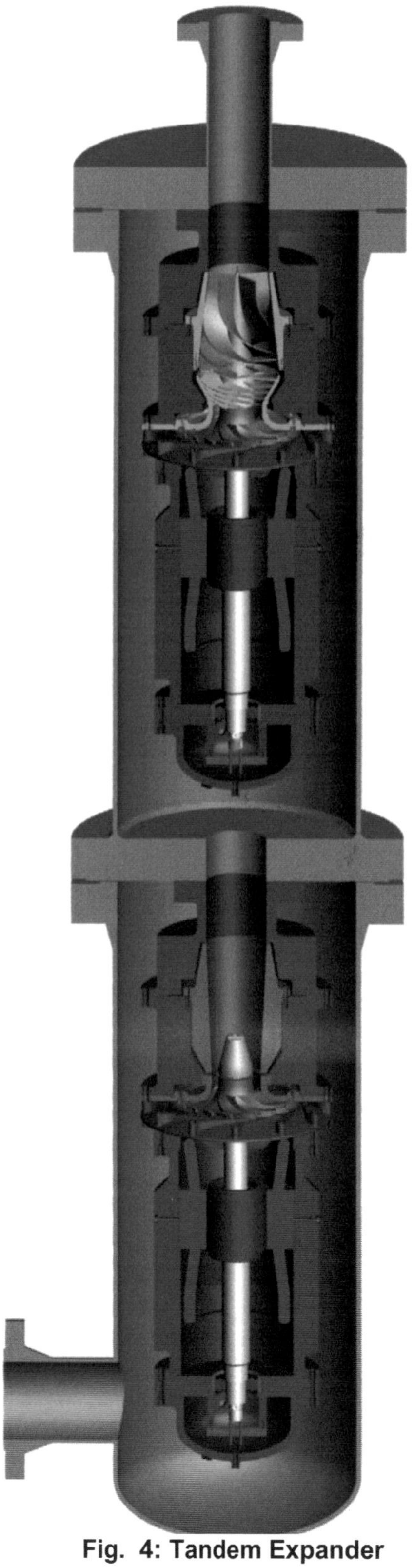

Fig. 4: Tandem Expander

Tandem Expander Advantages

The Tandem Expander design can accommodate higher head without increasing the size of the generators or the diameter of the vessel. The design also allows for multi-phase capacity and replaces both liquid and two-phase expansion by

Joule-Thomson (JT) valves. The design also improves efficiency by virtue of grater flexibility; each expander operates at a speed that maximizes efficiency. The variable speed for each expander allows for constant adjustment to match process changes without diverging from the best efficiency point (BEP). The design maintains typical vessel diameter; therefore, maintains a small footprint. Cost is saved in the machine design as well by maintaining a reasonable hydraulic diameter, saving material cost, and by maintaining generator diameter. To build an expander to achieve the same results as the Tandem Expander requires increasing the generator size which increases diameter, increasing cost.

Conclusions

The presented designs incorporate field proven machinery, but offer new, cost effective solutions to increased demands for larger flow , larger head, higher efficiency, more flexibility, compact size, cost effective solutions, and retro-fitting for older plants to maintain competiveness.

Biography of Speaker

Joel attended University at the United States Air Force Academy, University of Nevada and Pennsylvania State University, earning a Bachelor of Science Degree in Mechanical Engineering and a Master of Science Degree in Aerospace Engineering. Prior to joining Ebara, Joel worked for ten years as a research engineer for the Advanced Programs Department of the Rocketdyne Division of Rockwell International. During this time he was involved in analysis, development and testing of prototype hypersonic scramjet engines in support of NASA and U.S. Air Force Programs. Since joining Ebara International in 1995 he has worked as a Project Engineer, Project Manager, Chief Engineer, Chief Operating Office of the Cryodynamics Division and eventually Chief Executive Officer of the Corporation. Additional activities during this time have included new product development, hydraulic modeling and design, rotordynamic analysis, field service troubleshooting support, system design, sales support and customer coordination. jmadison@ebaraintl.com

Fundamentals of a single stage liquid expander for flashing liquids

Roger Dambach, Deputy Director Process & Shipboard
Josef Pozivil, Director Technology
Jonathan Marlier, Manager Aerodynamics

Cryostar SAS
Zone Industrielle
68220 Hésingue
FRANCE
www.cryostar.com

Prepared for submission to the AIChE Spring National Meeting
Tampa, Florida, 27 – 30 April 2009

ABSTRACT

The air separation industry has always taken the lead in developing cryogenic equipment that was later successfully applied into the context of the LNG industry. Not surprisingly liquid expanders were first operated in the air separation industry before they were designed and installed in the LNG industry. The present paper argues that the LNG industry needs to look again to the air separation community when considering the installation of flashing liquid expanders. It analyses the fundamental design features and design limits of a flashing liquid expander based on a radial inflow turbine. It further demonstrates that the design of flashing liquid expanders is entirely contained within the design envelope of existing turboexpanders. It finally shows why a single stage liquid expander based on the radial inflow turbine technology is the best suited machine for the purpose of flashing liquid expansion.

LNG PUMPS FOR FLOATING STORAGE AND REGAS UNIT

Fabien Wahl
Director of Marketing
Nikkiso Cryo Inc.
North Las Vegas, Nevada, USA

AIChE Spring Meeting, April 2009
9th Topical Conference on Natural Gas Utilization
T6a03 LNG Equipment – II
Tampa, Florida, April 26-30, 2009

Historically, liquefied natural gas (LNG) distribution has been accomplished through the use of land-based receiving terminals. In recent years we have experienced the rapid development of the Floating Storage and Regas Unit (or FSRU). The development of the FSRU has in part been fueled by lower capital costs, shorter on-line cycle times, operational flexibility, lesser demand for coastal real estate, and homeland security concerns. The challenge to beat in order to put a regasification unit on board a floating and seagoing vessel, was to develop materials that would perform regardless of the motion of the ocean.

FSRU Concept.

The FSRU concept is to combine a LNG carrier for the transport and storage of LNG, with a complete regasification unit for the distribution.

Typically, LNG is loaded on board the FSRU ship from a liquefaction plant. The loaded FSRU then sails to the distribution point. Once there, the regasification plant on board the ship is started and the gas is delivered to shore and onto the grid. After regasifying its cargo, the FSRU ship then sails to once again repeat the loading and distribution process.

Another application of the FSRU concept involves a static FSRU quasi permanently moored to its send out buoy. In this case, LNG is transferred to the FSRU ship from a traditional LNG carrier.

The challenge of shipboard regasification, is that of operating a facility with rotating equipment subjected to accelerations, often quite severe, as shown in Table 1.

Table 1. Operation environment

Accelerations [G]	Regas Operation	Transit Seagoing
Vertical	0.4	0.8

Transversal	0.2	0.5
Longitudinal	0.1	0.6

Table 1 shows that the pumps must be designed to address two rather different conditions: moderate acceleration values during the regasification phase when the pump is operating; important acceleration values while sailing at sea, when the pump is not operating.

The traditional cargo pumps on board a LNG carrier are designed strictly for offloading in an almost wave less, protected port environment. The deep sea regas operation environment is therefore more severe than what is known to the classic ship board industry. While non-operating conditions are the same for cryogenic pumps, the high-pressure send out pumps on board FSRU are much heavier than the cargo pumps, and that introduces a whole set of design considerations for the pump manufacturer.

The following illustrates how Nikkiso has addressed the regasification challenges of an FSRU.

LNG Pumps for FSRU.

Figure 1 shows the Nikkiso Cryo regasification ship-set which is used on board the Exmar fleet since 2004. On the left, is the emergency/in-tank pump. Lying on the Horizontal Assembly Stand ©, there is the send out pump. On the far right, is the line packing pump.

Figure 1. Nikkiso Cryo regasification ship-set.

The in-tank pumps are located inside the emergency wells of the traditional LNG carrier. Unlike the original emergency pumps part of a classical ship set, these in-tank pumps are permanently lowered in the column. These pumps feed the suction drum, and are rated for a flow of 620m3/h, with a differential head of 115m. The pump rated power is 200kW.

The send out pumps are located on the bow of the ship. They are mounted in their own suction vessel, themselves integrated onto skids. These pumps feed the vaporizers at a flow of 205m3/h, and with a differential head of 2370m. The pump rated power is 1100kW.

The line packing pumps are vessel mounted pumps also located on the bow of the ship. They deliver the same head as the send-out pumps, 2370m, but at a much smaller flow rate, 20m3/h. These 200kW pumps are operated during the start-up of the regasification unit, in order to pressurize the vaporizers and the downstream pipeline smoothly without any water hammer.

Technical challenges.

In-tank pumps.

Unlike the emergency pumps, the in-tank pumps [Figure 2] are lowered permanently in the emergency columns. Traditionally, an in-tank pump is lowered with cables and rests on the foot valve by gravity only. When not operating, and during severe sea conditions, the pump would be free to bounce vertically on the foot valve seat. Nikkiso has replaced the lifting cables by a segmented pushing pole. This pole keeps the pump on the foot valve seat and prevents any movement of the pump in the column. The design of the foot valve seat has been enhanced to a conical shape which offers a self centering and optimized contact surface.

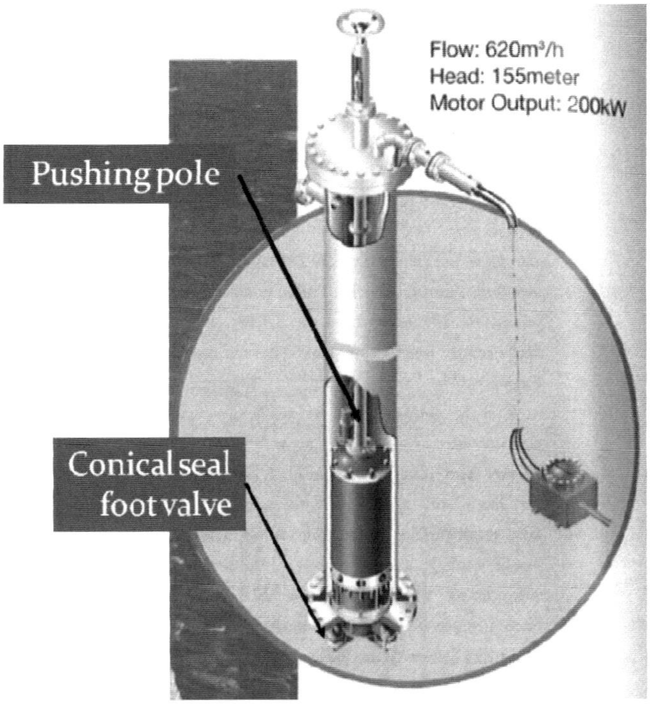

Figure 2. Emergency / In-tank pump.

Send out pumps.

The mass and the size of the send out pump [Figure 3], combined with the ship's motion, induce high mechanical stress issues. In traditional land-based applications, the send-out pump is hanging inside its suction vessel by its discharge manifold. By contrast, the ship is pitching and rolling, and the mass-inertia-induced lateral displacement of the pump would generate high stress loads on the bolt circle connecting the pump to the suction vessel. In order to prevent and relative motion of the pump with respect to the vessel, Nikkiso has added a pump stabilizing pin underneath the suction manifold of the pump. This brass male pin engages in the stainless steel female receiver at the bottom of the vessel. The design enables longitudinal translation to account for thermal contraction differences between the aluminum pump and the stainless steel vessel. By design, the pin prevents any lateral displacement, therefore suppressing any bending moment on the fanged connection between the pump and its vessel.

The same relative motion could occur externally between the suction vessel and the deck of the ship. In this case, the vessel supporting pods would be subject to an increased bending stress. Nikkiso has added the same pin concept between the bottom of the vessel and the deck of the ship.

This dual pin arrangement between the pump and vessel, and between the vessel and deck, creates a rigid arrangement with no lateral degree of freedom. This eliminates all detrimental stresses created by the ship's motion.

In the traditional design of a land-based application, and when the pump is not operating, the dead weight of the rotor is supported by only one radial bearing. The vertical acceleration experienced on a ship would lead the rotor to bounce on and off this radial bearing. The balls would brinnel the races and the bearing would self-destruct. Nikkiso has added a stowage piston located on the head plate. This piston reaches through the discharge manifold to the end of the shaft. This piston is actuated either with compressed nitrogen or by an electric motor. When the pump is not operating, the stowage piston lifts the shaft and relieves the supporting bearing from the rotor weight, hence preventing any damage. Just before pump operation, the rotor is lowered smoothly onto the support bearing.

Figure 3. Send out pump and stress control features.

The stabilizing pins and the stowage piston have been patented under Patent No US 7,063,512 B2 on June 20th, 2006.

Send out pumps are classically long multistage pump. In the case of the Neptune project, Nikkiso supplies a 15-stage pump. The lateral displacement of the shaft induced by the motion of the ship must be carefully addressed to prevent a rotor/stator contact during operation in rough weather conditions. Nikkiso has added three radial ball bearings along the pump shaft line. This unique construction adds stiffness to the rotor, and a better alignment across all contact points. Bushings disposed in all non-bearing-carrying stages, provide additional damping. This stiff and dampened rotordynamics enables a safe pump operation while the ship is pitching and rolling in the ocean.

During pump operation, the weight of the rotor and the hydraulic forces are compensated by the Nikkiso balance drum assembly. In a typical motionless land-based design, the balance drum only accounts for changes in hydraulic forces. In case of vertical acceleration changes, the weight force is also changing. The Nikkiso balance drum design is a separate function in the pumps, and is independent from the design of any other component of the pump. Therefore, the balance drum assembly is sized in order to meet the requirement of the axial thrust balancing, regardless of the main flow and head specifications.

Line packing pump.

The line packing pump specifications combine a rather deviated couple of low flow and high head. Traditional pump selection based on a 60Hz two pole motor would result in a slim 20-stage, 3.4m-long pump as shown in Figure 4. The impact of vertical and longitudinal acceleration forces at sea on a long and thin pump is far from ideal. Therefore Nikkiso has specifically developed a Small High Pressure pump, which takes advantage of a higher speed of rotation to reconcile low flow and high head. Nikkiso's line packing pump [Figure 4] uses a VFD driven motor at 120Hz. This compact Small High Pressure pump has only 8 stages and measures only 1.9m in length.

The high speed design was completed using a detailed rotordynamics analysis and impeller stress calculations.

Model	6286L20-P200F	60510L8-P200F
Diameter [mm]	520	540
Length [mm]	3430	1909
Speed [rpm]	3600 (sync)	7020 (actual)

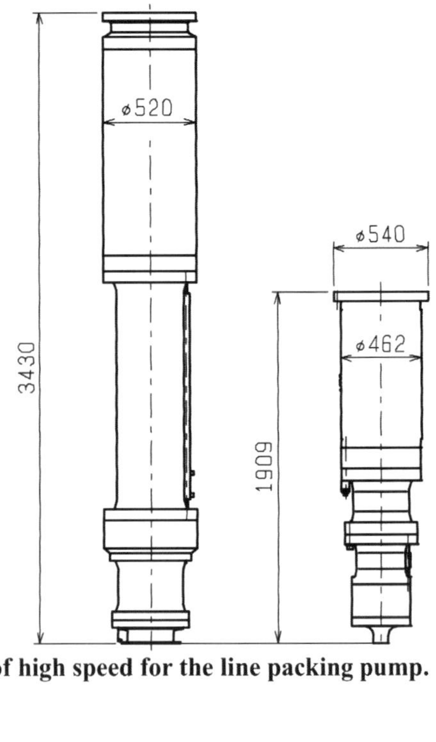

Figure 4. Benefit of high speed for the line packing pump.

Validations.

In 2004, El Paso constructed an entire shipboard regasification unit and mounted it on an articulated platform on the peak shaving site of Trussville, Alabama. The platform was then subjected to motions simulating the motion of the bow of the ship in 8 meter waves. The regasification was opearted at full capacity with LNG supplied by the adjacent Trussville peak shaver. Comparisons of data between the steady and articulated operations show that the send out pump service is insensitive to the motion, while matching the performance requirements.

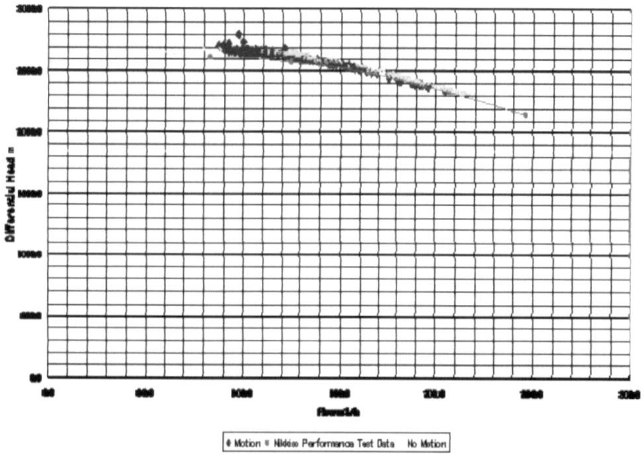

Figure 5. Send out performance curve with and without motion.

The Small High Pressure line packing pump was successfully tested on the Nikkiso test stand of North Las Vegas, Nevada. The pump was tested at 110Hz and 120Hz. The operational speed was then determined to be 116Hz. Impeller trim was unnecessary, because the final performance tune up is achieved by speed adjustment.

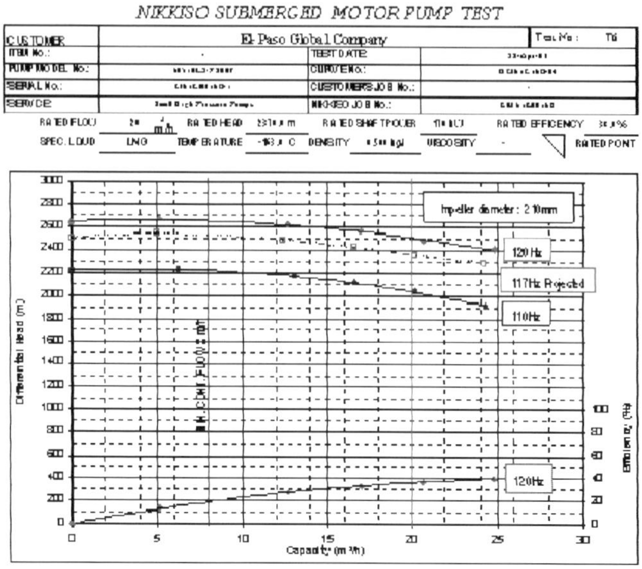

Figure 6. Line packing pump high speed validation.

Following the validation program, the Nikkiso regasification ship-set was certified by Bureau Veritas Marine and Det Norkse Veritas Maritime.

In 2005, the FSRU ship Excellence was sending out gas at 100% of capacity from her Energy Bridge boy in the Gulf of Mexico, in 4-meter waves, and just 60 miles from the eye of hurricane Katrina. This operation was successfully proving the robustness of the entire floating regasification concept during one of Mother Nature's worst storm conditions.

Conclusions.

Nikkiso successfully designed, tested and certified a pump ship-set for floating regasification operations. The pump set comprises an in-tank pump, a send out pump and a line packing pump. Original technologies addressing the specific needs of shipboard regasification have been developed, patented and successfully validated before shipboard use. This line of product is presently in proven service, or awaiting installation, on board all Exmar ships, Neptune ships, and the Golar Winter.

References

Biography of Speaker

Fabien Wahl graduated from the Ecole Centrale de Lyon, France and has his Master's of Engineering from Cornell University, USA. He started his career as a turbomachinery designer, for the European space industry, and participated in the design of the Snecma/JC Carter LNG cargo pump. Fabien Wahl joined Nikkiso Cryo Inc. in 2007 as director of marketing and support to R&D, promoting and joining in the continuous efforts of improving the product line.

fabien.wahl@nikkisocryo.com

Improved Methodology to handle Meteorological Data for LNG Projects

72d

Meredith Chapeaux and Stanley Huang
Chevron Energy Technology Company
Houston, Texas

AIChE Spring Meeting, April 2009
9th Topical Conference on Natural Gas Utilization
Tampa, Florida, April 26-29, 2009

INTRODUCTION

For an LNG plant, on schedule delivery of contractual quantities of LNG is of critical importance due to the possible severe penalties for late or non-delivery. Ambient temperature variation can significantly influence LNG production and therefore, during the design phase of the project, it is important to include sufficient margins to account for these variations. However, overly generous margins can result in an inefficient plant and negatively impact the overall project economics. Nowadays, every project is faced with this difficult challenge of interpreting site ambient temperature variations and determining the impact it will have on its facilities.

With advances in collection instruments for meteorological data and easy accessibility to large quantities of data through web-based communication systems, there are numerous sets of data available for any selected location. To make things more complicated, the meteorological data management has been more and more based on statistical approaches. Hence, the average and extreme temperatures of a location are not definite numbers. Rather, they are expressed in forms of probability functions.

This paper discusses an improved methodology in handling meteorological data, which are typically recorded on a daily or monthly basis. Usually, some types of statistical analyses are also presented for users' convenience. For example, the temperature data may be presented in daily average, monthly maximum, etc, based on calendar frames. However, the so-called "temperature distribution" is the real basis for calculating LNG plant capacities because capacity is a direct function of temperatures, but not calendar time frames. Converting from the meteorological data to temperature distributions demands serious considerations because perception that the "average annual temperature" is representative of a site's temperature data set may not be accurate.

TECHNICAL BACKGROUND

This section provides a foundation for numerical manipulation of meteorological data. The focus is on the significance of data conversions, for example, from monthly temperatures to temperature distributions, or so-called exceedance curves. This paper will not address probability issues, for example, if it is adequate to use P50 data to represent the average temperatures. These probability issues are directly related to risk tolerance of a project and are better discussed with other financial considerations.

Nonetheless, acquiring a feeling for the sensitivity of different process parameters will be helpful in making various project decisions.

Characterizing cyclic functions

Cyclic functions can be characterized by the following features: frequency, peak, valley, transition period, and returning to the initial point. For climate data, by default the frequency is fixed to a one year period. To facilitate discussions, four analytical functions are created and each one with a range from +10 °C to -10 °C. Table 1 provides the specifics of each function.

Table 1 – Example of cyclic functions of different characteristics

	Function	Description	Inverse function
Model 1	Linear	Temp profile is evenly distributed	Linear
Model 2	Sine x	Commonly observed in nature	Arc sine
Model 3	(Sine)^3	Gradual peak-valley transition	Reverse operations
Model 4	(Sine)^1/3	Flat peaks and valleys, rapid transition	Reverse operations

The first function is based on a linear function while the other three functions are derived from a sine wave function. Reasons for the choice are elaborated as follows:

1. The linear function represents an evenly distributed function. It is intuitive and concept can be grasped easily.
2. Sine functions are commonly used to express naturally-occurring phenomena. They are familiar to most people.
3. Analytical functions provide good platforms where exact solutions eliminate complications due to numerical handling.
4. All listed functions are symmetrical with respect to the extreme values. This provides a good check point for accuracies of all numerical manipulations.

Figure 1 shows the four functions in a twelve month period. The four functions share the same periodic features, namely, the peak occurs at Month 3 and valley at Month 9. The peak and valley are separated by 6 months, or half of a year. Also, the average annual temperature is 0 °C.

The main differences among the four functions reside in their peak-to-valley transition behaviors. The linear function (Model 1) depicts a monotone transition, whereas the sine function (Model 2) paints a more realistic a picture. Model 3 is a sine function raised to the third power. The function represents short summer and winter seasons with relatively long and smooth transition periods. In contrast, Model 4 is a sine function raised to the power of 1/3. The function represents flat summer and winter seasons with relatively short and sharp transition periods. Each function has its own characteristic temperature distribution, which will be elaborated upon in the next section.

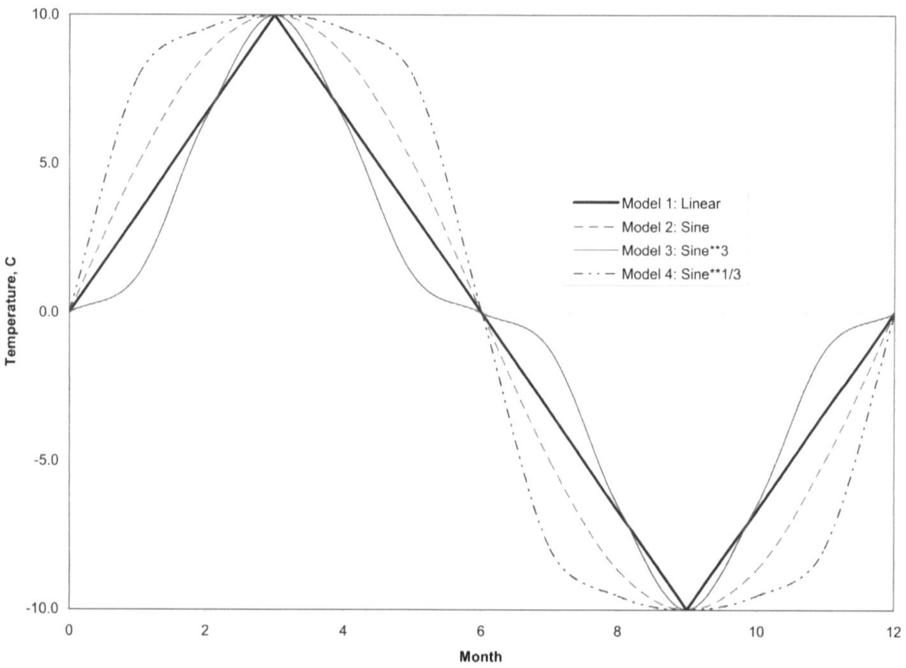

Figure 1 – Temperature models with various peak-valley transitions

Temperature distributions

The mathematical representation of Table 1 or Figure 1 can be written in the following form:

$$T = f(t) \tag{1}$$

where T is the temperature in °C, and t is the time frame in days or months. To derive the accumulated temperature distribution as a function of time, the following equation should be used [Stewart, 2008]:

$$t = \int_{T\min}^{T} \frac{d}{dT}[t] \ dT = \int_{T\min}^{T} \frac{d}{dT}\left[f^{-1}(T)\right] \ dT \tag{2}$$

The derivative of the inverse function warrants special discussion. It is apparent that each horizontal portion in Figure 1 would cause difficulties of singularity during the integration. Physically, this represents a time period when the temperature remains stable and specifying the temperature does not map to a specific time. The reason for adopting analytical functions in this work is for illustration that the inverse functions of linear and sinusoidal type functions can be defined with infinite accuracy except at one point in each temperature extreme [Chapra, 2008]. In the following discussions, the inverse function mapping is conducted analytically, whereas the integration is performed numerically using a spreadsheet. Unless stated otherwise, a resolution of 1 °C is found adequate in all numerical manipulations of this work.

For actual weather data, the numerical differentiation and integration of Equation (2) can be quite delicate, above all if the yearly temperature variations are small (i.e. tropical region). The method requires the actual data to be curve fitted by a polynomial function which is then integrated numerically. However, integration around temperature extremes can become quite sensitive to the temperature scale that is being used. This results in an approximation to the actual number of days spent at that specific temperature. Moreover, the sharper the peak around one temperature extreme, the more difficult it is to accurately approximate the number of days. Therefore, the total numbers of days after integration may not be 365 days exactly. However, the discrepancy in this work fell within 80-85%.

PRESENTATION OF THE NEW METHODOLOGY

A new method is proposed in this section to simplify the calculation of temperature distributions based on the seasonal data. Two of the aforementioned four models will be used as examples for illustrations.

The temperature distribution within each month is assumed to be represented by a normal distribution with its mean value being equal to the averaged temperature and sigma (or standard deviation) to be determined later.

The immediate advantage of this approach is obvious because the calculation of annual temperature distribution becomes a straightforward exercise of summing contributions from each month. This simplifies the calculation procedures and, more significantly, bypasses the handling of singular points corresponding to the horizontal portions in Figure 1.

Using as an example the sine and sin^3 functions from the cyclic models presented in the previous section and implementing the above described calculation method results in the following.

Model 2 – Sine function

Figure 2 shows the temperature distributions acquired by different methods. The upper-left quarter of the figure is acquired by the arc sine function. There are two sharp horns at the two extreme (reversing) temperatures, corresponding to the horizontal portions in Figure 1. In the inverse function mappings, the horizontal portions in the very vicinity of the temperature extremes become almost singular. Consequently, these horns may be difficult to duplicate using analytical functions, such as normal distribution functions. Nonetheless, the upper-right quarter using a sigma value of 2 °C is a reasonable representation. The other quarters show the impact of increasing the sigma value from 2 to 10 °C. As the value increases the resultant figures lose foci and their characteristic features.

Physically, this observation indicates that the average temperature of 0 °C appears with the least frequency. The function is poorly represented by this value. Instead, the function would be better characterized by the two extreme values. Notice that large sigma values would totally change the true shape of the distribution function.

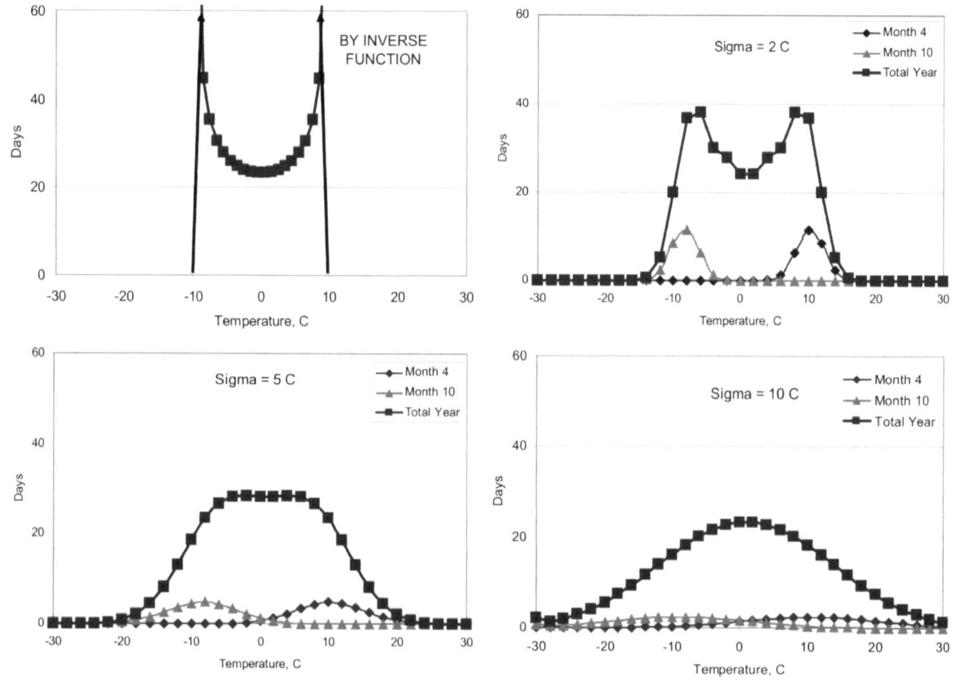

Figure 2 - Temperature Distributions for a Sine Function (Model 2)

Figure 3 shows the accumulated days of this function.

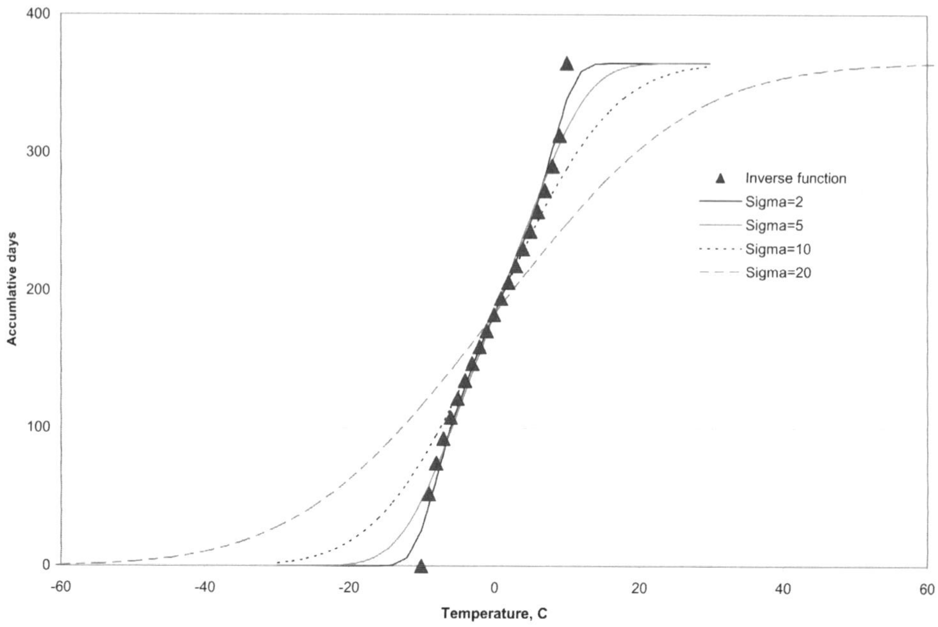

Figure 3 – Cumulative Temperature Distributions for a Sine Function (Model 2)

As expected, there are two points that jump at the two extreme temperatures of +10 °C and -10 °C. This curve is better approximated when the sigma value of the monthly distributions decreases from 10 to 2 °C.

Model 3 – Sin^3 function

Figure 4 shows the temperature distributions acquired by different methods.

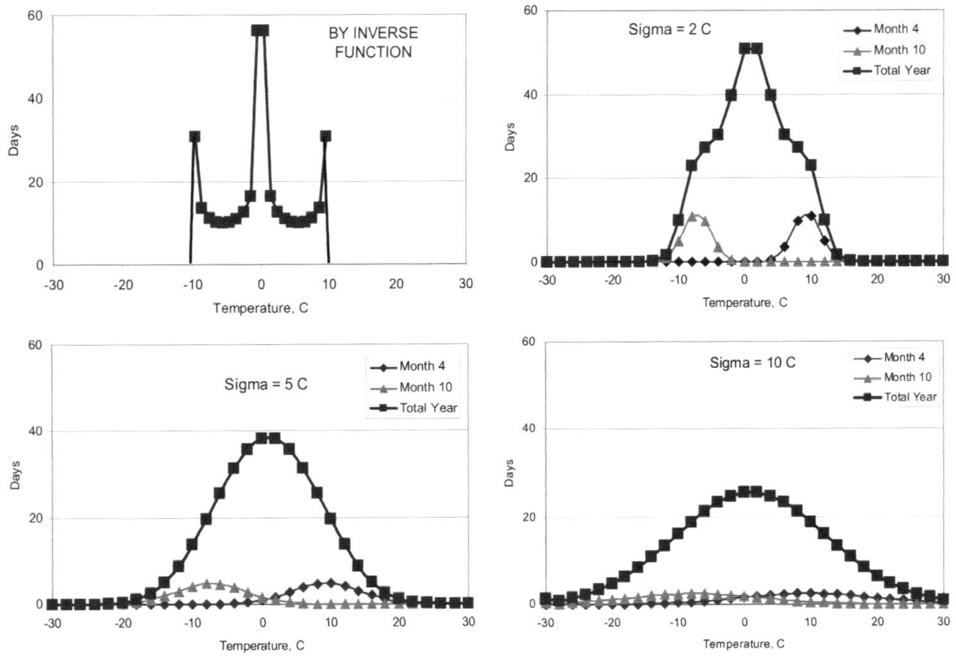

Figure 4 - Temperature Distributions for a Sin^3 Function (Model 3)

The upper-left quarter of the figure is acquired by the inverse operations of sine and power functions. There are two sharp horns at the two extreme (reversing) temperatures, corresponding to the horizontal portions in Figure 1. Additionally, there is a third horn in the central part corresponding to the flat portion around month 6. Although the distribution pattern is complex, the upper-right quarter using a sigma value of 2 °C is a reasonable representation. Again, the other quarters show the impacts by increasing the sigma values from 2 to 10 °C. As the value increases the resultant figures lose foci and their characteristic features.

Physically, this observation indicates that the average temperature of 0 °C appears with a high frequency. The function is well represented by this value. This is typical for functions with long transition periods between extreme values. Notice that large sigma values would not totally distort the true shape of the distribution function.

Figure 5 shows the accumulated days of this function.

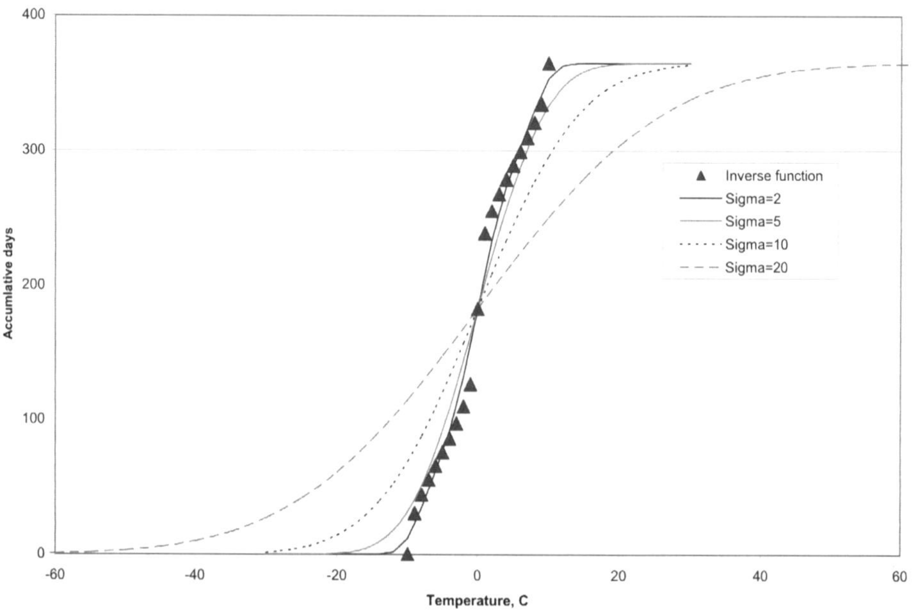

Figure 5 – Cumulative Temperature Distributions for a Sin^3 Function (Model 3)

As expected, there are three points that jump at the two extreme temperatures of +10 °C and -10 °C as well as 0 °C. This curve is better approximated when the sigma value of the monthly distributions decreases from 10 to 2 °C.

Comments on Sigma Values

It was mentioned earlier that the only parameter left to be determined is the proper value of sigma in this new methodology. As is presented above, decreasing the sigma values sharpens the composite temperature distribution functions.

With larger values of sigma, e.g. 10 °C, all resultant distributions appear to be broad and normally distributed. However, it has already been shown that this can be improper for some cases. As the sigma values reduces to 2 °C, the resultant temperature distribution best represents the numerical results acquired from inverse functions. However, if the sigma value continues to decrease, e.g., 1 °C, the resultant distributions start to show irregularities due to numerical processing. In other words, the monthly distribution becomes so sharp and discreet that features of continuous functions are lost. The actual value of sigma that starts generating irregularities in the distribution is highly dependant on the original data set and the temperature variations within that set as will be shown in the following sections.

INTERPRETATION OF METEOROLOGICAL DATA

The new methodology will be applied to interpret actual meteorological data. The compiled data include existing and planned major LNG projects which have been described in the open domain. It is worth mentioning that all major LNG projects, including export terminals, are in the vicinity of oceans and therefore certain moderations of the climate patterns by the oceans may be anticipated.

This section is divided into four subsections for clarity. The first two subsections describe general patterns of meteorological data and certain adaptations of the methodology which were required for handling real data compared to the derivation of the methodology when based on mathematical functions. The next two sections present the results.

Data Source and Climate Patterns

Unless specified otherwise, the climate data for this work are extracted from the following website: http://www.climate-charts.com/. The author of this site claimed that "All of the World Charts and about 10% of the USA charts are based on data from the World Meteorological Organization" and the website does not require user registration before using the data. Data of selected locations were compared with known values collected from other reliable project data and confirmed.

Climate data associated with LNG plants can mostly be divided into three regions: tropical, arctic and desert. Typical climate data for the tropical region show high average temperature with small temperature swings. For the arctic region, the typical climate data will show low average temperatures with large temperature swings. Finally, for desert region climate data, the average temperature will also be high but with large temperature swings.

Figure 6 shows the average temperatures versus latitudes for locations of existing or planned LNG plants.

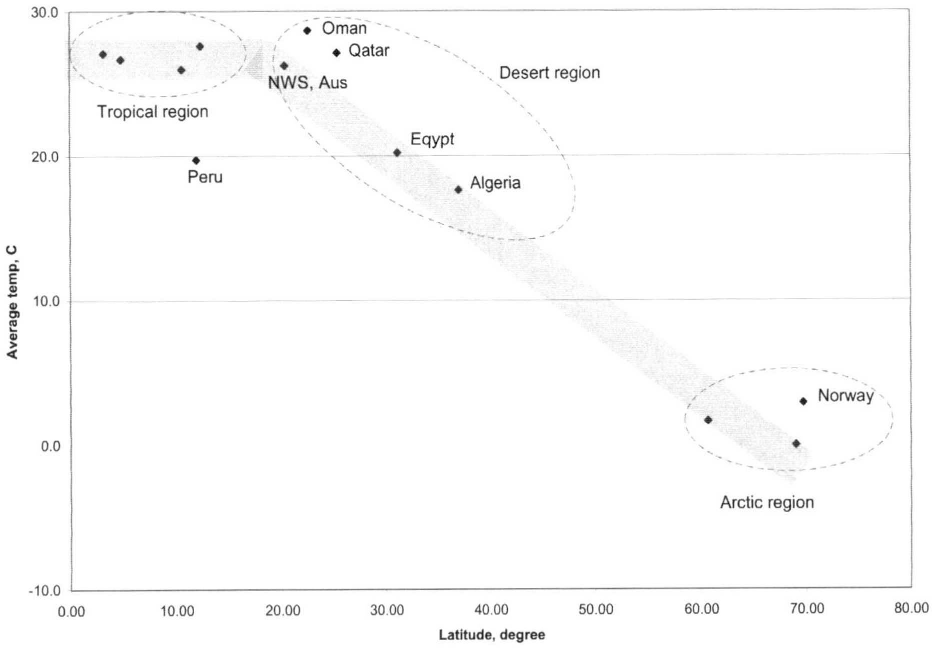

Figure 6 – Global LNG plants clustering in three climate regions

The three climate regions are marked and LNG plants are grouped accordingly. The following observations can be made:

- There is definitely a correlation between latitude and average temperature. It should be pointed out that all LNG plants are located close to the oceans. Therefore, certain levels of climatic moderation are anticipated.
- Impacts of local factors are visible. For example, Peru LNG is chilly among locations of the same latitude due to the Pacific cold current. Conversely, Statoil LNG of Norway is relatively warm thanks to the Atlantic warm current.

Figure 7 plots the temperature swings versus the average temperature for the same data.

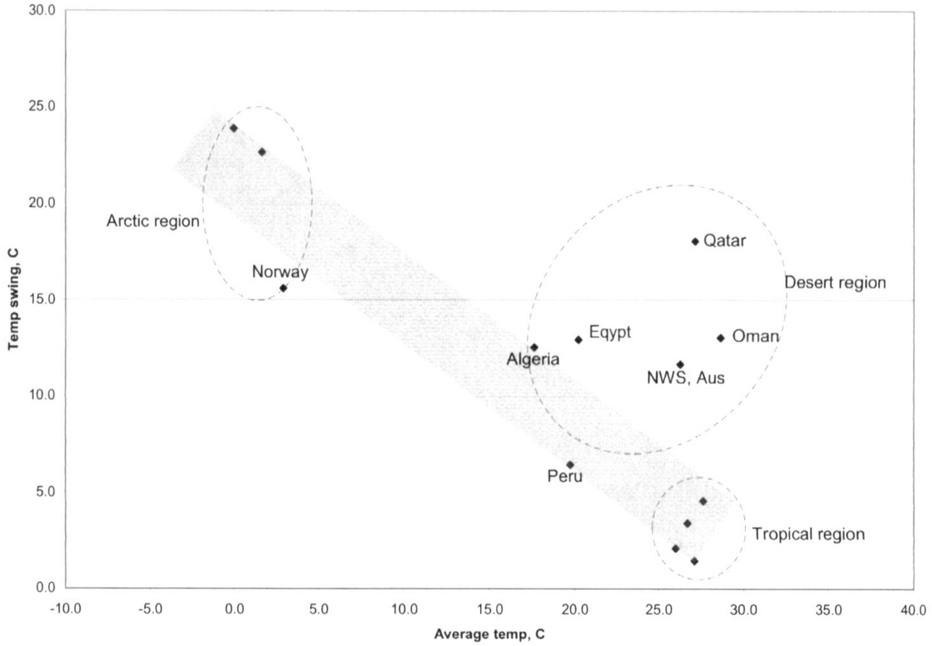

Figure 7 – Global LNG plants clustering in three climate regions

Again, a trend can be observed. Generally, the higher the average temperature, the narrower the seasonal temperature swings. Hence, plants in the arctic region require serious consideration of temperature impacts, but not in tropical regions.

However, the desert climate may distort this relation significantly. As shown in Figure 7, the temperature swings in the desert area are higher than normally anticipated. Unlike tropical regions, where temperature swings are within 5 °C, the temperature swings in the desert region are in the range of 15 °C for similar average temperatures.

The data of five locations will be presented in this study: Murmansk, Russia; Tromso, Norway; Doha, Qatar; Lagos, Nigeria and Piarco Int'l Trinidad, Trinidad and Tobago. The first two are in the arctic region, Qatar is in the desert region and the last two are in the tropical region. The first three locations show relatively large temperature swings while the last two have very small temperature swings. Table 2 summarizes the annual average temperatures and yearly temperature swings for each location while Figure 8 shows their detailed monthly average temperature.

Table 2 – Annual average temperature and temperature swing for each location

Location	Annual Average Temp. (°C)	Temperature Swing (°C)
Murmansk, Russia	-0.3	25
Tromso, Norway	2.9	16
Doha, Qatar	26.7	18
Lagos, Nigeria	26.8	3.5
Piarco Int'l Trinidad, Trinidad	26	1.7

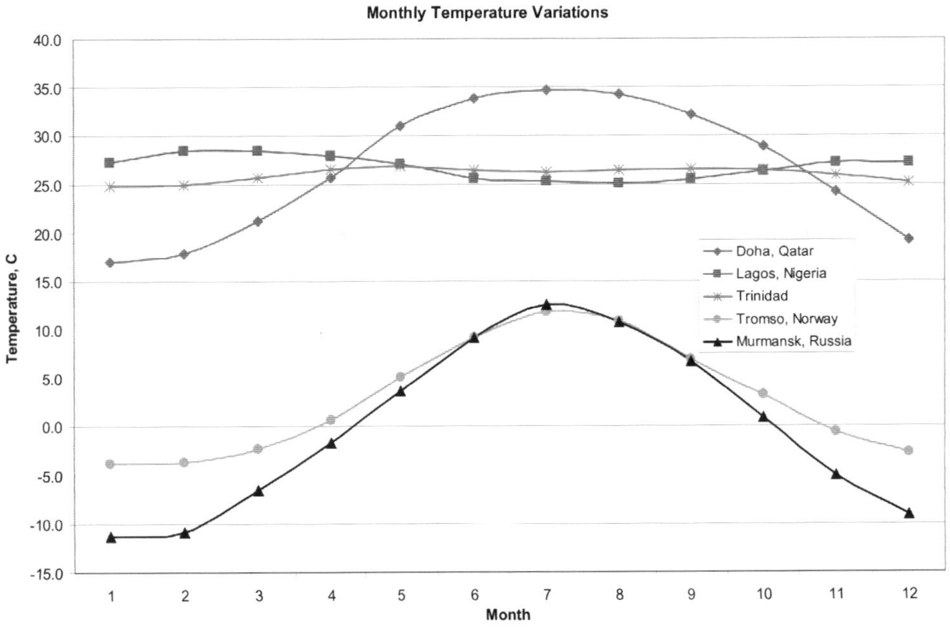

Figure 8 - Monthly temperature variations for the five sites studied

Numerical Methods

The suitability of the proposed new method for converting climate data to temperature distributions will be demonstrated. There are two major differences in the real data and mathematical models:

1. The shapes of mathematical functions are symmetric. No such regularity can be said about real climate data.
2. Unlike mathematical models, the analytical function to represent the climate data has to be derived on a case by case basis. This can be done conveniently using polynomial functions. Therefore, the inverse operation can be approximated by numerical solutions. However, the singularities resulted from horizontal portions in the climate data cannot be handled elegantly by polynomial functions which will add difficulty to the representation of tropical data.

For each of the location selected, the monthly average temperature data curves were used to determine the inverse function. Using polynomials functions, the data was approximated and then integrated numerically. Following the method presented in the previous section, the annual average temperature of each location was used in parallel to derive the inverse function analytically by using normal distribution functions.

Determining the value of sigma is also important in this work. A large value of sigma will result in the data set being represented by a normal distribution which will not approximate the inverse function correctly. Too small a sigma results in irregularities in the data and therefore is not helpful either. Different data set require different sigma values and from the discussions below, it seems that the sigma value is linked to the data temperature swing. For data with high temperature swings, a sigma of 2 °C is appropriate however, for data with smaller temperature swings a sigma of less than 1 °C seems to be required. However no correlations appear clearly at this time.

Cases of Wide Temperature Swings

Figures 9 through 13 show the results of each location's analysis. The results are presented in two categories of wide versus narrow temperature swings. This subsection contains data of wide temperature swings.

Figure 9 is the representation of the data for Murmansk.

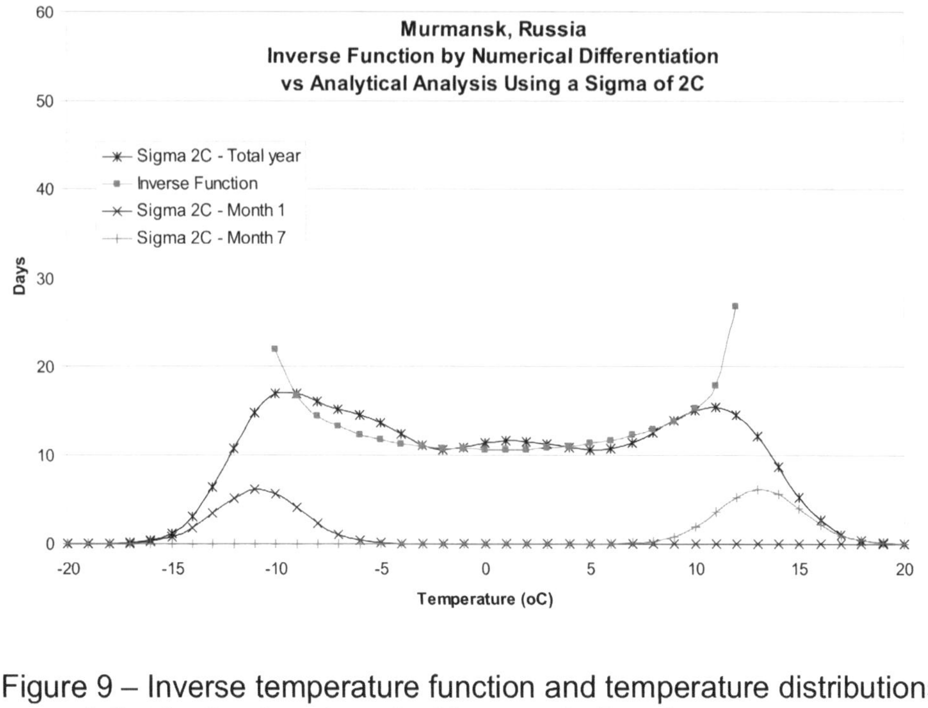

Figure 9 – Inverse temperature function and temperature distributions acquired through normal distribution functions for Murmansk, Russia

It shows that the analytical representation with a sigma value of 2 °C is a reasonable approximation of the inverse function. The inverse function, approximated with polynomial functions, shows two sharp, asymmetric horns at the two extreme temperatures, corresponding to the horizontal portions in Figure 8. In the inverse function mappings, the horizontal portions become singular. Consequently, these horns are difficult to duplicate using the analytical functions. However, the approximation by normal distributions is still reasonable. This analysis was done at different values of sigma and it shows the same trend as for the cyclic functions: increasing the value of sigma results in the data losing foci and characteristic features; reducing the sigma value below 2 °C results in distortion in the calculated distributions.

The above description is typical for data sets with large temperature swings, as can also be seen in Figures 10 and 11 for the average temperature data of Tromso and Doha.

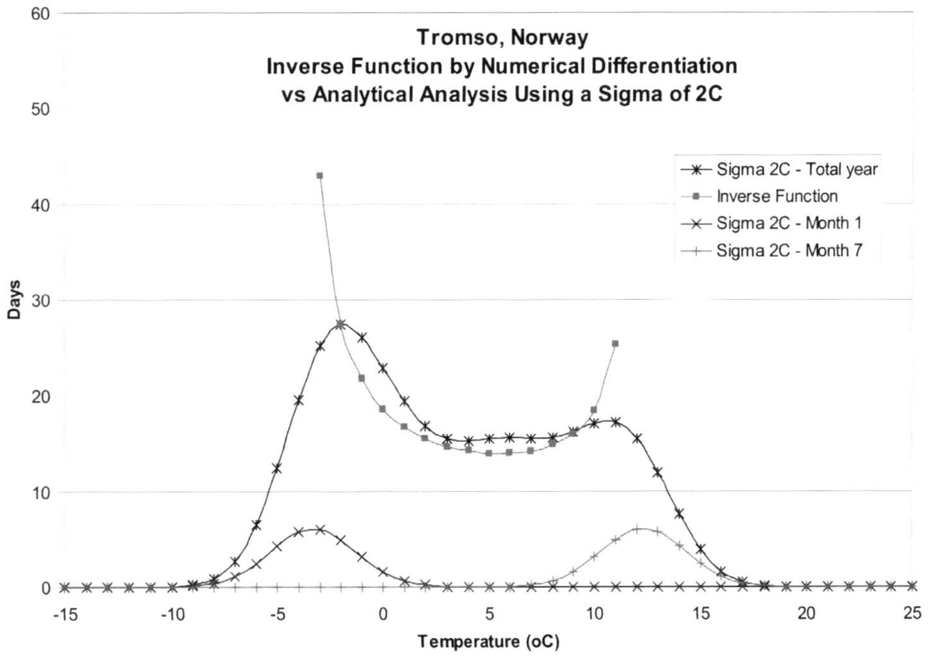

Figure 10 - Inverse temperature function and temperature distributions acquired through normal distribution functions for Tromso, Norway

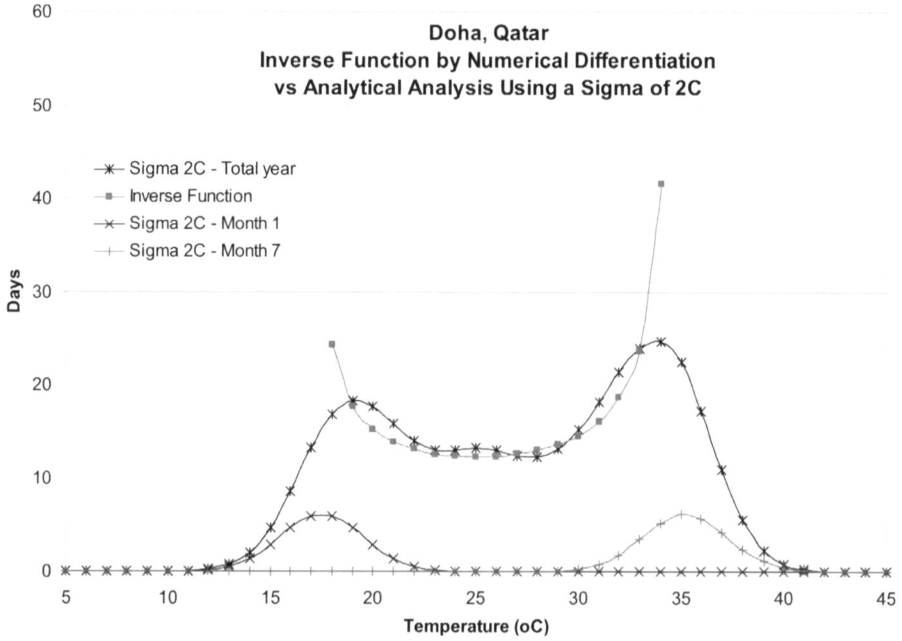

Figure 11 - Inverse temperature function and temperature distributions acquired through normal distribution functions for Doha, Qatar

Cases of Narrow Temperature Swings

However, the analysis becomes more tedious when the temperature swing is small as will be shown with the last two data series. To be able to determine the inverse function, the first problem to resolve is to approximate the function with polynomials. Even though the cyclic function can be approximated by sinusoidal functions, the break point between the upswing and downswing sections is definitively not as clear as for the data in arctic and desert regions. First of all, the climate data shows two small peaks compared to the one large peak for the previous models. This makes the approximation using polynomial functions difficult. Secondly, with such small temperature variations, it is clear that there will not be large peaks at the two temperature extremes (small peak to valley transition) but rather a large peak close to the average value. Due to the aforementioned reasons, the resolution used in numerical manipulations is reduced from 1 °C to 0.1 °C in this section. As a result, the ordinates in Figures 12 and 13 are different in scale from the others.

Figure 12 shows the resultant inverse function and its approximation using analytical function with a sigma of 0.5 °C for Lagos.

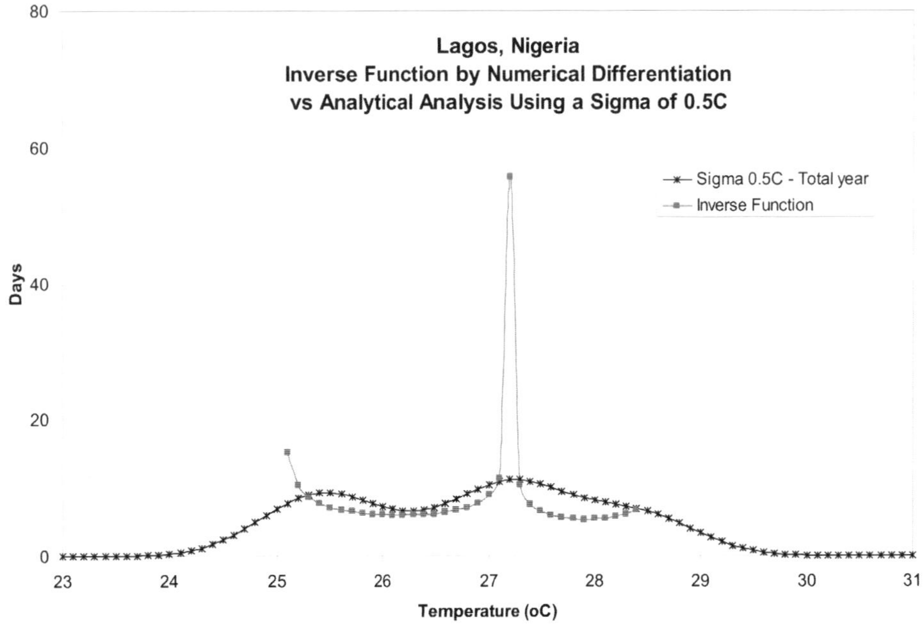

Figure 12 - Inverse temperature function and temperature distributions acquired through normal distribution functions for Lagos, Nigeria

The inverse function clearly shows that there is one sharp peak in the middle (close to the average temperature) with two smaller peaks at the two extreme temperatures, corresponding to the horizontal portions in Figure 8. In the inverse function mappings, the horizontal portions become singular. Consequently, these horns, particularly the large central peak becomes difficult to duplicate using analytical functions. As is shown in Figure 12, the representation of the inverse function by analytical methods in this case is not as accurate as for the previous sets of data. Trying to adjust the sigma value, by either increasing or reducing it does not yield any better results.

The above conclusion also applies to the data in Figure 13, representing the information around the average temperature distribution in Trinidad. The approximation in this case uses a sigma value of 0.2 °C and is a little more accurate than the Lagos data. However, the intensity of the large average peak is again not met by the analytical approximation.

Figure 14 summarizes this section. The results of Murmansk, Doha, and Lagos are plotted together to provide an overall perspective. The Lagos data in Figure 12 are re-grouped to the common base of 1 °C resolution. This is to provide a leveled ground for comparison. Some details of the distribution function are lost during the process. However, the messages remain the same.

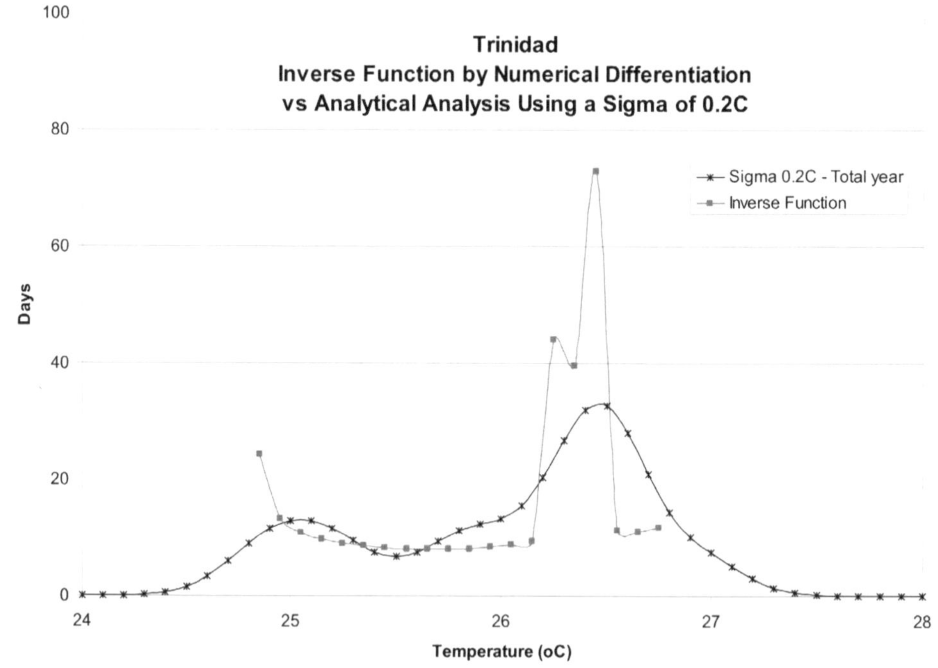

Figure 13 - Inverse temperature function and temperature distributions acquired through normal distribution functions for Piarco Int'l Trinidad, Trinidad and Tobago

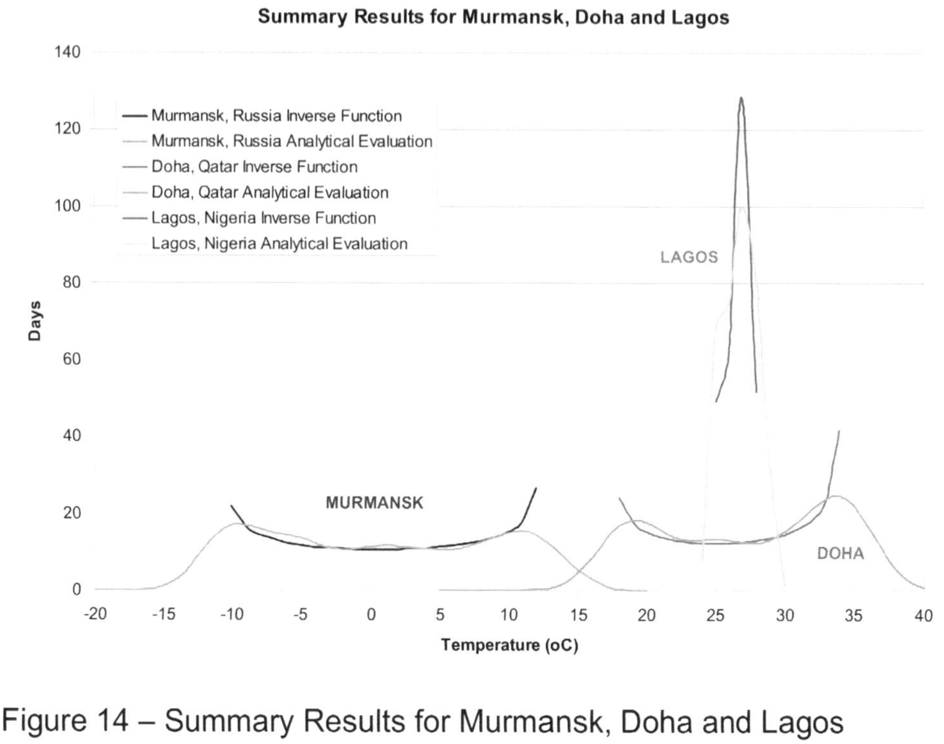

Figure 14 – Summary Results for Murmansk, Doha and Lagos

This figure clearly indicates that the proposed methodology presented in this paper applies with significant advantages to location with large temperature swings where there will be a high impact from the extreme temperatures. In a tropical climate, where the temperature swings are really small, the average temperature is as good a representation of the data set as the methodology proposed above and therefore there are no clear advantages of going through detailed analysis.

CONCLUSIONS

The conclusions of this work are listed as follows:

- A new method for handling meteorological data is introduced. The method disapproves current misconception that the temperature distribution function is a normal distribution (Gaussian function) with the annual average temperature being the central point. Instead, the temperature extremes should have the highest probability of appearance because they represent the turning of seasons. In comparison, the "annual average temperature" represents a transitional season, which should pass fairly rapidly.
- This has the greatest impact in cases where the temperature swing is the greatest, i.e. for plants in the arctic or desert regions. For plants in tropical climate, the small temperature swing results in small variation to the currently widely used method and the annual average temperature is a fair representation of the data set.
- Choosing the appropriate temperature scale to perform the numerical integration and the right value of sigma in the analytical analysis will help greatly with the accuracy of the results. Sigma values of about $2^{o}C$ seem to apply to temperature data set with a temperature swing of more than $15^{o}C$ whereas for data set with temperature swings of less than $5^{o}C$, a sigma value of less than $1^{o}C$ seems more accurate.
- Specific project risks/benefits analysis need to be performed to evaluate the cost reward impact that such a methodology could bring to the life cycle cost of such a project.

REFERENCES

- Chapra, S. C., "Applied Numerical Methods with Matlab ", 2nd Edition, McGraw-Hill, New York (2008)
- Stewart, J., "Calculus", 6th Edition, Thomson Publishing, CA (2008)

Biographies

Meredith Chapeaux is a process engineer for Chevron Energy Technology Company. She graduated in 1999 from the University of Texas at Dallas with a BS degree in Mathematics and Chemistry and received her Master's degree in Chemical Engineering in 2001 from Texas A&M University. Meredith has been in the industry for ten years focusing on LNG and Gas Processing.
E-mail: Meredith.Chapeaux@Chevron.com

Dr. Stanley Huang has a specialty area in cryogenic applications, particularly in LNG and gas processing. He graduated from National Taiwan University with a B.S. degree and attended Purdue University in 1981. He earned his Master and Ph.D. degrees there, all in Chemical Engineering. Additionally, he also acquired a Master of Science in physics. Stanley has 20 years of industrial experiences and is a Registered Professional Engineer in Texas. Currently he is a Staff LNG Process Engineer of Chevron.
E-mail: Shhuang@Chevron.com

72e Strategic Model of LNG Arbitrage: Analysis of LNG Trade in Atlantic Basin

Svetlana Ikonnikova, Dmitry Volkov, Gürcan Gülen, and Ruzanna Makaryan. Center for Energy Economics, University of Texas, Austin, 1801 Allen Parkway, Houston, TX 77019

Introduction

With the increase in oil and gas prices over the last decade, the role of LNG in natural gas markets has become more prominent. According to the BP Statistical Review of World Energy (2003-2008), the total world LNG trade has increased by more than 40% over the last five years, with an average growth rate about 7%. Imports of North America have risen by about 50%from 2002 to 2007. This dynamics attracts growing attention to the LNG markets.

Historically, LNG contracts have been signed for 20-25 years and had destination clauses. Over the last decade the structure of the LNG markets has transformed. Most of the new contracts have a much shortened duration of 10-12 year. Destination clauses have not been completely remove, but relaxed allowing buyers to redirect their ships to extract a higher rent, which is to be shared with the seller. As the world market was growing a spot markets emerged. Now exporters sell LNG without contracts competing on spot markets directly.

The study, based on a comprehensive analytical model of the LNG market, aims to answer the following questions:

• How will LNG trade develop: will the LNG market continue to grow?

• Will the relation between markets get stronger and prices converge? If so, could it lead to decoupling of NG prices in general and LNG in particular, from crude oil/oil product prices?

• Will the arbitrage continue to exist? If yes, in what scale?

Preliminary Conclusions

Based on our analytical model, we have shown that it is beneficial for an LNG importer as well as an exporter to enter long-term contracts, if other market players do so. The intuition behind this result is similar to one in Allaz & Villa (1992). We have derived the shadow price of LNG export capacities. Based on its value we calculate: how much LNG-exporting countries should invest to benefit from the LNG spot trade; how much LNG-importers are willing to pay for have flexible destination clauses and hence, be able to resell LNG; and how much LNG importers should invest in regasification facilities given the expected market conditions. Our results allow to explain observed price dispersion in the

Atlantic Basin. Moreover, we show that given the capacity constraints (and/or high capacity costs) and information asymmetry the price difference may continue to exist even if the market is global (central). Though the exact figures are subject to change with respect to willingness to pay in different countries as well as marginal costs of supply and capacities, the overall incentives and effects obtained are robust.

References Cited

1 Friedmand, Daniel, "A simple testable model of double auction markets", Journal of Economic Behavior and Organization 15 (1991), pp. 47-70

2 Hubert, Franz; Ikonnikova, Svetlana, "Investment Options and Bargaining Power in the Eurasian Supply Chain for Natural Gas", 2005 North American Winter Meeting of the Econometric Society, Philadelphia, PA

3 Allaz, B., J.- Vila, L., "Cournot competition, forward markets and efficiency", Journal of Economic Theory 59, pp.1-16.

Economic and Technical Challenges in Lined Rock Cavern LNG Storage System

89a

Ho-Yeong Kim, Ph. D., *Vice President*
Sang-woo Woo, *Senior Manager*
Dae-Hyuck Lee, Ph. D., *General Manager*
Joseph H. Cho, Ph. D, P.E. *Technology Director*
SK Engineering & Construction Co., Ltd.

AIChE Spring Meeting, April 2009
9th Topical Conference on Natural Gas Utilization
Tampa, Florida, April 26-30, 2009

ABSTRACT

This paper describes the economic advantages of the lined rock carven system against conventional LNG storage tank. The lined rock carven system, basically similar to the membrane LNG storage system, shows more economical when its capacity is beyond 300,000 m^3.

This paper also discusses the challenges of the lined rock carven system based on pilot storage system with liquid nitrogen. Impact of rock freezing and hydrogeology on the design and operation will also be discussed.

INTRODUCTION

At LNG terminals, conventional aboveground tank or in-ground tank type have been successfully used worldwide to store LNG. Nevertheless, these types of storage tank may have some problems in case that there requires large land area for storage, large reclamation of seaside and industries have already developed and free remaining area is small and expensive.

Based on experience of underground storage for crude oil and various types of hydrocarbons, underground storage system was thought to be more economic way to store LNG regardless of the above-mentioned conditions.

With this idea, some attempts have been tried to develop the LNG underground storage system but were not successful due to unlined containment and failure of controlling the temperature and groundwater. From the past trials, it is known that the thermal stress and freezing of groundwater nearby cavern generate cracks in the host rock mass. Generated thermal cracks contribute to induce gas leakage and to increase in exchanging the heat flux between stored LNG and rockmass. Eventually, this causes excess of boil-off rate from the stored LNG.

New concept of storing LNG in a lined rock cavern with containment system which can overcome these problems has been developed by Geostock, SKEC and SN Technigaz.

To demonstrate the technical feasibility of this concept, a pilot plant was constructed in Daejeon Science Complex in 2003, which had been under operation for storing LN2 (Boiling Temperature: -196°C) since January 2004, and now been decommissioned.

The objectives of this paper are to introduce lined rock cavern LNG storage system as a new technology, to evaluate and compare the construction cost in comparison with conventional aboveground and in-ground tank storage based on the Korean reference, and to examine the real-scale applicability of a lined underground rock storage system which has been verified by a successful operation of the Daejeon LNG pilot plant.

CONCEPT OF LINED ROCK CAVERN LNG STORAGE SYSTEM

The basic concept for the development of an LNG storage in membrane Lined Rock Cavern is a combination of well-proven technologies and a new concept named as "formation of ice ring (Fig. 1) :

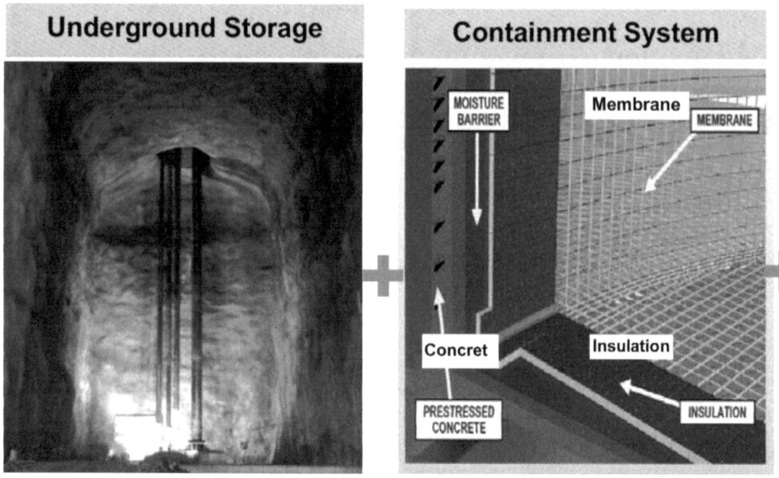

Fig. 1 - Basic Concept of Lined Rock Cavern LNG Storage System

Underground Storage of Crude Oil and Hydrocarbons

The first underground storage one brought into service at the beginning of 20th century. Since this date, this technology, which proved reliable, experienced a strong development related to its intrinsic qualities but also on the innovations and technical progress from which it regularly profited. Currently, large capacity of storage cavern units with more than several hundred of thousand of cubic meter, for a broad range type of hydrocarbon products and crude oil, are constructed and operated successfully especially in Korea.

Membrane Containment for LNG Product

The membrane containment system, which was introduced in 1962 on a prototype ship, has since been successfully used in several LNG carriers and storage tanks. This membrane system provides the proper thermal protection to the surrounding rock mass, preventing excessive stresses and crack formation and reducing boil-off to level comparable with conventional LNG storage tanks.

Formation of Ice Ring

Formation of ice ring as an impervious second barrier against the leakage of contained LNG can be explained by the each stage of construction of lined rock cavern (See Fig. 2). Groundwater is temporarily removed from the rock surrounding the cavern during the first phase of construction. This preliminary de-saturation of the host rock mass aims at preventing unacceptable hydrostatic pore pressure and ice formation behind the cavern lining. After excavation of cavern and installation of containment system are completed, LNG will be stored in the cavern and then the cold front starts propagation from the cavern at the same time. When the cold front has advanced far enough from the cavern wall, drainage can be stopped to allow groundwater progressively to rise up and quickly form a thick ring of ice around the cavern. After ice ring is completely formed, operation of the drainage system is stopped. The drainage period during LNG storage will last several months or years depending on the thermal properties of rock masses and hydro-geological characteristics on the site.

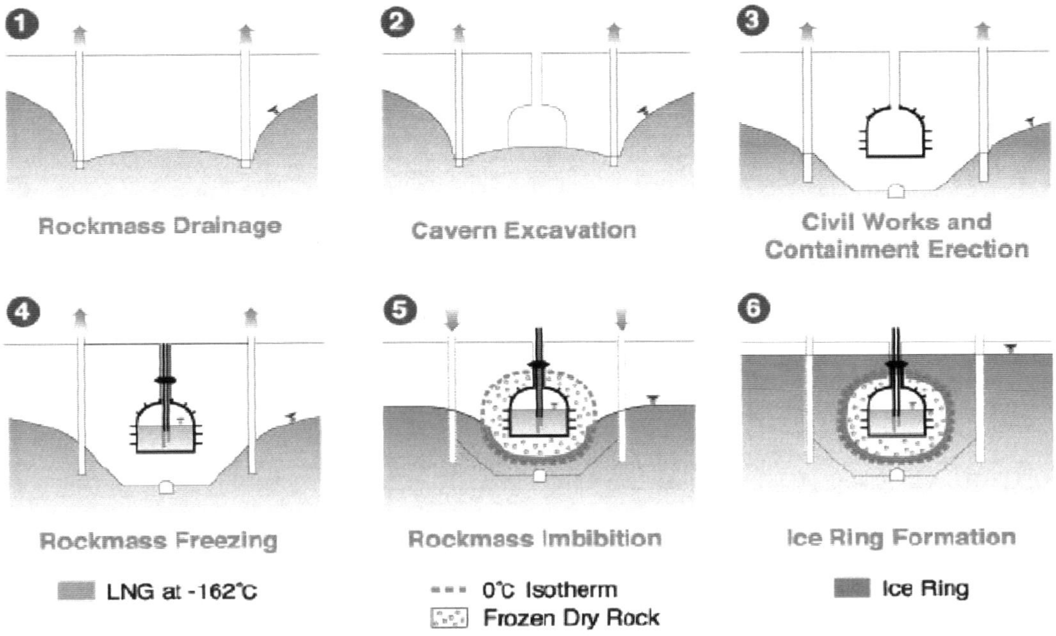

Fig. 2 - Schematic Diagram of Formation of Ice Ring

This new concept of LNG storage in membrane lined rock cavern is valid for any type of export or receiving terminal, but also for peak-shaving or large capacity stockpile storage.

Typical Underground Storage Layout

The cavern layout geometry should be adapted to local geological and hydro-geological conditions of the LNG terminal facilities. A wide variety of different geometries can be considered. The general cross section of storage gallery is comparable to unlined underground storage technology, with horse shape galleries.

The following Figs. 3 and 4 show respectively typical sections of underground works and galleries arrangement that can be considered in good rock mass conditions.

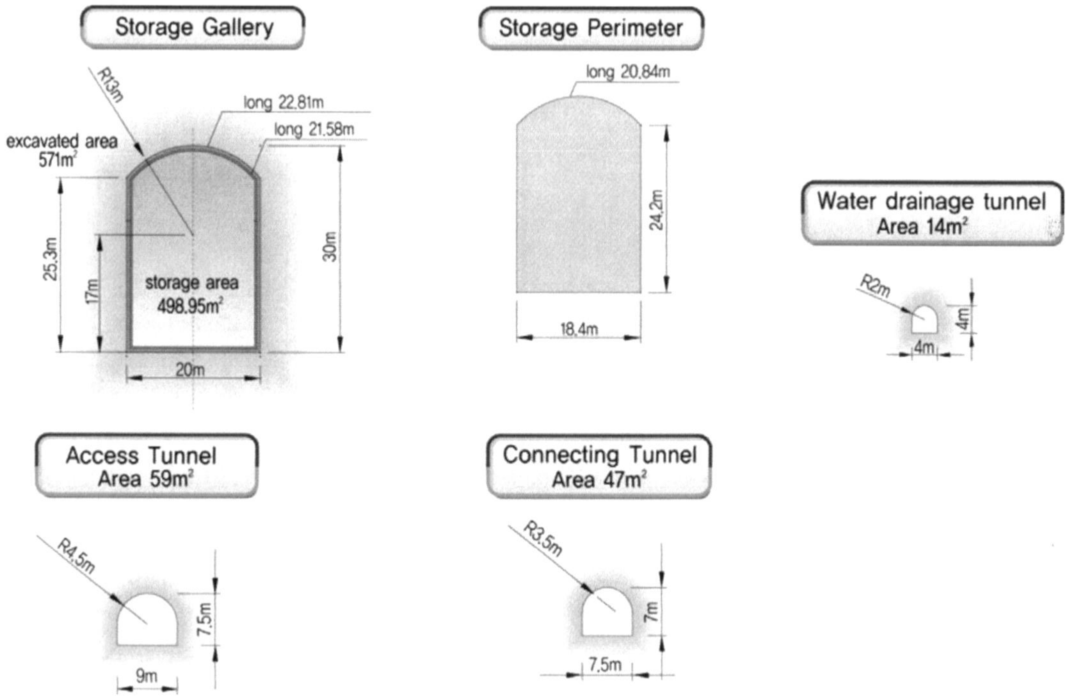

Fig. 3 - Typical Cross Section of Underground Galleries

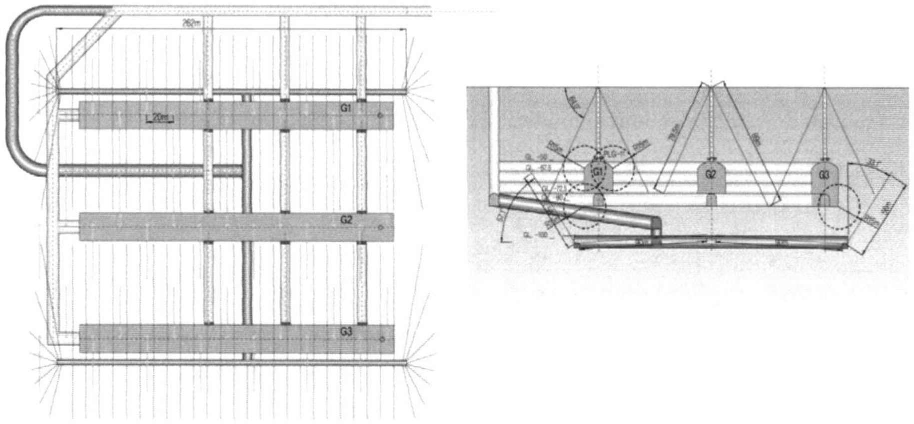

Figure 4.a. Plan view   Figure4.b. Vertical cross section

Fig. 4 - Typical Layout of Underground Storage (320 000 m^3)

Equipment, Shaft and Pipework Arangement

The storage cavern equipment is similar to the ones of a conventional LNG terminal, allowing it to operate in exactly the same way as an aboveground tank (see Fig. 5).

Internal piping arrangement is also similar to the one of a membrane LNG storage tank. All pipes are embedded in a concrete structure near the bottom of the operation shaft. Inside the cavern the pipes are joined together by braces forming a pipe tower designed to permit differential pipe contraction.

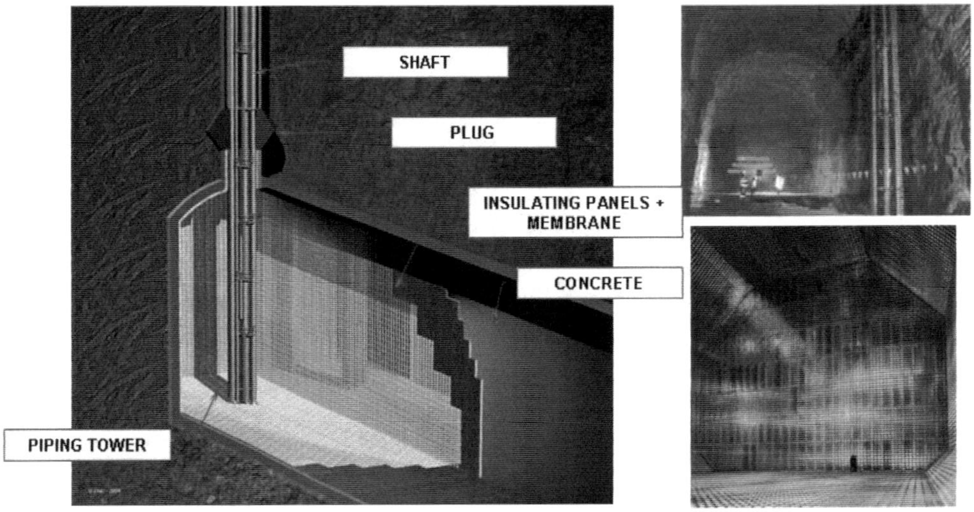

Fig. 5. Typical Cavern Equipment Arrangement.

(with pictures of LPG cavern storage and membrane tank of LNG carrier)

ADVANTAGES AND ECONOMICS OF UNDERGROUND STORAGE

There are many advantages of underground storage in terms of safety, security and environmental acceptability compared with aboveground tanks and in-ground tanks.

Underground storage is much safer in consideration of fire on plant or decreased potential damages in case of industrial accident nearby due to its multi-component barrier with liner and ice ring, and lower vulnerability against earthquake and typhoon.

In security aspects, underground storage can easily prevent from sabotage or terrorism acts. Because of no need of large reclaimed areas and less earthworks at ground level, underground storage is environmentally more friendly, and eventually it is better acceptable by people located nearby.

Nevertheless, some other advantages should be also considered because of the minimization of total space required for the LNG terminal (see Fig. 6). This is due to the fact that LNG storage is about 50 m underground, this can represent a huge cost saving especially in seashore areas where industries are already developed and free remaining areas are small and expensive. It is also the case in areas whose topography needs expensive reclaimed land.

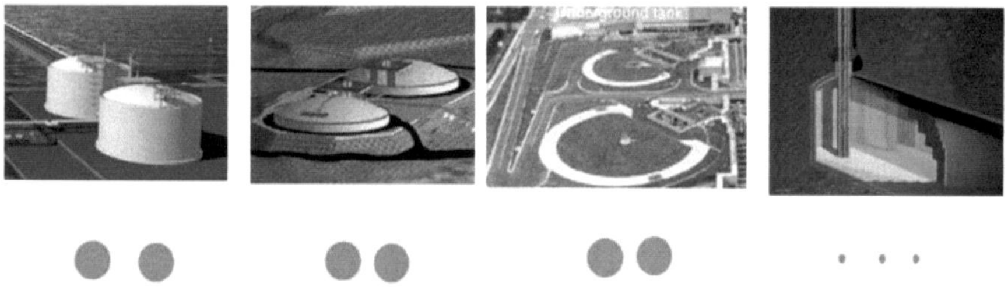

Fig. 6 - Required Area at Grade Depending on Storage Type (for 320 000 m^3)

Small galleries should be avoided wherever possible by reason of their poorer capacity/area ratio. Underground storage in the form of a gallery of around 20m width by 30m height cross section (see Fig. 3) is the most favorable which has been studied geometry in cost terms versus rock behavior.

Moreover, mining technologies and membrane containment system have such flexibility that unit storage capacity has no limit. Crude oil caverns are up to 4,500,000 m^3 in Korea and LNG lined caverns can be increased till such value.

This comparative cost estimate between aboveground and cavern storage concerns only the storage itself and its equipment. It does not take into account the substantial cost saving which could be made, in the case of the cavern storage, for the safety equipment (impounding basin, peripheral retention wall, fire fighting systems, etc.) and possibly for the piping length and terminal surface area reduction. Moreover, operation costs, including maintenance are very attractive.

In 2008, a national forum for cost comparison among conventional aboveground, in-ground LNG tank and underground storage was held in Korea where consisted of Ministry of Knowledge and Economics (MKE), Korea Gas Corp. (KOGAS), Korea National Oil Corp. (KNOC), Korea Institute of Geoscience and Mineral Resources (KIGAM), experts of engineering consultants and prestigious professors steered by Congress committee. Costs of varying stored volume from 200,000m^3 to 1,000,000 m^3 with increment of 200,000m^3 were evaluated and compared with reference price as of March 2006 in Korea. In all cases, underground storage is the most economic over the stored volume of 300,000m^3, and at 400,000m^3, cost for underground storage can be economical by 8% and 34% compared to aboveground and in-ground tank respectively (see Fig. 7).

Moreover, operation cost for underground storage units are highly competitive as compared to aboveground and in-ground tanks as systems like slab heating or fire water are not necessary or can be tremendously reduced. Based on Korean reference which has been implemented on crude oil storage by Korea National Oil Company, operation cost of underground storage is 63% less than that of aboveground one.

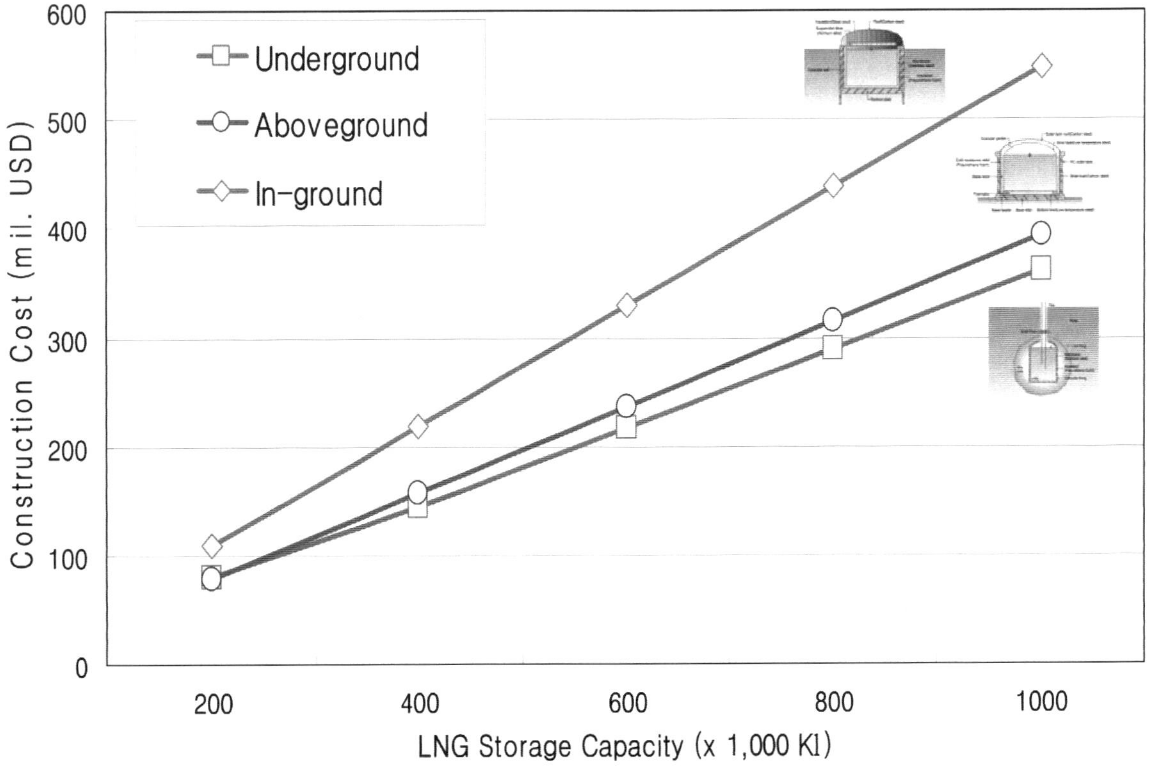

Fig. 7 - CAPEX Evaluation and Comparison (South Korean Context)

THE LNG PILOT CAVERN

The objectives of this pilot project are: to demonstrate the feasibility of the technology, to check adequacy between the results of previous computations and laboratory tests regarding in-situ measurements, to test the overall performance of the storage and to have an efficient return of experience to improve design and construction of industrial scale projects.

The LNG pilot cavern is located in Daejeon, about 200 km south of Seoul, in an existing cavern implemented within KIGAM research centre area. The experimental cavern roof lies at a depth of about 20 m below the ground surface. In order to have the containment system completed, a concrete wall closes the entrance of the cavern. Additionally, a platform above the entrance of the cavern is made to install instruments, manhole and piping. The internal dimensions of the completed pilot plant have a sectional dimension of 3.5m x 3.5 m, a length of 10 m, amounting to a working volume of 110 m^3 (see Fig. 8).

The containment system, which is used for underground lined rock caverns, is similar to the one used and improved by SN Technigaz since 1962 for the membrane type LNG storage tanks and LNG carriers. The modular structure of the containment system makes it very flexible, improving construction and adaptation to cavern geometry. The thickness of the insulating panels can be chosen as per the requirement of the thermal efficiency and the nearly unstressed membrane permits its usage in very large scale future projects.

The containment system is composed of several layers, from rock to LNG (see Fig. 9) as follows. A concrete structure ensures the transition between the containment system and the rock. It is made of reinforced cast in place concrete. Insulating panels made of foam are sandwiched between plywood sheets and are bonded on the concrete using load-bearing mastic. Thickness of the insulation panel is 300 mm so that the rock temperature will not fall below –50°C after 30 years and the boil-off rate will stay at acceptable limit. Finally, a 1.2 mm thick stainless steel corrugated membrane attached to the insulation panel provides gas tightness at low temperature. All the surfaces are covered with concrete lining, insulating panels and the stainless steel membrane sheets.

Fig. 8 - A bird's-Eye-view and Cross Section of Pilot Cavern

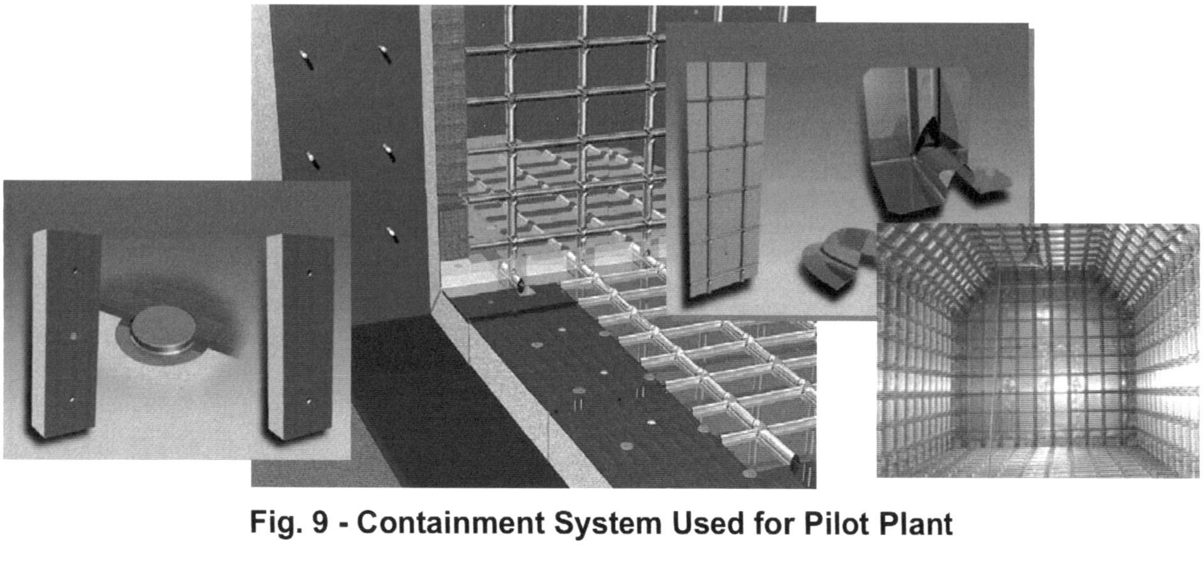

Fig. 9 - Containment System Used for Pilot Plant

Process and Pilot Instrumentation

For safety and practical reasons, liquefied nitrogen was used instead of LNG. The cryogenic pilot is operated from a laboratory room located in the access tunnel. It houses the Data Control System (DCS) for the process of the cryogenic pilot and for the geotechnical monitoring. An LN2 plant is installed outside in a fenced site, containing one LN2 storage tank, a vaporizer and associated control systems in order to inert, cool down the cryogenic pilot, fill it completely with LN2 and compensate for the gaseous N2 boil-off (see Fig. 10).

These operations use two insulated pipes from this dedicated site to the cryogenic pilot through the access tunnel and the special local enlargement in the cavern roof to go over the plug and penetrate inside the cavern. The additional line is necessary to release overpressures if they happen. Some instruments installed in the containment system and cavern permit to follow the main parameters (inner pressure and temperature, LN2 level, containment deformation) and to operate the pilot.

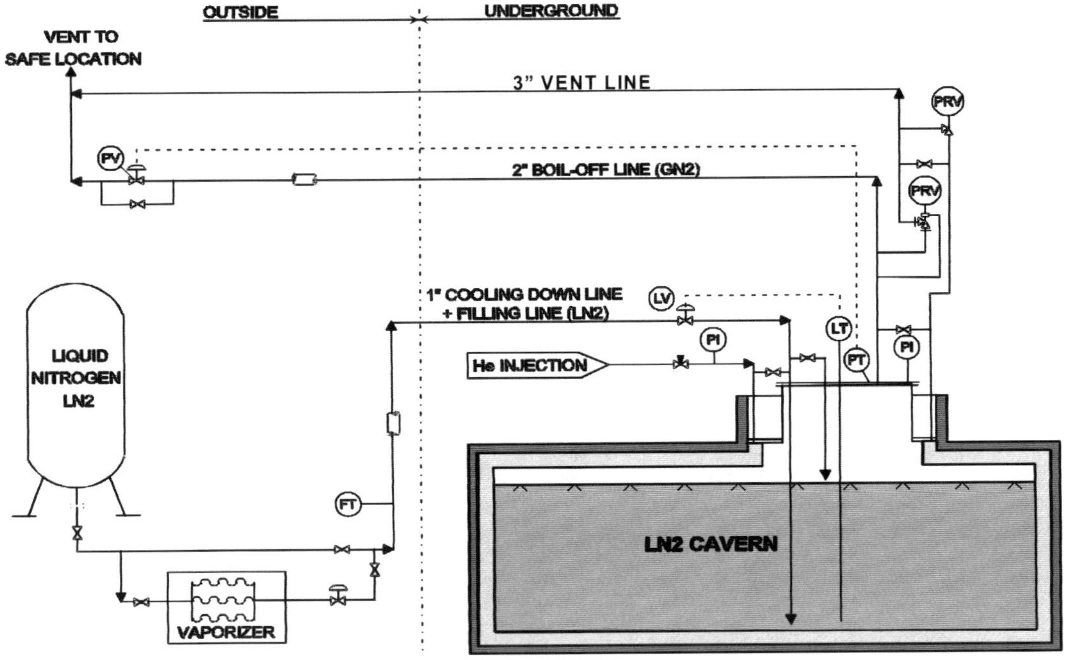

Fig. 10 - Schematic Process Flow Diagram of Pilot Plant

Pilot Test

A comprehensive monitoring system was provided for measuring temperature, thermo-mechanical and groundwater responses of the rock and concrete during pilot test. Also numerous parameters such as LN2 level, temperatures, pressure and boil off rate for the containment system were monitored during the operation. The behaviour of surrounding rocks and the containment system was recorded during three successive phases of operation: a) first six months duration during which a full level is maintained by filling with LN2 in order to compensate for loss of boil-off, b) second six months during which no more filling is performed allowing the cavern to empty naturally, and c) third six months during which the empty cavern is heated up till the ambient temperature is reached and the pilot cavern is decommissioned.

The results from the pilot test confirmed that both construction and operation of underground LNG storage in lined rock caverns are technically feasible. Major conclusions of the test can be summarized as follows:

a) The 20 boreholes drainage system has proved its efficiency during the successive construction and operation periods and allowed maintaining the rock mass de-saturated during the cooling down period. Efficiency tests performed at this stage showed that almost 99% of the natural water flows towards the pilot cavern could be drained back by the drainage system. The re-saturation of the rock mass was performed by stopping progressively the drainage. Finally, the drainage system allowed draining again the host rockmass during the thawing progress of the rockmass.

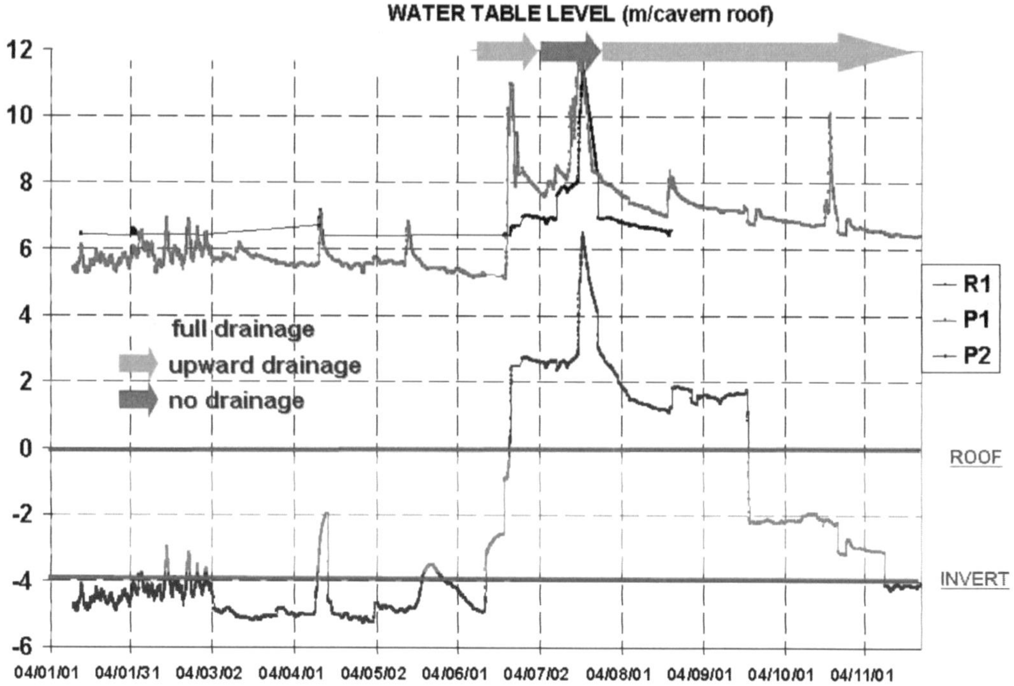

Fig. 11 - Piezometric Variations by Operation of Drainage System

b) Thermal responses of the rock mass under very low temperature of about -30°C around the rock cavern could be well predicted by numerical models.

c) Thermal stress-induced displacements occurred toward inner rock mass, which is favorable to stability aspects of the cavern. Displacements are of relatively low amplitude within 3~5 mm corresponding to about 0.2% of cavern radius. Data recorded at the end of thawing period shows that small displacement(about 1mm) is expected due to local rearrangement of the rockmass. And conventional rock reinforcements such as rock bolt and shotcrete remain effective at very low temperature during the pilot operation.

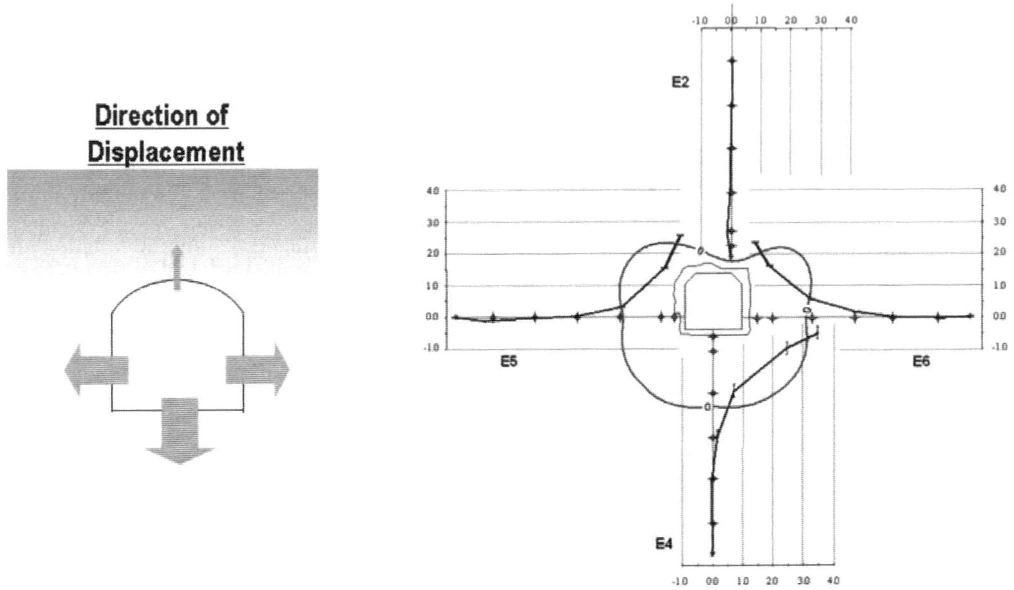

Fig. 12- Displacement of Rock Mass around Pilot Cavern

d) From the results of hydro-thermal analysis, ice ring is formed with the assumption that average distance of zero degree isotherm reaches 3 or 4 meter from the cavern wall in rock mass with the hydraulic conductivity of 10^{-7} to 10^{-6} m/s by controlling groundwater drainage system (see Fig. 13).

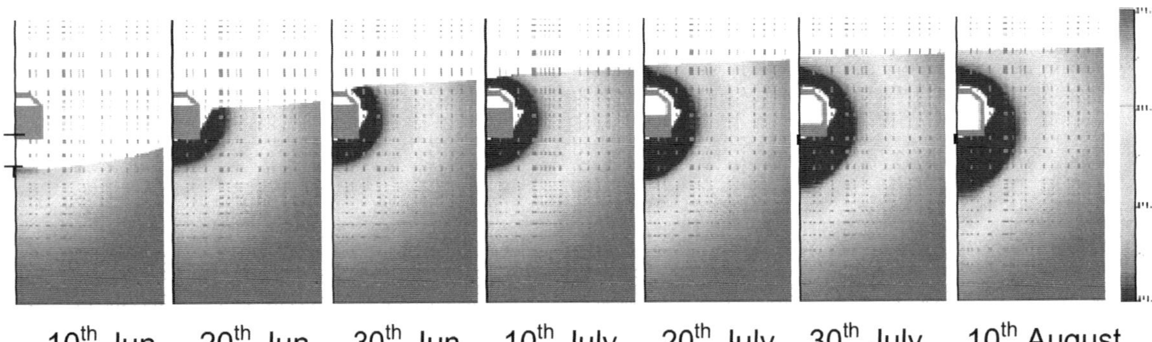

Fig. 13- Simulation Results for Ice Ring Formation Around the Pilot Cavern

e) By comparing boil off gas ratio (BOR) occurred during operation of the pilot with estimated ones, BOR from underground LNG storage cavern can be estimated by numerical and theoretical methods (see Fig. 14).

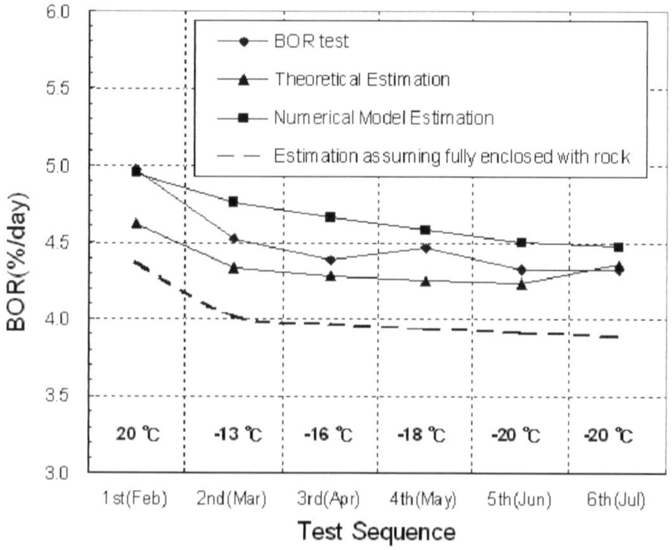

Fig. 14 - Boil-off Gas Ratio Occurred during Operation of the Pilot Plant

CONCLUSION

As compared with the conventional aboveground and in-ground storage tanks, the use of lined rock cavern LNG storage system at the LNG terminals can be more economic way in the aspects of CAPEX and OPEX. In addition, it has also the advantage of safety, security and environmental acceptability against the conventional tanks.

From construction and operation of Daejeon Pilot plant, technical feasibility of underground lined rock storage system can be well proved and the real scale applicability of it has been evaluated by the results from a successful operation of the pilot plant.

Lined rock cavern LNG storage system can be realized in due course at some countries which have suffered from the shortage of storage capacity of LNG and seasonal extreme variation of domestic demand, and of which industries are already developed and free remaining areas are small and expensive.

REFERENCES

Amantini, E., Chanfreau, E., "Development and construction of a pilot lined cavern for LNG underground storage", 14th International conferences & Exhibition on Liquefied Natural Gas, Qatar, PO-33., 2004.

Chanfreau, E., Amantini, E., Lee, D.H., "Assesing the use of underground storage tanks at LNG terminals", LNG Journal, June 2006.

Chung, S.K., "Thermo-mechanical behavior of rock masses around underground LNG storage cavern", Keynote lectures in Rock Mechanics in Underground Construction – 4th Asian Rock Mechanics Symposium, Singapore, pp. 19-28., November 2006.

Chung, S.K., Park, E.S., Han, K.C., "Feasibility study of underground LNG storage system in rock cavern", 11th ACUUS Conference, Athens, Greece, pp. 501~506, September 2007.

Jeon, Y.S., Park, E.S., Chung, S.K., Lee, D.H., Kim, H.Y., "Numerical simulation of fracture mechanisms for rock masses under low temperature conditions", Tunnelling and Underground Space Technology, Vol.21, Issues 3-4, pp.470-471., 2006.

Jeong, W.C., Woo, S.W., Lee, H.S., Lee, D.H., Lee, J.M., Choi, Y.T., Jung, Y.B. Park, E.S. and Jeong, S.K., "Coupled Heat Transfer and Water Flow Analysis on Ice Ring Formation around a Underground LNG Storage Cavern", Proceedings of the GEOPROC 2006 International Symposium, Weiya Xu (ed). Nanjing China, p.602-625., May 2006.

Lee, D.H., Kim H.Y., Gatelier, N., Amantini, E., "Numerical Study on the Estimation of the Temperature Profile and Thermo-mechanical Behaviour in Rock around the Taejon LNG Pilot Cavern", International Symposium on the Fusion Tech. of Geosystem Eng., Rock Eng. And Geophys. Exploration, Seoul, Korea, p.233-237., November 2003.

Lee, D.H., "Thermo-geomechanical monitoring & results", Proceedings of International Symposium on LNG Storage in Lined Rock Cavern, Energies, Saint Quentin en Yvelines, France., December 2004

Lee, D.H., Lee, H.S., Kim, H.Y and Gatelier, N., "Measurements and Analysis of Rock Mass Responses around a Pilot Lined Rock Cavern for LNG Underground Storage", Proceedings of EUROCK 2005, Koecny (ed), Brno, Czech, p.287-292., May, 2005.

Park, E.S., Thermo-mechanical consideration for the cryogenic storage, International workshop on the Underground Storage Facilities in conjunction with the 4th ARMS, Singapore, pp.81-98., November 2006

Park, E.S., Chung, S.K., Lee, D.H. and Kim, H.Y., Experience of operating an underground LNG storage pilot cavern, 11th ACUUS Conference, Athens, Greece, pp. 157~162, September 2007

SKEC, Geostock, SN Technigaz, 2004. Proceedings of International Symposium on LNG Storage in Lined Rock Caverns, Seoul, Korea., June 2004

Author Biography

First Author:
Name: Joseph H. Cho
Position: Technology Director
Company: SK Engineering & Construction

Biography:
Joseph Cho is a Technology Director in the LNG Division at SKEC. He is a registered professional engineer and has been in the LNG industry for twenty three years. He was involved in process design of LNG liquefaction plants and import terminals for various regions as a senior engineer at top tier EPC companies. His scope of work includes managing LNG/Oil/Gas technology, feasibility and optimization studies, economic evaluations, computer simulations, process equipment specifications, control philosophy and review of output from general engineering. He graduated from the University of Oklahoma with a Ph.D. degree in Petroleum Engineering.

LNG Transfer in Harsh Environments - Introduction of a New Tandem Mooring Concept

89b

Sven Hoog (PhD), Hartmut Koch and Roland Huhn
IMPaC Offshore Engineering, Hamburg, Germany

Christian Frohne (PhD)
Nexans Deutschland, Hannover, Germany

Jörn Homann
Brugg Pipesystems, Wunstorf, Germany

Günther F. Clauss (Prof., PhD), Florian Sprenger, and Daniel Testa
Technical University Berlin, Berlin, Germany

AIChE Spring Meeting, April 2009
9th Topical Conference on Natural Gas Utilization
Tampa, Florida, April 26-30, 2009

Abstract

For loading LNG shuttle carriers at an offshore terminal, the technical issues of the vessel mooring configuration (side-by-side or tandem configuration) as well as the approach and handling system for the transfer lines (either composite hoses or flexible cryogenic pipes) are influenced and limited by weather conditions, especially by sea state.

This paper introduces the new innovative offshore LNG loading system "Maritime Pipe Loading System 20 (MPLS20)", proposed by the project partners Nexans and Brugg, leading manufacturers of vacuum insulated, flexible cryogenic transfer pipes, IMPaC, an innovative engineering company that has been involved in many projects for the international oil and gas industry for 25 years and the Technical University Berlin, Department of Land- and Sea Transportation Systems, with great expertise in numerical analyses and model tests.

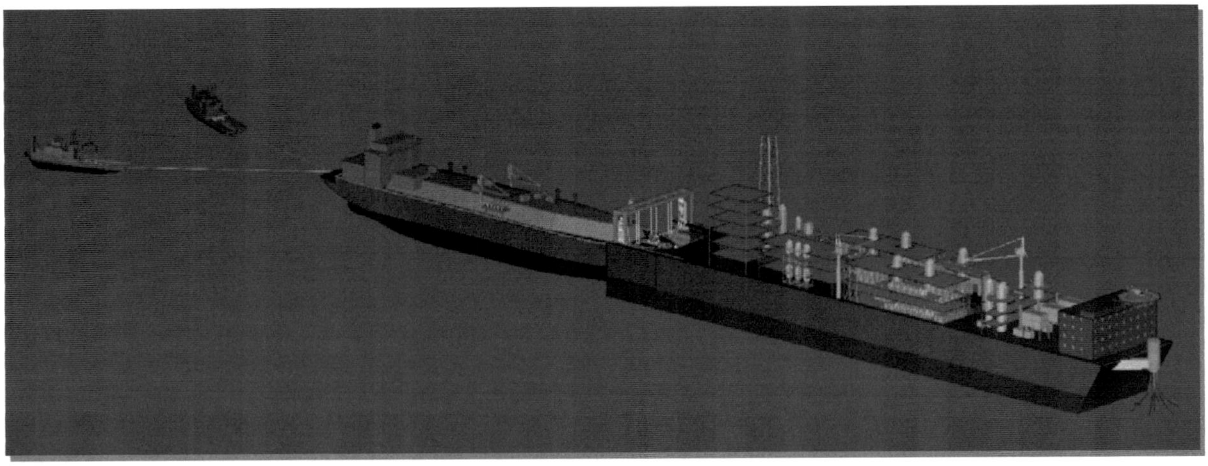

Figure 1: Illustration of the new proposed offshore LNG transfer system with the LNGC berthed in tandem configuration to the 'mooring bay' at the LNG terminal

The new concept is based on a unique tandem mooring configuration (Figure 1). In comparison to standard operations applied in the oil business for about 40 years, the concept introduces a mooring bay for shuttle tankers. Extensive numerical simulations are conducted to determine the working envelope of motions and mooring forces.

As the Nexans/Brugg corrugated metal pipes provide a flexible double containment system all relevant safety issues are well addressed, as required by EN1474-2/-3. Thus, LNG transfer can take place even under severe environmental conditions which makes this new concept superior to other approaches such as side-by-side configurations using composite hoses.

Introduction

For several decades, natural gas was merely a byproduct of oil production. Today, its importance as energy source is still growing. Most natural gas is produced from offshore fields, with associated problems of transportation to further onshore processing. The deployment of LNG tank ships (Liquefied Natural Gas) is an alternative to pipelines with increasing significance (Ref. 6). In order to achieve economically reasonable transportation, the natural gas (mostly methane) is cooled down to -162 degrees C, whereby it is liquefied and reduced to 1/600th of its original volume. For safe storage of LNG, specially insulated loading pipes and tanks are required.

State-of the art technology allows loading/offloading procedures in still water and moderate sea states with 10.5" pipes. The increasing loading capacity of LNG carriers (up to 260.000 m^3) creates a new market for fast and safe loading/offloading concepts — i.e. larger pipe diameters and operations in rough seas.

Based on existing techniques used for offshore crude oil transfer, a number of studies on LNG transfer have been carried out, resulting in potential vessel mooring configurations and designs for suitable transfer techniques. It is important to point out that no 'show stoppers' have been identified within these studies, neither in terms of safety issues nor in terms of LNG production or re-gasification in marine environments.

Two mooring configurations have been identified to work satisfactorily for the connection of shuttle tankers (either LNGC or crude oil tanker) and terminals (for LNG or oil production) for offshore cargo transfer:

The first is the side-by-side configuration, which is based on the emergency unloading technology and proved to work in smooth and moderate weather conditions (Ref. 8).

The second is the tandem configuration which is today's standard for crude oil transfer between FPSOs and shuttle tankers, especially in harsh environmental conditions like the North Sea (Figure 2). With dozens of installations and hundreds of operations every year, this configuration has shown a very good tracking record in terms of safety and reliability for decades.

The turnaround time of the shuttle tankers depends on the production rate and storage capacity of the terminal vessel. According to current rules and regulations for mooring of crude oil shuttle tankers to terminals in tandem configuration, a distance between the terminal stern and the shuttle tanker bow in the range of 50 to 90 m should be achieved (e.g. Ref. 9).

Figure 2: Tandem mooring configuration for standard offshore crude oil transfer, using floating and submerged hoses; fishtailing movements of the shuttle tankers moored to FPSOs (both left); bow loading systems (right) (Sources: Kongsberg, APL)

Depending on environmental constraints at the specific operational location, the terminal can be spread moored or turret moored to the seabed, the latter allowing the coupled floating vessels to weathervane in 360 degrees around the mooring center. Once the shuttle tanker is moored to the terminal (constituting coupled multi-body system), a phenomenon called fishtailing may occur, where the carrier periodically swings behind the stern of the terminal (Figure 2). To prevent fishtailing, modern shuttle tankers are often equipped with dynamic positioning systems (DP2 or higher) and/or offshore tugs are employed to support station keeping and alignment to the terminal vessel.

Due to the specific thermodynamical properties of the liquid crude oil and the required transfer rate, the transfer line can be designed as a single floating or submerged hose with a length of up to 100 m or more. Here, the coupling of the hose flange to the shuttle tanker receiving flange is realized by means of a specialized and well approved pull-in device located amidships or at the bow of the tanker (Figure 2). Operations with the transfer line in aerial mode (no water contact of the hose or flexible pipe) are rare.

In conclusion, none of the 'conventional' vessel mooring configurations and transfer techniques can easily be adapted to meet the requirements of LNG transfer operations, especially when designed for harsh environmental conditions with significant wave heights up to e.g. 5.5 m, zero-up-crossing periods between 8 and 12 seconds as well as significant wind and current loads for transfer durations of 18-24 hours.

It is evident that the choice of the mooring configuration is one major key to the successful development of a suitable offshore LNG transfer system.

Such a new concept should ideally combine the main advantages of the proven configurations – i.e. side-by-side and tandem: A short free span distance of the transfer lines (as featured for SbS configurations) as well as minimum relative motions between carrier and terminal and tolerable mooring forces (as featured for tandem configurations) allowing save and reliable handling, approach and coupling of the transfer lines.

The new offshore transfer system introduced in this paper meets those requirements: The system, which is developed by the project consortium MPLS20, is based on Nexans' and Brugg's newly developed CRYOFLEX flexible LNG transfer pipe with ID 16-inch. The unique design of the corrugated and super-insulated pipes with their seamless manufacturing principle provides a flexible and monitorable double containment system for most types of cryogenic fluids.

The overall design of the transfer system is further based on IMPaC's newly developed and patented offshore 'mooring bay' concept. By means of a 'loading bridge', it allows simultaneous handling and approach of up to four flexible LNG transfer pipes in aerial mode, providing safe, robust and efficient operations.

In the following sections, the new proposed tandem mooring concept will be presented in detail. An outlook and a brief view on development perspectives conclude the paper.

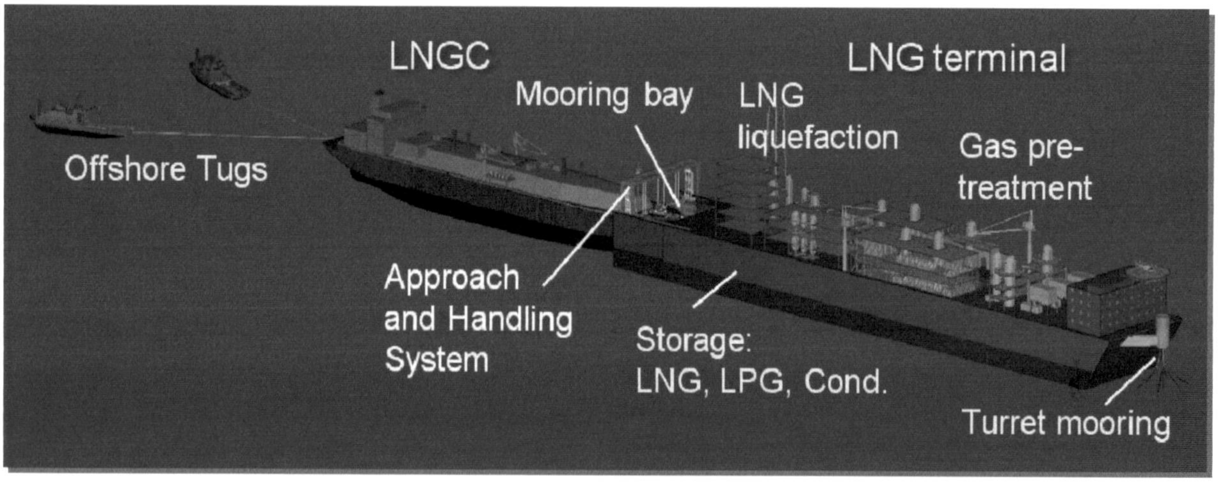

Figure 3: Main components of the new Offshore LNG transfer system: Offshore tugs, LNGC, LNG terminal with turret mooring, mooring bay and approach and handling system

A New Offshore LNG Transfer System

The proposed transfer system features a generic LNG terminal design with the new mooring bay concept, a modified standard LNGC and the approach and handling system for the newly developed transfer pipes. The main components of the system are shown in Figure 3.

As the system design is based on a generic approach, no specific operational location is considered. Nevertheless, the LNG terminal is exemplarily moored at a water depth of 100 m, which is representative e.g. for the North Sea with the associated JONSWAP spectrum. The main dimensions of the terminal and the carrier are listed in Table 1.

Parameter	LNG Terminal	LNG Carrier
Length over all	360 m (+40 m mooring bay)	285 m
Breadth	65 m	42 m
Draught	12 m	12 m
Height	33 m	26 m
Displacement	275.087 m³	103.921 m³
LNG storage cap.	280.000 m³	138.000 m³

Table 1: Main dimensions of LNG terminal and LNG carrier

The LNG terminal hull used in this concept is of barge type with a wave flattening bow, providing an optimum cargo loading capacity of up to 280.000 m^3 LNG in five independent SPB tanks (Self-supporting, Prismatic, IMO Type B) which are sloshing-proof and allow a flat deck (Figure 3). With this LNG buffer storage, even today's largest LNGC with transport capacities up to 265.000 m^3 can be handled. Additional storage capacity for the gas byproducts LPG (Liquefied Petrol Gases) and hydrocarbon condensate of 25.000 m^3 each is also provided. The storage capacity as well as the required deck area of the LNGC is mainly driven by numerous equipment for gas pre-treatment and liquefaction as follows (Ref. 4):

- Columns
- Exchangers
- Compressors / blowers
- Pumps
- Filters
- Tanks
- Vessels / drums
- Boilers / furnaces / flares
- Misc.

Other relevant contributions to the weight and volume magnitude of the terminal are resulting from standard marine equipment like cranes, power generators, ballast control system etc. as well as from accommodations and life saving equipment for the personnel.

Due to active ballasting, the barge is outlined to work with a nearly constant draught of abt. 12 m, which equals the LNGC draught.

The barge is permanently moored to one location by means of a passive 12-point external turret mooring system, allowing the terminal and the coupled multi-body system to weathervane in 360 degrees. It should be noted that it is not intended to design an optimum turret mooring system within the framework of the MPLS20 project.

The aft end deck area is dedicated to accommodate the loading bridge and the transfer pipes with the handling header when in standby or during maintenance works. Due to the LNG and methane gas specifications, it must be checked whether the related deck area has to be classified as explosion-protection-zone (if ex-zone 1 or 2 is not yet defined) so that all equipment mounted to that area complies with related rules and regulations.

Regarding the LNGC, the storage concept is independent from tank types, so that spherical Moss or SPB can be used within the LNGC. In the following investigations, the tanker is equipped with four membrane tanks whose main dimensions and cargo capacities are given in Table 1. Due to active ballasting, the LNGC also operates at a constant draught of abt. 12 m, the same as for the LNG terminal barge.

The carrier must be slightly modified compared to today's standards, as one additional and specially designed receiving manifold is placed at the deck bow area. This bow manifold completely enters the mooring bay at the aft end of the terminal when the LNGC is moored for cargo transfer, significantly reducing the free span lengths of the transfer pipes compared to crude oil transfer techniques.

The bow deck area accommodates standard anchor winches and – if necessary – chain stoppers as well as Quick Release Hooks (QRH) for mooring to the mooring wings.

Due to the LNG and methane gas specification, the LNGC bow deck area around the manifold also has to be classified as ex-zone (if zone 1 or 2 is not yet defined) so that all equipment mounted to that area complies with related rules and regulations.

Bridging the Gap: Tandem mooring with the new 'Mooring Bay' Concept

The new mooring system features a mooring bay, framed by two steel structures, the so called 'mooring wings' (Figure 4), which are fixed to the terminal's aft end at starboard and port side, respectively.

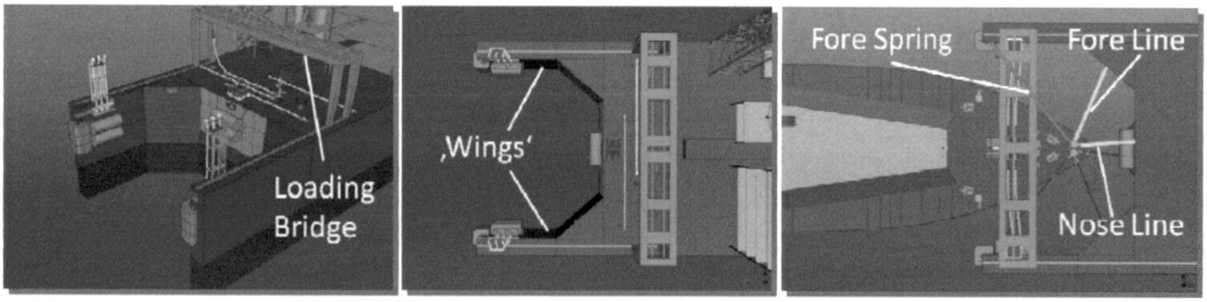

Figure 4: The mooring bay concept features two side ,wings' with six moorings in a symmetrical arrangement (right)

Each wing provides mooring hawsers for the LNGC resulting in a symmetrical arrangement of four moorings (two 'fore springs' and two 'fore lines').
Together with two additional 'nose lines' reaching from the aft end of the terminal to the bow center of the LNGC, all moorings are fixed to QRHs at the LNGC deck (Figure 4, right).

The mooring lines are operated by load adequate winches and heave compensation systems, reducing line peak loads.

As the LNGC is actively pulled into the mooring bay, the wings with the mooring arrangement provide a unique solution to stop the incoming vessel in a controlled manner at the required position right below the loading bridge.

A number of foam fenders are installed at each inner wall of the mooring bay to prevent hard impacts if the LNGC exceeds the tolerable working envelope inside the mooring bay (Figure 4).

The cargo is transferred via rigid pipes from the terminal LNG tanks through the wings into three separate export flanges at each wing. These rigid pipe flanges are located high above the barge's weather deck so that the handling as well as draining and purging of the flexible transfer pipes can be proceeded in a safe, efficient and reliable way (see paragraph 'Operating Phases').

Flexible LNG Transfer Pipes

Vacuum insulated pipe systems are common in all kind of cryogenic applications. Due to the double containment system, the 'Pipe in Pipe' technology has advantages in terms of leak detection and risk assessment of the installation with best thermal insulation properties.

The LNG industries' demand for vacuum insulated pipe concepts based on the experiences from cryogenic applications is increasing.

Nexans in cooperation with Brugg is developing an adaption of the well known flexible vacuum insulated pipe system CRYOFLEX for offshore LNG loading applications (Figure 5).

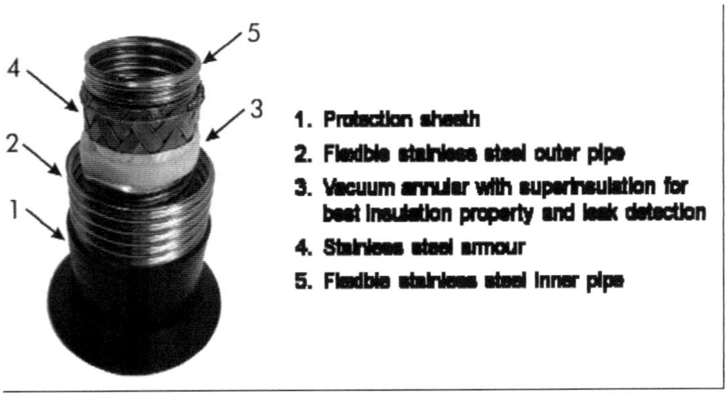

Figure 5: The LNG transfer pipe system CRYOFLEX offers a newly developed design with 16-inch inner diameter

A corrugated inner pipe (5) of 316L, surrounded by another corrugated stainless steel pipe (2) of 316L is insulated by the vacuum space and super insulation (3) between these two pipes. A stainless steel braiding (4) on the inner pipe bears the end cap load of the pipe. For mechanical protection an outer sheath of PE or PA can be applied.

The vacuum does not need any active pumping during operation for more than 15 years. The pipe structure offers leak detection in the vacuum space, which gives a signal to the system, if the vacuum degrades due to a leak in either the inner or the outer pipe. Nevertheless, in case of a leak, the gas cannot escape to the environment because of the double containment system.

This pipe is widely used in cryogenic applications. Amongst others the ARIANE space rocket test fields use an installation with this technology for more than ten years without any maintenance.

The pipe design in Figure 5 is a special development for LNG offshore loading for lengths up to 30 m. With this technology, the MPLS20 requirement for reasonable short length and high flexibility can be achieved with good fatigue properties.

Other LNG offshore systems require longer pipes with a more robust and stiff pipe design. Tandem loading configuration with connection to the standard amidships manifold can be achieved by either floating or submerged pipe configuration. For these scenarios, a similar pipe system with heavier steel armor in a riser-like design will be tested according to EN1474-2.

Loading Crane and Header for safe handling of Transfer Pipes

When moored to the mooring bay, a rail mounted moveable gantry crane (the so called loading bridge) bridges the bay from one wing to the other (see Figure 4).

The loading bridge is designed to handle all four transfer pipes simultaneously by means of a header structure. For coupling of the flexible transfer pipes to the receiving flanges of the LNGC, standard components for Quick Connect/Disconnect Couplers (QCDC) and Emergency Release Couplers (ERC) can be used.

At the terminal side, the transfer pipes are connected to the rigid conductor pipes by manual couplers. Here, two out of three available rigid pipes can be used to for LNG transfer; one flange per wing is spare. Three pipes are dedicated for LNG transfer, one pipe for vapor return.

Motion analyses show tolerable torsion in the pipes during operations, so that swivels are not required in the transfer line assembly (refer paragraph 'Motion and Mooring Analysis').

The two part header structure combines the following active functionality (Figure 6):

- simultaneous support and operation of all four flexible pipes with related QCDC and ERC
- winch driven fine approach, alignment and landing at the LNGC receiving manifold aided by guide posts
- damping of the touch down at the manifold by means of hydraulic dampeners
- operation (closing and disconnection) of all four QCDCs
- operation (disconnection) of all four ERCs in an ESD situation
- remote controlled departing of both header parts (and subsequent lifting of the upper part to a safe position by means of pre-tensioned wires suspended from the loading bridge) in an ESD situation

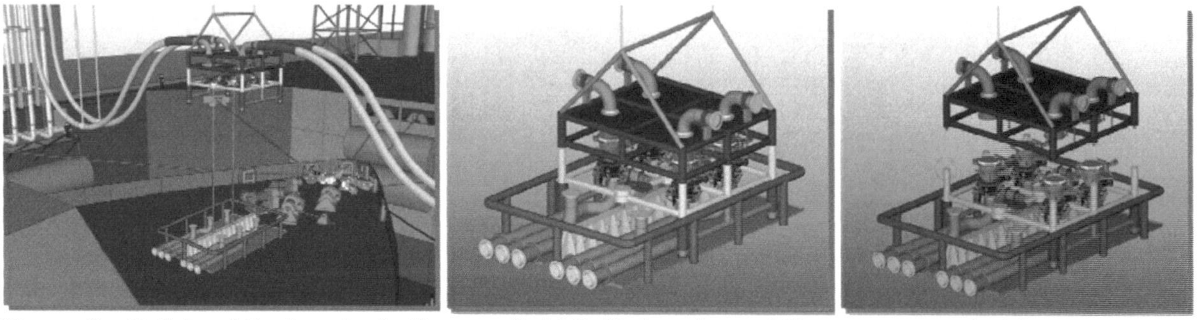

Figure 6: Pull-in of the header to the manifold (left); header and flanges connected for cargo transfer (middle); ESD situation (right): ERCs disconnected, header departed and upper part lifted off the lower part to a safe position

The flexible transfer pipes are mounted to rigid pipe elbows fixed to the upper part of the header. A bending stiffener is mounted to the endings of each pipe to prevent excessive bending moments. In addition, pre-tensioned wires reaching from air-winches mounted on traveling trolleys at the loading bridge support the header as well as the free spanning section of the pipes (Figure 6, left).

During standby or maintenance, the loading bridge can be moved on rails and secured onto the service deck area at the very aft end of the terminal (see Figure 4, left and middle). At the current design status, a deck area of 10-15 m × 65 m will be necessary for this purpose. This region will also be part of the transfer system ex-zone declaration.

In analogy to the terminal flanges, the LNGC receiving manifold provides two rows of three flanges, offering flexibility in case of connectivity problems. The flange openings are horizontally oriented, but in case of minor spillages of LNG after decoupling of the QCDCs or ERCs, collecting trays and a sheathing permanently flooded with water are mounted on the deck below and besides the manifold area (Ref. 7).

Operating Phases

Based on the new 16-inch ID transfer pipes as well as the new concepts for the mooring bay and the approach and handling system for the transfer lines, the following specific operating phases can be identified (
Table 2), an illustration is shown in Figure 7:

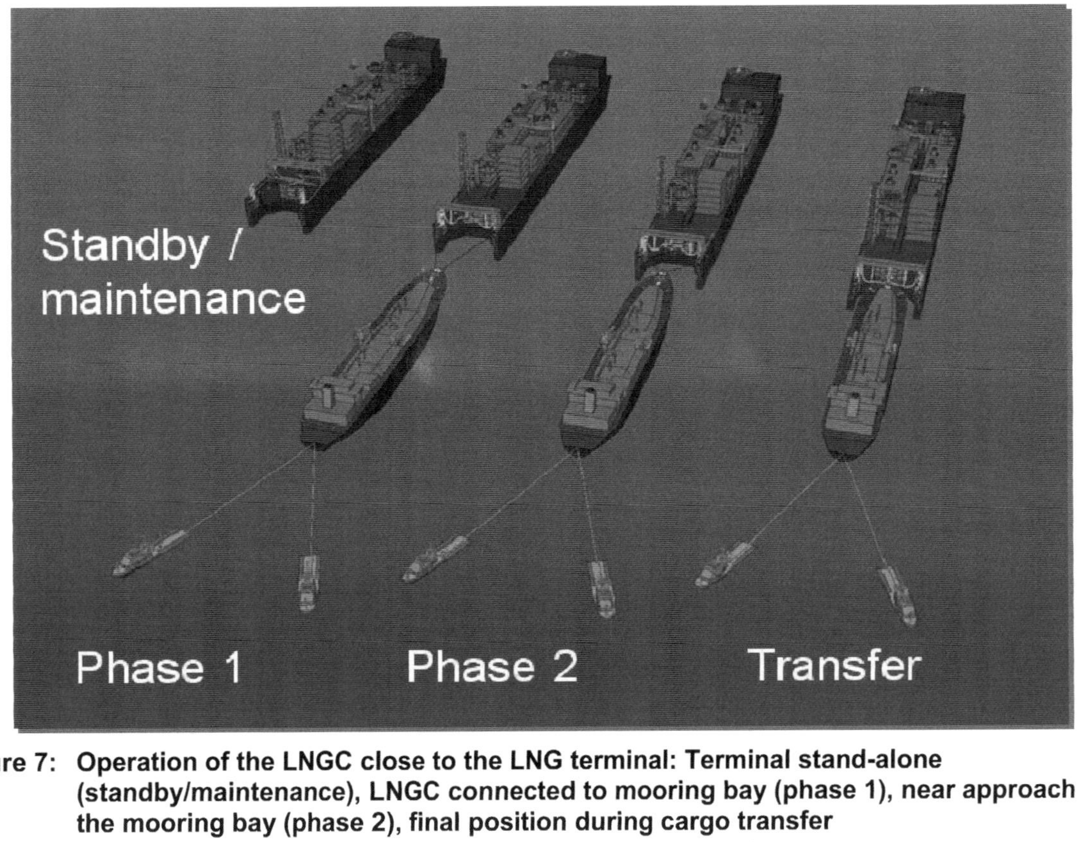

Figure 7: Operation of the LNGC close to the LNG terminal: Terminal stand-alone (standby/maintenance), LNGC connected to mooring bay (phase 1), near approach to the mooring bay (phase 2), final position during cargo transfer

Distance between LNGC and mooring bay entrance	Operations	Operating Phase
> 100 m	The LNG terminal stand-alone: Transfer system standby, waiting for next LNGC or ready for carrying out maintenance works	Standby / maintenance
100 to 80 m 	Shoot over of mooring pilot lines from the terminal to the LNGC; start of the terminal winches for slow controlled pull-in of the LNGC to the terminal aft end; LNGC thrusters slowly backwards; tug(s) assisting at heading alignment	Phase 1
50 to 30 m 	slow motion of winches; employment of next moorings; final alignment of the vessels aided by LNGC thruster and (two) offshore tug(s); further slow pull-in with winches; LNGC thrusters slowly backwards	Phase 2
0 to -20 m	very slow pull-in of the LNGC via bow mooring lines; loading bridge is in standby	
-20 to -30 m	Full stop of pull-in; stopping aided by backward pulling moorings reaching from mooring bay entrance; tightening of all six mooring lines in a symmetrical arrangement; tension control of mooring lines	
~ -30 m 	Loading header coupled to the LNGC bow manifold; transfer pipes cooled down and start of cargo transfer; mooring winches and heave compensation systems act for reduction of peak loads in the hawsers	Cargo transfer phase
~ -30 m 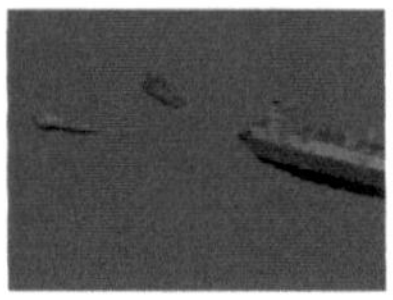	Pumps stopped; transfer lines drained and purged with nitrogen; QCDCs decoupled; header lifted from the manifold; moorings partly decoupled, partly remaining for controlled cast off of the LNGC from the terminal; LNGC backward leaving the mooring bay by own thrusters, assisted by pulling tugs	Normal stop of cargo transfer
~ -30 m	Emergency Situation: Cargo pumps stopped; valves closed in less than 30 s; disconnection of the ERCs; parting and lifting of the upper header part; activation of QRHs and cast off of the LNGC from the mooring bay, assisted by tugs	ESD 1 / ESD 2

Table 2: Main operating phases of the tandem loading system: LNGC and LNG terminal with 'mooring bay'

Motion and Mooring Analyses

For ensuring save loading and offloading procedures, detailed knowledge on the motion characteristics of the carrier and the terminal in tandem configuration is required. Due to the turret mooring of the terminal, the entire system is weathervaning according to the angle of attack β of the superimposed environmental loads from waves, current and wind (Figure 8).

As an idealized case, head waves (β = 180 degrees) are exclusively considered in the investigations. The focus lies on the transfer configuration with completely filled cargo tanks and the respective distance of 10 to 12 m between terminal stern and carrier bow is applied for all computations.

In the next step of the project, differing angles of attack resulting from incoming waves, wind and currents will also be considered.

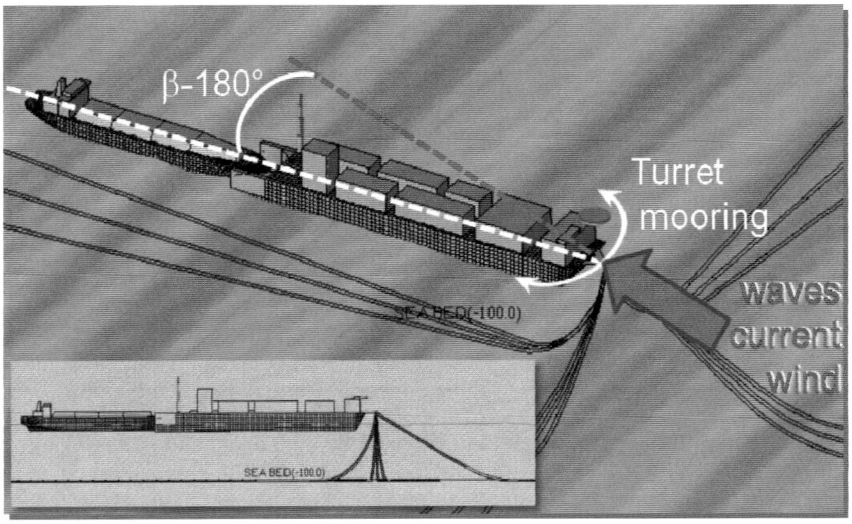

Figure 8: Environmental loads acting on the new proposed offshore LNG transfer system with the LNGC berthed to the 'mooring bay' at the LNG terminal in tandem configuration

The design criteria are resulting from the following environmental conditions, which are assumed to be maximum values for the analyses:

Environmental Design Parameter	Design Values
significant Wave Height, H_s	5.5 m
range of zero up-crossing Periods, T_0	8-12 s
Wave Spectrum, peakedness factor γ	JONSWAP, $\gamma = 3.3$
Wind speed, v_w	30 m/s
Current speed, v_c	1.0 m/s
angle of attack for all loads (current stage of analysis), β	180°
Water depth, d	100 m

Table 3: Generic design criteria for the new offshore LNG transfer system

The two software packages ANSYS AQWA and WAMIT have been applied independently by the MPLS20 partners IMPaC and TUB to determine the motions and reaction forces in the moorings of the multi-body system (Ref. 1, Ref. 5).

Both programs are based on potential theory and good agreement for the results of frequency domain analyses have been achieved (see Ref. 2 for details). Due to different modeling capabilities, comparisons of time domain results have been carried out exemplarily for the same significant wave heights and ranges of wave periods. Additional analyses for the combined set of environmental loads listed in Table 3 as well as time domain analyses for a standard three hour short-time statistical approach are carried out with ANSYS AQWA exclusively.

Note that the given environmental loads are defined to be maximum values in which regular decoupling procedures of the transfer system as well as ESD operations must be possible in a safe and reliable manner. Normal cargo transfer should be possible up to these values. It is assumed that the assisting offshore tugs are not the limiting factor for the operations, since they do not come into direct contact with the LNGC hull.

Due to the early stage of the project analyses results must be treated as preliminary. Nevertheless, it turns out that the relative motions in x- and z-direction as well as the maximum loads in the mooring lines between LNGC and LNG terminal for the multi-body system (LNGC, LNG terminal barge, turret mooring) exposed to extreme head seas resulting from second-order wave (drift) forces, can be accommodated with the mooring bay concept. More detailed analyses will follow and results will be presented in a later stage.

Compared to second-order forces, first-order forces resulting from linear sea state induced body motions are significantly higher and cannot be absorbed by the mooring lines. Thus, the moorings have to be veered and hauled up by individual heave compensation systems and/or active winches in order to reduce excessive peak loads and prevent damages in the mooring lines.

Conclusion and Outlook

An innovative offshore LNG transfer system is introduced, where the shuttle carrier is towed into a mooring bay at the stern of the LNG terminal. By using six hawsers in a symmetrical arrangement with the unique geometry of the mooring bay, the LNGC can be stopped in a distance of abt. 10 to 12 m behind the terminal, right below a standby loading crane. The loading crane is founded on the so called mooring wings and bridges the mooring bay. By means of pre-tensioned wires reaching from the crane, four flexible transfer pipes with an inner diameter of 16-inch (or more) are operated via a dedicated header structure, which provides all required functionality - even in ESD situations.

The transfer pipes introduced are newly developed by Nexans and Brugg and feature 16-inch ID. They provide a double containment system for the cryogenic cargo, allowing monitoring of the vacuum super-insulation for more than 15 years without intervention.

First results from numerical motion and mooring analyses for the newly developed system prove that LNG transfer can take place in open seas even at harsh environmental conditions.

Within the joint project MPLS20, the focus lies on the development of a new offshore LNG transfer system. This system may require new operational procedures for safe loading and unloading of cargo at sea. One issue to be investigated in detail is the influence of partly filled tanks with associated free fluid surfaces on the motion behavior of the carrier vessel during loading and unloading operations.

Since the superposition of waves, wind and currents from different directions can lead to a range of headings from 150 to 210 degrees, further investigations have to be conducted. For these cases, the influence of the roll motion is no longer negligible and requires detailed knowledge of viscous damping coefficients in order to obtain reasonable results.

Extensive tank tests will be carried out in the next project phase aimed at the determination of the damping coefficients as well as on the verification of the numerical analyses. It is planned to model and visualize the main operations in a Ship-Handling-Simulator; results will be discussed with experienced mooring masters.

The project developments are accompanied from the beginning by a 'Failure Mode and Effects Analysis' (FMEA) carried out by Germanischer Lloyd. The objective is to consider critical feedback in the early design phase to receive an 'Approval in Principle' for the offshore LNG transfer system and its related operations.

Acknowledgement

The authors want to express their gratitude to the German Federal Ministry of Economics and Technology (BMWi) and the Projektträger Jülich (PTJ) for (partly) funding and supervision of the project 'MPLS20 — Maritime Pipeloading System 20' (FKZ 03SX240).

The authors want to thank all colleagues from Nexans, Brugg, IMPaC and the Technical University Berlin who are contributing to the successful project MPLS20. Thanks to colleagues from Deloitte (Hamburg) for the administrative support.

Special thanks to the specialists from Germanischer Lloyd, Hamburg, for carrying out the FMEA for the new offshore LNG transfer system and for the very helpful feedback.

References

1. ANSYS, Inc.: Software ANSYS AQWA, user guide, Canonsburg, USA, 2008
2. Clauss, G.F., Sprenger, F., Testa, D., Hoog, S., Huhn, R.: "Motion Behaviour of a new Offshore LNG Transfer System at Harsh Operational Conditions", OMAE 2009-79391, to be published at Proc. of OMAE Conference 2009, May 31-June 05, Honolulu, USA, 2009
3. Frohne, C., Harten, F., Schippl, K., Stehen, K. E., Haakonsen, R., Eide, J., Høvik, J.: "Innovative Pipe System for Offshore LNG Transfer", OTC 19239, Proc. Offshore Technology Conf., Houston, USA, 2008
4. Gas Technology Institute, "Understanding LNG Terminals and Terminal Operations: A Critical Link in the LNG Chain", training course hand out, Savannah, Georgia, USA, March 3-6, 2008
5. Massachusetts Institute of Technology: "WAMIT Version 5.1 — A Radiation-Diffraction Panel Program for Wave-Body Interactions", user guide, 1994
6. Royal Institute of Naval Architects (RINA): "Gas Ships: Trends and Technology", 2006
7. SIGTTO / OCIMF: "Recommendations for Manifolds for Refrigerated Liquefied Natural Gas Carriers (LNG)", Second Edition, Witherby & Co. Ltd., 1994

8. SIGTTO / OCIMF: "Ship to Ship Transfer Guide (Liquefied Gases)", Second Edition, Witherby & Co. Ltd., 1995
9. UKOOA FPSO Committee: "Tandem Loading Guidelines, Vol. 1, FPSO / Tanker Risk Control During Offtake", Rev. C, 2002

Biography of Speaker

Since 2004 Sven Hoog is Project Manager for Naval and other Marine Systems for IMPaC Offshore Engineering in Hamburg, Germany, a company providing engineering services for the Oil and Gas Industry, both offshore and onshore.

Sven Hoog studied Naval Architecture and Ocean Engineering at the Technical University of Berlin, where he also received his Doctoral degree in ocean engineering.

He is involved in numerous projects for the oil and gas industry, the scope ranging from developments for floating drilling structures to production systems for ultra-deepwater.

Since several years he is also engaged in the development of LNG transfer systems, for onshore as well as for offshore application.

He is head of the civil, structural and naval department at IMPaC and he leads the research activities. hoog at impac.de

89c

Modeling the Spreading of Large LNG Spills on Water

Nicolas Ponchaut, Melissa Chernovsky, Filippo Gavelli, and **Harri K. Kytomaa**.
Exponent, Inc., 9 Strathmore Rd, Natick, MA 01760

This paper reviews and summarizes the models that are currently used to describe the evolution of a pool of spilled liquid that is spreading on water. In particular, the paper focuses on the spread of Liquid Natural Gas (LNG) on water in the event of a large spill, which is a critical element in the risk and safety analyses for LNG terminals. This paper first revisits the history of the early experimental work and the models that were derived from it, identifying their contributions and limitations. The models calculate the local height and speed evolution of the pool, along with the trajectory of its front. This paper then presents a time-dependent one-dimensional framework that models the transient behavior of a spill that is then used to compare the different model assumptions for large circular and semi-circular pools. The paper uses the GASP model developed by Webber as a baseline to describe the impact of different friction models between the spreading pool and the water. Friction becomes an important effect for very large spills, but it has garnered limited attention so far largely due to the design of early experiments, in which friction was not important compared to other effects. This paper demonstrates the importance of friction, the lack of validation of the current friction models with experiments and the fact that current models do not capture the expected variation in friction under the footprint of the spill associated with thermal and flow variations between the source and the leading edge of the spill.

Absorption Characteristic of LNG as A Function of IR Wavelength

89c

Carolina Herrera, Ray Mentzer, and M. Sam Mannan
Mary Kay O'Connor Process Safety Center
Artie McFerrin Department of Chemical Engineering
Texas A&M University

AIChE Spring Meeting, April 2009
9th Topical Conference on Natural Gas Utilization
Tampa, Florida, April 26-30, 2009

In the past, LNG research has focused on the study of dispersion models and mitigation systems, where mass rate of evaporation is one of the key parameters used in the models. Previous investigators have reported an overestimation of the mass vaporized from a LNG boiling pool compared with experimental results when calorimeters were used as an indirect measurement. Several reasons have been proposed to explain this discrepancy; however, no final explanation has been provided. Therefore, an investigation into this area is necessary. This current effort is aimed at the estimation of the evaporation rate due to the radiant heat introduced to the LNG pool by the heat flux from the fire. This procedure will allow determining the amount of heat lost by either reflection or absorption of the LNG itself, as well as the amount of heat that is being used to evaporate the liquid.

Previous work and Motivation

Small and large scale experiments either on water or land have been performed in the past in order to study and model the thermal hazards arising from hydrocarbon pool fires starting in 1960s. As a result, further understanding of LNG spills and fires has been achieved. However, gaps still remain when modeling LNG pool fires, particularly the theoretical evaluation of burning rates due to radiant heat from the fire has been difficult to develop due to the complexity of the combustion.

Mass burning rates (mass evaporated per surface area per unit time) have been measured and evaluated in field and laboratory experiments. Most small-scale field tests derive mass burning rates from data obtained from liquid regression rates or linear regression rates (volumetric loss of mass per surface area per unit time) along with the observed burning pool surface. On the other hand, burning rates obtained from small laboratory tests are characterized by the uncertainty of the data for predicting consequences when scaling up.

A summary containing some of the burning rate data from different tests on land is shown in Table 1 and further discussed below.

Table 1. *Reviewed experimental burning rate data of hydrocarbon pool fires[1]*

Test	Diameter (m)	Substrate	Regression rate (m/s)	Burning rate (Kg/m^2 s)
Esso	18 (equivalent)	Trench	$1.6*10^{-4}$	0.195
Shell	20	Concrete dike	$2.37*10^{-4}$	0.106
US Bureau of Mines	0.38	Ground dike	$1.94*10^{-4}$	0.083
AGA	1.8	Ground dike	$5.2*10^{-5}$	
	6.1	Ground dike	$1.5*10^{-4}$	
	24	Ground dike	$2.3*10^{-4}$ (estimated)	
Gaz de France	35	Concrete dike	$3.11*10^{-4}$	0.14, 0.33 (measured, calculated)

- Esso: these tests differ from others in that the area of the spill was irregular and the level of the burning pool was maintained constant by metered continuous LNG flow. As a result, the consumption of LNG due to the evaporation was known. In addition, the calculation of the liquid regression rate (m/s) was not possible since the surface area of the pool was not accurately known. To overcome this, data of regression rates reported elsewhere were used to calculate the surface area and therefore report the burning rate as a function of the pool area. Calculations of radiant thermal hazard distances were made based on the estimated regression rate and other measured data[2].

- Shell: consisted of a series of LNG, LPG, and kerosene spills on concrete. For the 20 m diameter pool, the registered burning time for methane was about 240s. The burning rate was measured and found to be reasonably constant through the entire burning period.

- American Gas Association (AGA): this series of tests were performed on land. Liquid regression rates were measured by depth gages, which measured the change in the pool level with time. These regression rates were found to be influenced by heat transfer from the ground and overall to be quasi-steady[3].

- Gaz de France: the largest series of tests conducted on a concrete dike. To understand the characteristics of the fire, a number of instruments were used to measure different aspects of steady state pool fires. In this, the burning rate of the fuel is driven solely by the radiation input from the flame since the pool has achieved a steady state heat flow with the ground. To obtain this type of information, calorimeters were located on the surface of the pool to measure the heat input from the fire, depth gages were used to measure the liquid regression rate, and thermocouples were used to measure the heat input from the ground. This allowed for the indirect and direct calculation of mean burning rates.

The indirect calculation was done using the values of the column labeled "heat flux feedback from fire to the pool" as latent heats along with the heat of evaporation for LNG of 510 KJ/Kg. The results of this calculation are shown in Table 2. Also included in this table are the values obtained for the direct measurement of the pool depth variation. These values averaged 0.14 Kg/m^2 s, which differed by a factor of almost 2.5 from those calculated using the data of heat feedback from the fire. Several reasons were proposed to explain this discrepancy such as absorption and scattering of methane vapor and liquid droplets on the pool surface, reflection of the liquid in the pool, and absorption of low temperature soot at the bottom of the fire [4, 5].

Table 2. *Summary of Montoir test results [5]*

Test	Steady-state (methane burning period) (s)	Heat flux feedback from fire to the pool (kW/m^2)	Liquid (Kg/m^2 s)	
			Calculated from measured heat feedback	Helium purged dip tube data
1	30-170	168.4±45.7	0.330±0.09	0.146
2	35-200	188.6±55.4	0.370±0.11	0.142
3	57-160	167.4±44.7	0.328±0.09	0.134

Although the review of previous experiments on land is not extensive, it does show the potential limitations of the data obtained: burning rates seem to be dependant to the conditions of the tests, they contain uncertainties that are not always reported in the literature, and they can be miscalculated and therefore propagate errors if further calculations are performed using these values.

The major hazard arising from pool fires is the radiant heat from the flame reaching the surroundings. Semi-empirical and field models (computational fluids models) are used to predict this thermal hazard distance. The available semi-empirical models are the point source and solid flame model, and as semi-empirical models, they require experimental data of burning rates at different conditions to be input as known parameters. However, this data is not always available and, moreover, relying on this limited data has the potential to result in poor performance of the model causing an over prediction [6] of the thermal hazard distances from pool fires. In addition, experimental data includes inaccuracies of the measurements and unstudied phenomenon occurring in large LNG fires as described in the Montoir test. Both models have as a major limitation the extrapolation of the calculations outside the limited range of conditions over which they were validated (ref).

The limitations encountered when predicting hazard distances with the available models are the incentive for this work, which consists of the search for an explanation for the incorrect estimation of burning rates when heat feedback data from the fire is used and the development of a correlation that allows the calculation of the burning rate as function of known parameters

Methodology

This research project will focus on the use of an infrared spectrometer with a cryogenic cell cooled to 110K to detect the absorption characteristics of LNG at different wavelengths. The final goal is to estimate the energy absorbed by the liquid and the losses due to reflection and scattering and correlate the results to be used in the modeling of pool fires.

The overall acquisition of the data will be accomplished by: i) cooling the cell, ii) obtaining spectra of the background (empty cell), iii) filling the cell with LNG, iv) stabilizing the equipment for data acquisition, and v) spectrum recording at different temperatures. The spectrum obtained from the measurements will be integrated to find the total energy absorbed and therefore a correlation to predict the burning rate of LNG will be developed. The hypothetical results are illustrated in figure 1.

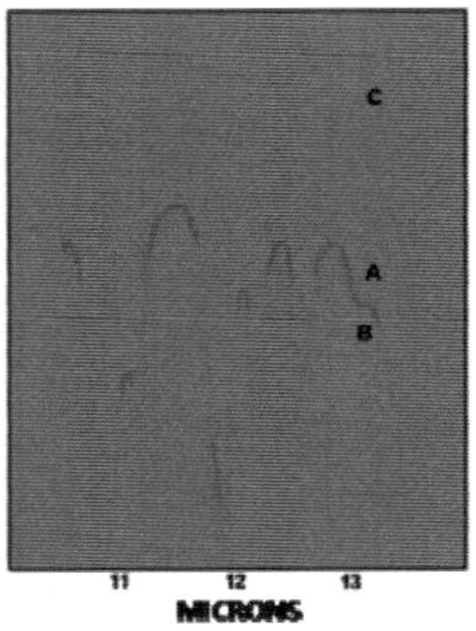

Figure 1. *Hypothetical absorption spectra for the background (A), background and LNG (B) and LNG (C) at 110K in the infrared region[7].*

This project is divided in two phases as shown in Figure 2. The first phase will be accomplished by experimental work under controlled conditions, in which IR spectroscopy and radiometry techniques will be applied. The anticipated objectives to be attained in this phase are:

- Determine the absorption characteristics for LNG in liquid phase at low temperatures.
- Determine the percentage of infrared radiation absorbed by LNG with the variation of the IR wavelength and cell length.
- Obtain a thermal absorption equation as function of wavelength, path length, and temperature.

The second phase of this project consists of the application of the results obtained in phase I and the interpretation of the existing data to aid in pool fire modeling. The following are the objectives to be achieved in this phase:

- Validate the use of calorimeters to calculate evaporation rates of LNG pool fires
- Study of the relation between mass burning rates and heat losses due to optic effects.

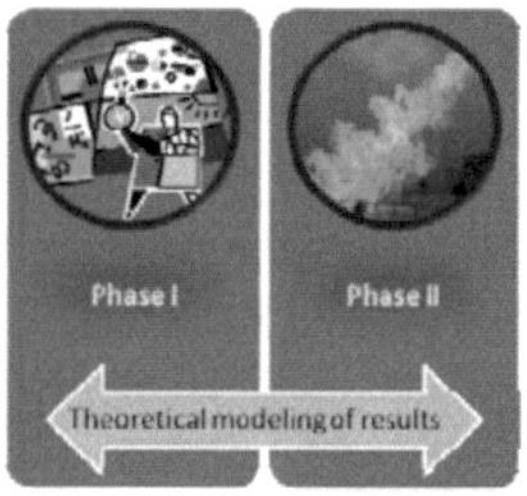

Figure 2. *Phases of research plan: phase I carried out under controlled conditions, and phase II to correlate the data*

Both phases will be interconnected by the theoretical correlation of the results observed in the experiments. The integration of energy balance with the theory and experimental optical radiation measurements will provide a plausible explanation for the phenomenon observed in Montoir tests.

As describe previously, the overestimation of mass of LNG evaporated was predicted by the following assumption and the combination of mass and energy balances:

Energy balance on the pool surface where $T_{liquid} = T_{boiling\ point}$:

$$Q_{evap} = Q_{conv} + Q_{rad} + Q_{cond} - Q_{loss},$$

where Q_{evap} is the heat exiting the system as mass evaporated, Q_{conv} is the convective heat, Q_{cond} is the conductive heat, Q_{rad} is the radiation heat entering the pool, and Q_{loss} represents the heat loss.

At steady-state, Q_{cond} and Q_{conv} can be neglected and the energy balance is reduced to:

$$Q_{evap} = Q_{rad} - Q_{loss}$$

Q_{loss} is assumed to be very small ($Q_{evap} = Q_{rad}$) and from this, the mass evaporated is calculated as:

$$m = \frac{Q_{evap}}{\Delta H_{evap}}, \qquad \Delta H_{evap}\text{: heat of evaporation}$$

Clearly, neglecting Q_{loss} is at best an approximation with growing magnitude as the size of the fire increases, since the mass burning rate resulting is larger than that obtained by direct estimation. In other words,

$$Q_{evap} = Q_{rad} - \lfloor Q_{rf} + Q_{tr} \rfloor,$$

where Q_{evap} is in reality Q_{abs} and $Q_{rf} + Q_{tr} = Q_{loss}$

Q_{abs} in the energy absorbed by the system and used to evaporate the LNG, Q_{rf} is the energy reflected by the pool and Q_{abs} s the energy transmitted (not absorbed) by the LNG to the surrounding.

Conceptual design

The initial setup was conceived according to the existing cryogenic equipment used to perform similar measurements on other liquids. Infrared measurements of absorption spectrum at temperatures as low as 20K for gas and liquid phases of nitrogen[8], methane[9, 10], and hydrogen[11] were found through a literature review and contributed to the conceptual design in the present research. At the present stage, the key components involve a cryostat in which the sample (LNG) will be pumped from a reservoir cooled by LNG. This cryostat will have the advantage to fit commercially available Fourier transform infrared spectrometers (FTIR).

The procedure to acquire measurements will encompass the initial cooling of the cell, followed by the recording of the spectrum background, loading of the cell, and achievement of complete stable conditions of the liquid sample at the desired temperature. Once the temperature is reached, the radiation source within the FTIR spectrometer will radiate the sample and once the beam of light has been through the sample, it will reach the detector recording the energy absorbed by the sample. This signal will be processed by the computer connected to the system.

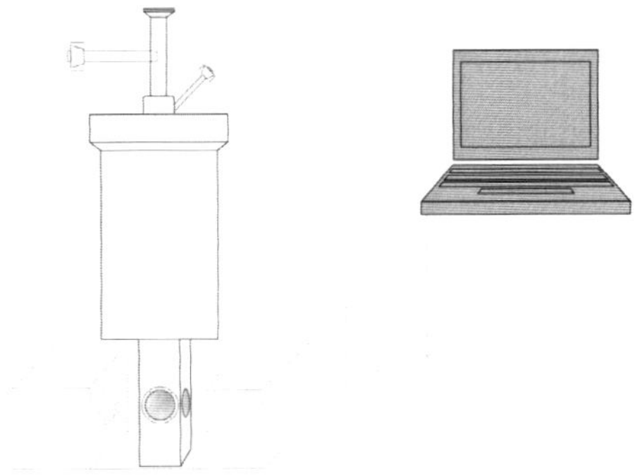

Figure 3. *Conceptual design of the key components of the system. Cryostat is displayed inside the IR spectrometer chamber*

Conclusion and Future Work

This research focuses on providing an explanation for the over prediction of the evaporation rate of LNG due to heat feedback from the pool fire. This explanation will contribute to the understanding of large-scale fire scenarios and will offer a tool for an improved prediction of the hazards arising when modeling pool fires.

The final experimental setup will be built keeping in mind the objectives to be accomplished. Further detail of the key components will be incorporated into the initial setup, including: material of construction, range of operation temperature, containing the LNG, radiation source, and detector type, etc.

References

1. Raj, P.K., *LNG fires: A review of experimental results, models and hazard prediction challenges (vol 140, pg 444, 2007).* Journal of Hazardous Materials, 2007. **143**(1-2): p. 603-603.
2. May, W.G. and W. McQueen, *RADIATION FROM LARGE LIQUEFIED NATURAL-GAS FIRES.* Combustion Science and Technology, 1973. **7**(2): p. 51-56.
3. Raj, P.K., Moussa, N.A., Aravamudan, K. , *Experiments involving pool and vapor fires from spills of liquefied natural gas on water*. 1979. p. 257.
4. Malvos, H., *Details of 35 m diameter LNG fire tests conducted in Montoir , France in 1987, and analysis of fire spectral and other data*, in *AIChE Spring National Meeting*. 2006, AICHE: Orlando, FL, United States. p. 1-20.
5. Hugues Malvos, P.K.R., *Thermal emission and other characteristics of large liquefied natural gas fires.* Process Safety Progress, 2007. **26**(3): p. 237-247.
6. Luketa-Hanlin, A., *A review of large-scale LNG spills: Experiments and modeling.* Journal of Hazardous Materials, 2006. **132**(2-3): p. 119-140.
7. Miller, R.G.J., *Laboratory methods in infrared spectroscopy.* 1965: Philadelphia, Sadtler Research Laboratories.
8. Wishnow, E.H., A. Leung, and H.P. Gush, *Cryogenic multiple reflection absorption cell and Fourier transform spectrometer system for the far infrared.* Review of Scientific Instruments, 1999. **70**(1): p. 23-31.
9. Kassi, S., et al., *The near-infrared (1.30-1.70 mu m) absorption spectrum of methane down to 77 K.* Physical Chemistry Chemical Physics, 2008. **10**(30): p. 4410-4419.
10. Birnbaum, G., et al., *Far infrared absorption in liquid methane: experiment and theory.* Molecular Physics, 1998. **93**(4): p. 573-581.
11. Buontempo, U., et al., *FAR INFRARED-ABSORPTION SPECTRUM OF N2 IN GAS AND LIQUID-PHASES.* Journal of Chemical Physics, 1975. **63**(6): p. 2570-2576.

Study on the Heat Transfer of Expansion Foam on LNG Pool

89d

Geunwoong Yun, Yuyan Guo, M. Sam Mannan
Mary Kay O'Connor Process Safety Center
Artie McFerrin Department of Chemical Engineering
Texas A&M University System, 200 Jack E. Brown Engineering Bldg,
College Station, Texas 77843-3122, USA

AIChE Spring Meeting, April 2009
9th Topical Conference on Natural Gas Utilization
Tampa, Florida, April 26-30, 2009

1. Introduction

The Liquefied Natural Gas (LNG) is one of the fastest growing energy sources in U.S. to diversify the energy portfolio and fulfill the increasing energy demands. In order to meet the increasing LNG demand, many LNG facilities including LNG importation terminals are operating currently. Moreover, many projects for LNG terminals have been proposed to fill the gap between supply and demand of LNG in North America. However, LNG is a highly flammable gas which may result in fire or explosion when it is discharged into undesired conditions. In order to secure LNG safety, several safety guidelines, such as NFPA 59 A and 49 CFR, are available. For controlling spilled LNG vapors and suppressing LNG fires, these guidelines recommend several safety measures, such as expansion foam application.

Expansion foam is one of the important safety measures to control vapor dispersion and LNG pool fires. The expansion foam is also used for fire-extinguishing systems for interior or exterior hazards in other industries such as refineries and petrochemicals. Fire-fighting foam, within the scope of this research, is an aggregation of small bubbles of lower density than LNG or oil that can cover horizontal surfaces of LNG. Air foam, which is generally used in LNG facilities, is made by mixing air into a water solution, containing a foam concentrate, by foam generators. The expansion foam flows freely over a volatile liquid or burning liquid surface and forms a continuous blanket that seals volatile flammable vapors from access to air. It is also capable of resisting disruption from wind or flame attack so as to keep continuous blanketing. According to NFPA 11 [1], foam can be subdivided into three ranges of expansion having foam-to-solution volume ratios: (1) Low-expansion foam – expansion up to 20; (2) Medium-expansion foam – expansion from 20 to 200; (3) High-expansion foam – expansion from 200 to approximately 1000.

It has been recognized that the effectiveness of expansion foam may be achieved by: (1) reducing the LNG vapor density to make it positively buoyant by increasing its temperature adding heat from water in the foam to LNG vapors while penetrating through the foam; (2) reducing radiant and convective heat input from surroundings to the LNG pool surface; (3) reducing the fire growth by lowering the oxygen concentration in the vicinity of the burning liquid. In other words, heat transfer between water in the foam and LNG vapors can play an important role to disperse LNG vapors and suppress LNG fires.

The effectiveness of expansion foam for vapor dispersion control and fire suppression is very dependent on the thickness of foam to cover LNG pool, application rate which is closely related with required time to cover overall surfaces of released LNG and reach minimum foam depth, and expansion ratio. If the foam thickness is relatively small, it may not provide enough heat to LNG vapors up to reach positive buoyancy and also sufficient blanketing for radiant heat from surroundings or fire. The minimum effective foam depth might be different with incident scenarios between with and without fire, because fire will break down expansion foam relatively quickly and provide significantly more heat to the foam so that water in the foam can evaporate. In other words, the phenomena of heat transfer of foam with or with fire are different. In the LNG facilities, the possible incident scenarios might involve bigger releases and/or fires due to its huge storage and processing capacity. Therefore, it is very important to identify minimum effective foam thickness for vapor dispersion control as well as fire suppression. Consequently, heat transfer study associated with field experiments is very necessary to obtain the required expansion foam depth for the safety of LNG or other flammable fuels.

Therefore, in this paper, heat transfer study between LNG and expansion foam is addressed and its results will be compared with experimental results achieved by Mary Kay O'Connor Process Safety Center (MKOPSC) to validate the model. Finally, this generalized heat transfer study will allow us to estimate the effective foam depth for LNG pool scenarios. Moreover, this paper will provide the basis for the estimation of minimum effective foam depth in case of foam application on fire, which is more complicated than the vapor control case.

2. Literature reviews

Takeno [2] studied effects of high expansion foam dispersed onto leaked LNG on the atmospheric diffusion of vaporized gas. He performed two kinds of experiments. One of them was to identify the effects of expansion foam application on the temperature increase of vaporized gas, using a small scale liquid pool to simulate an actual scale as shown in Figure 1. However, he used liquid nitrogen (LN_2) instead of LNG to consider experimental safety. In the experimental apparatus, 14 stationary thermocouples were arranged along a pole installed vertically to verify the liquid level and measure the temperature profile in and above the expansion foam. The other was to measure the change of evaporation rate after applied expansion foam meets with the LN_2 as shown in Figure 2. The evaporation rate which can be an input for overall heat balance was measured by a laboratory scale apparatus that included a load cell, which converts weight into voltage. However, it was a laboratory scale experiment to get the evaporation rate of LN_2, not field scale test using LNG. Thus, the resulted evaporation rate may not reflect the atmospheric outdoor conditions (e.g. wind effects, solar radiation, humidity, etc.) and possible real size of LNG release in the LNG facilities. Currently, two methods of direct measurement can be used to obtain the evaporation rate of LNG. Number one is to use thermocouples, which can also be called as "thermocouple trees". The thermocouples can detect the liquid level of LNG in terms of temperature measurements. Thus, with the given surface area of LNG and its level, the volumetric evaporation rate can be obtained and it can be converted into mass evaporation rate using specific gravity of LNG. Number two is to use the differential pressure transmitter, which can be called as "bubble meters". It measures the pressure difference between the atmospheric pressure and the hydrostatic pressure of LNG at the bottom of pit. The hydrostatic pressure of LNG differs as LNG liquid level decreases. The pressure difference can be converted into the level of LNG decrease so as to obtain evaporation rate. In addition, the differential pressure transmitter includes nitrogen line at the bottom end of the pipe to avoid any back flow. However, Takeno [2] did not use those direct measurements for evaporation rate.

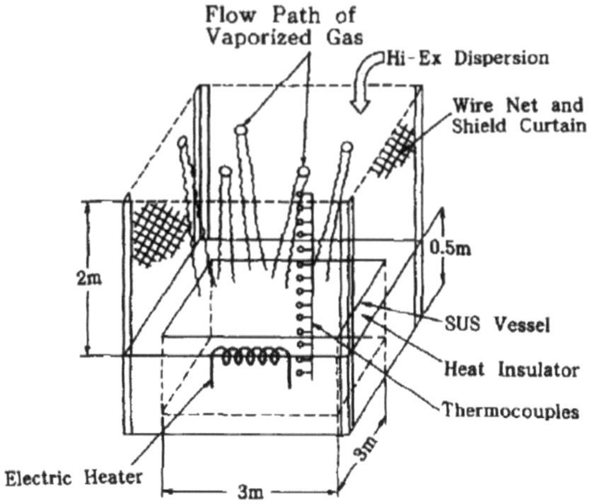

Figure 1 Experimental apparatus of expansion foam using LN_2 [2]

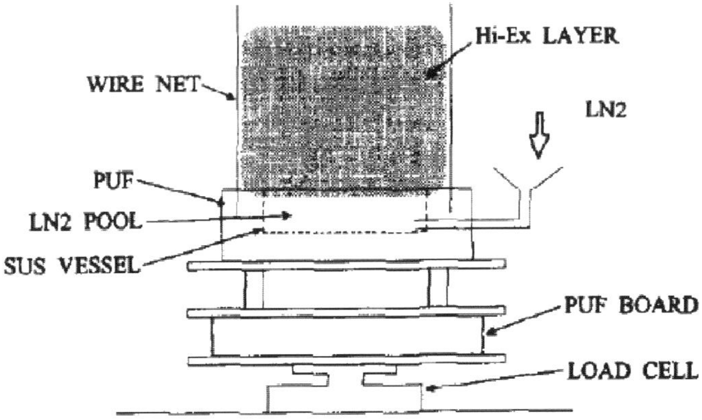

Figure 2 Experimental apparatus for evaporation rate [2]

Takeno's paper showed the discussion of heat balance of expansion foam layer, LN_2, and surroundings. The paper indicated three layers of expansion foam as shown in Figure 3 and 4; frozen layer, no-frozen layer, and dissipated layer. It considered two modes of heat transfer: (1) convection of water traveling downwards caused by dissipation of expansion foam; (2) thermal conduction through the expansion foam layer. In the paper, the total heat balance was expressed in the following equation; $Q_B + Q_D + Q_E = Q_A + Q_C$. However, it didn't include thermal convection from concrete dike or wall which is one of requirements for LNG storage tank systems and facilities. The paper also assumed that the radiation from the sun, Q_E, is zero since the experimental apparatus had the shield curtain. However, the assumption may not be realistic in the LNG or other facilities because released flammable liquid from the containments is recommended not to be enclosed. Moreover, in case of foam application to fire, the heat transfer is different with that without fire, but the paper did not address the issue with fire. In case of LNG pool fire, heat flux and radiant heat from fire to expansion foam as well as the change of latent heat due to vaporization of water should be considered additionally.

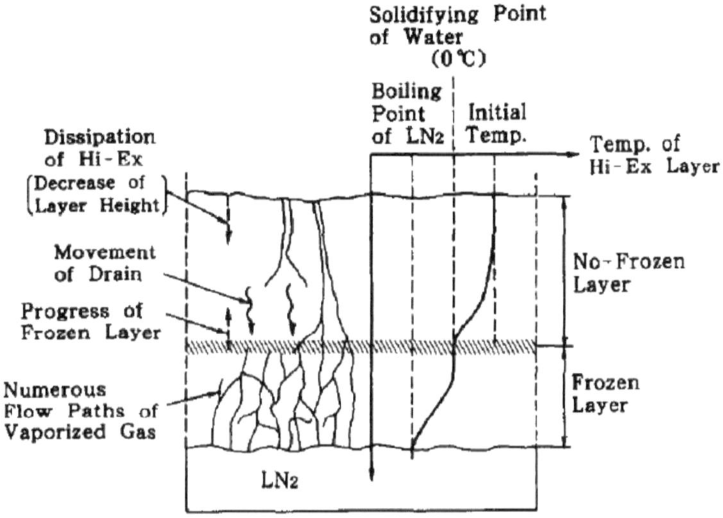

Figure 3 Model of heat transfer of expansion foam using LN_2 [2]

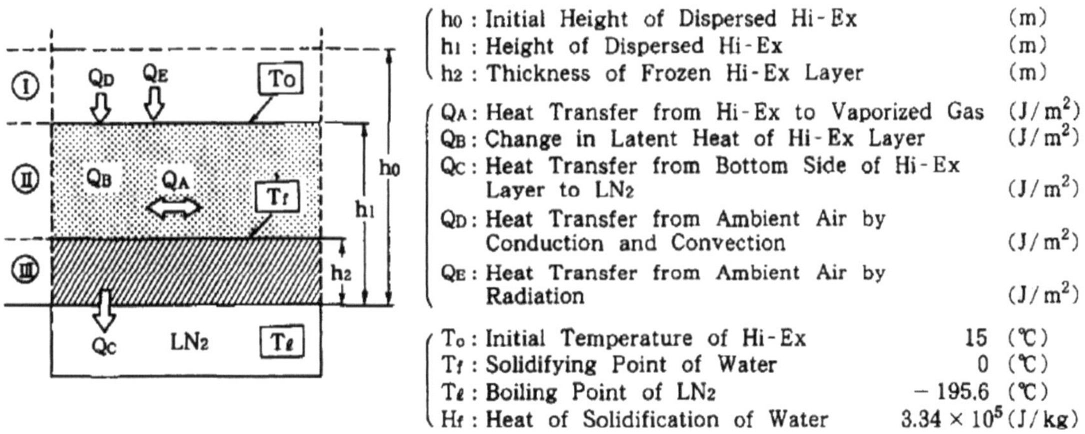

Figure 4 Heat balance in the whole expansion foam system using LN_2 [2]

Based on literatures and previous experiments, research needs and gaps can be summarized as shown in Figure 5.

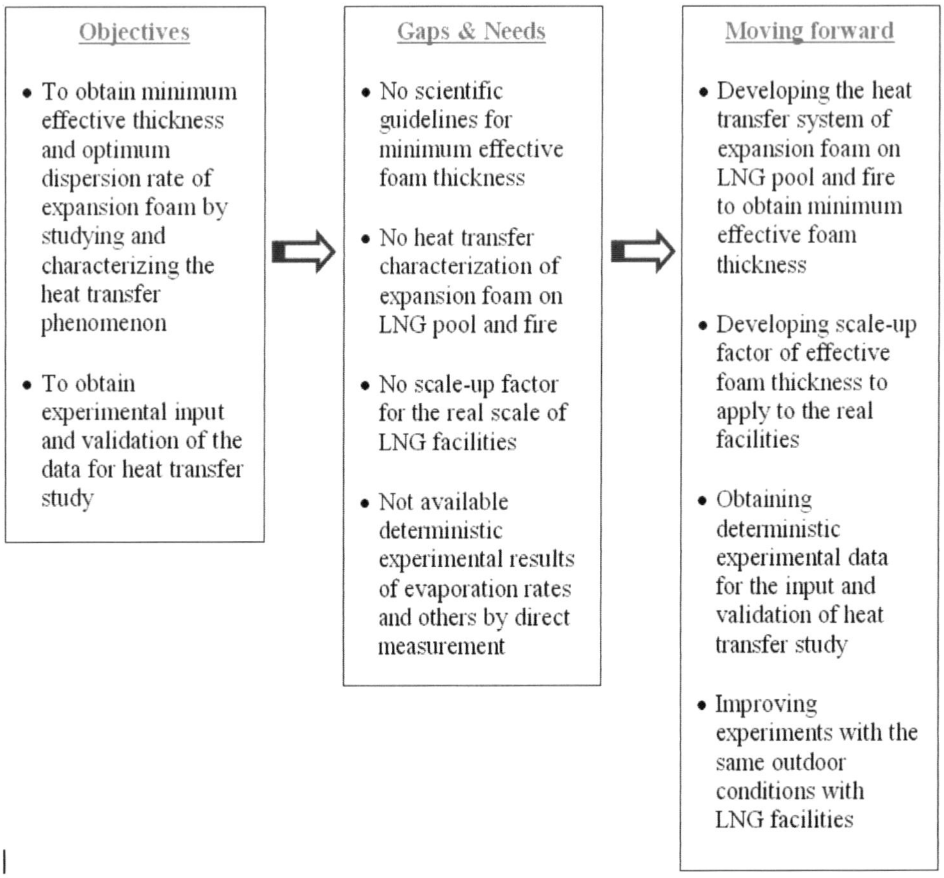

Figure 5 Summarized research needs and gaps

3. Research overview

Theoretical study of heat transfer

Theoretical study will focus on the real incident scenarios which might be occurred in the LNG facilities. NFPA 59A [3] requires impounding area with dikes or walls around the LNG container as shown in Figure 6. According to the standard, expansion foam equipment should be installed to the dike or impounding wall. If LNG releases from the container, it will be contained in the impounding area. The released LNG will touch the surface of concrete wall of a container, concrete dike, and concrete bottom. Moreover, the LNG will be exposed to solar radiation because the impounding area does not have a roof to block sunlight. Based on Takeno's paper [2], the total heat balance of expansion foam and surrounding including LNG should additionally include following factors as shown in Figures 7 and 8: (1) heat transfer from concrete wall and bottom by convection (Q_F) for the vapor control; (2) heat transfer from fire by convection and radiation (Q_G) for the fire suppression; (3) change in latent heat of foam layer due to the evaporation of water (Q_H) for the fire suppression. From the heat balance, the required amount of expansion foam to keep raising the temperature of LNG vapors can be estimated. Furthermore, it will be used not only to obtain scale-up factor for real cases of LNG facilities, but also to provide general heat transfer balance which can be applied to other flammable liquid facilities such as liquid ethylene and hydrogen.

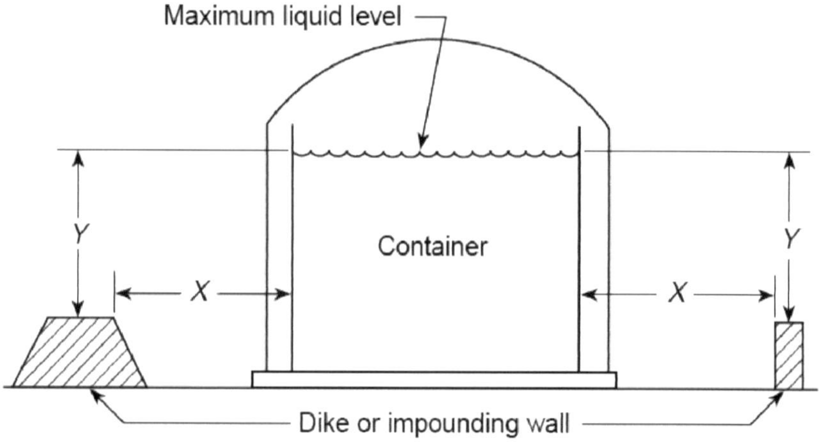

Figure 6 Dike or impounding wall proximity to containers [3]

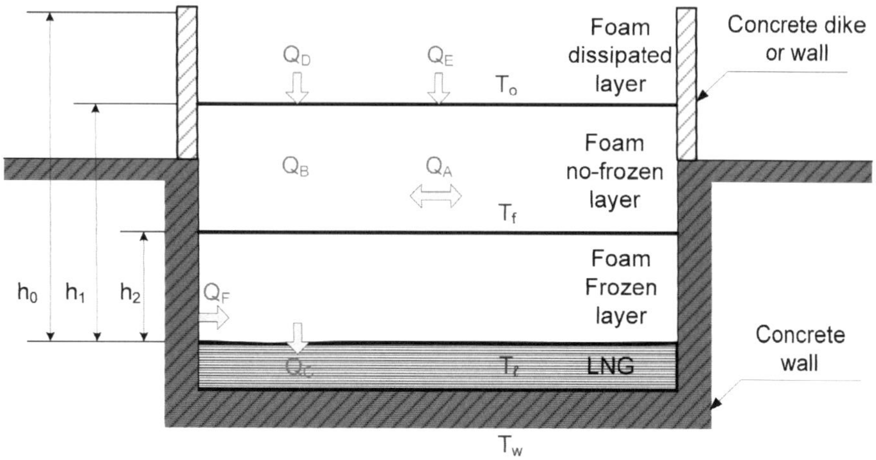

Figure 7 Heat balance during vapor control by expansion foam

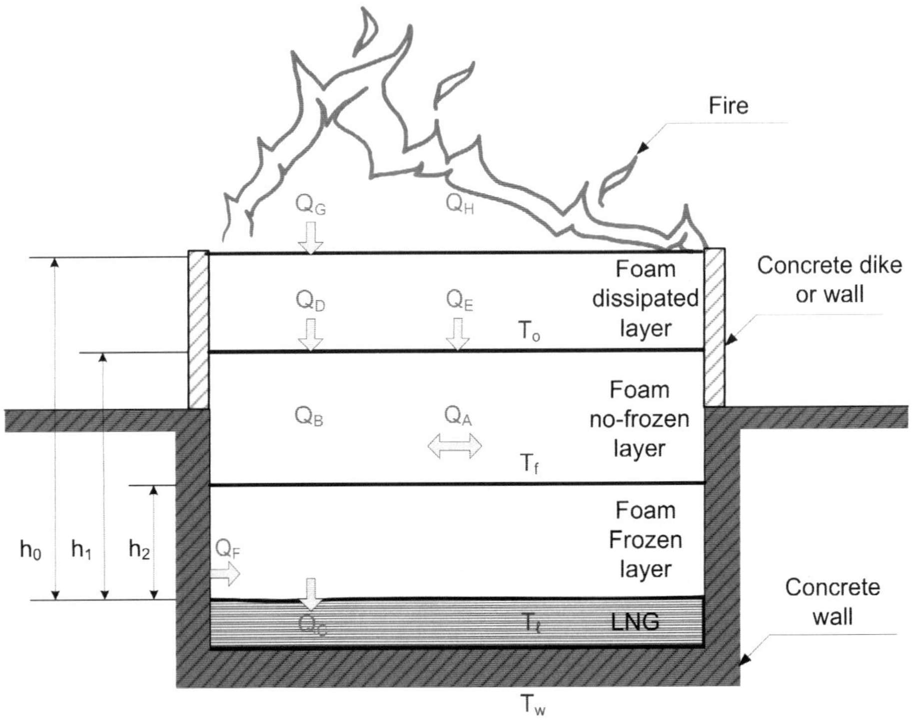

h_0: initial height of applied expansion foam (m)

h_1: height of applied expansion foam after having dissipation of foam (m)

h_2: thickness of frozen expansion foam (m)

Q_A: heat transfer from expansion foam to vaporized LNG (J/m^2)

Q_B: change in latent heat of expansion foam due to freezing of water (J/m^2)

Q_C: heat transfer from bottom side of expansion layer to LNG (J/m^2)

Q_D: heat transfer from ambient air by conduction and convection (J/m^2)

Q_E: heat transfer from ambient air by radiation (J/m^2)

Q_F: heat transfer from concrete wall by convection (J/m^2)

Q_G: heat transfer from fire from fire by convection and radiation, when fire (J/m^2)

Q_H: change in latent heat of foam layer due to boiling of water (J/m^2)

T_o: initial temperature of expansion foam (℃)

T_f: freezing temperature of water (℃)

T_ℓ: boiling point of water (℃)

T_w: initial temperature of concrete wall (℃)

Figure 8 Heat balance during fire suppression by expansion foam

Experiments

In order to validate the results of theoretical heat transfer study, following test will be conducted in the Brayton Fire Training Field in College Station. For the test of expansion foam on top of LNG, LNG will be released on concrete into a pit. The test will be composed of four steps as shown in Figure 9.

- Step 1: LNG release on concrete into a pit

  LNG will be released into a concrete pit where thermocouples are installed.

- Step 2: Application of expansion foam by four feet depth on top of LNG pool on concrete (vapor dispersion control)

  Expansion foam will be applied on the surface of LNG pool into a concrete pit. Through this step 2, temperature profile inside expansion foam and LNG vapor temperature above the foam to determine buoyancy will be obtained by thermocouples. In addition, to identify the wall convection effect to the foam and LNG vapors, three thermocouples will be installed very next to the concrete wall on the same location with other thermocouples inside the foam and these results will be compared with the one of other thermocouples. From some of the thermocouples and differential pressure meters, the evaporation rate will be obtained for the input of overall heat transfer balance. These data can be used for heat transfer study of expansion foam and LNG.

- Step 3: Application of expansion foam by eight feet depth on top of LNG pool on concrete (vapor dispersion control)

  Additionally, expansion foam will be filled up to the level of wall (eight feet). Then, temperature profile inside expansion foam and above the foam and evaporation rate will be obtained by thermocouples and its results will be compared with those of step 2 to determine the effects of foam thickness changes on LNG vapor temperature.

- Step 4: Ignition of the LNG pool applied expansion foam (fire suppression)

After having some certain time from step 3 and removing the additional walls, the LNG will be ignited. Through this step 4, temperature profile inside expansion foam and temperature of fire will be measured. LNG pool fire will create heat flux and radiant heat to the objects including expansion foam. The minimum effective foam depth might be different with the one of the vapor dispersion control since foam will be heated by fire and its temperature will be increased. If the foam temperature is over 100 ℃, the water in the foam will be vaporized and will break the foam down. Therefore, minimum foam depth should be determined by temperature profile and heat transfer study.

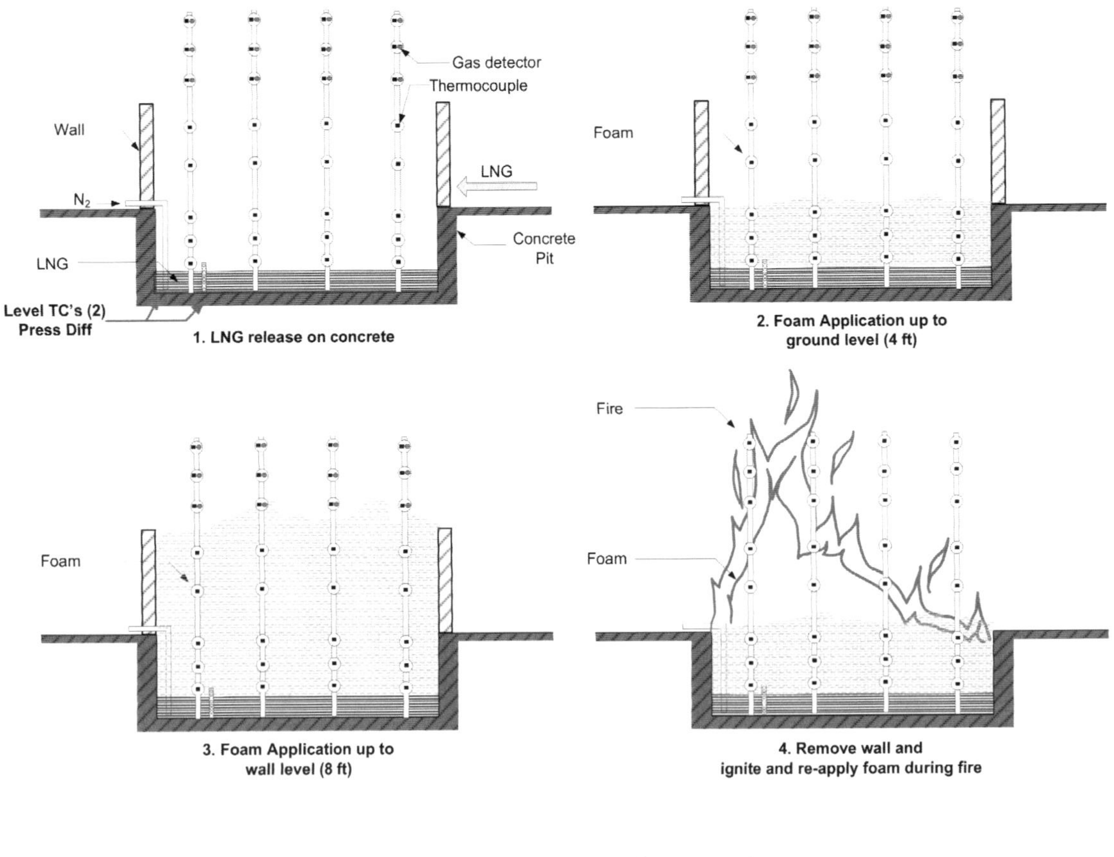

Figure 9 Experimental procedure

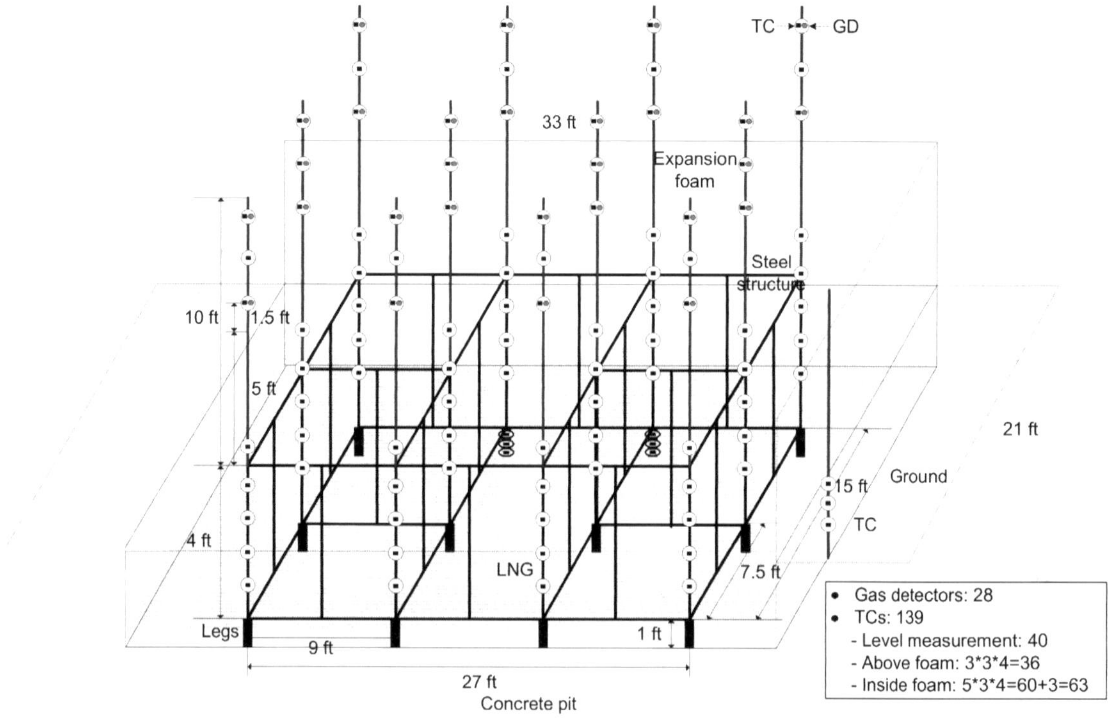

Figure 10 Detail experimental setup inside pit

Additionally, transparent foam box which has height measurement will be used for identifying foam changes when it contacts with LNG. This setup may provide ideas to determine minimum effective foam depth by investigating frozen layers of foam. Intuitively, it may be indicated that frozen layers may not provide enough heat to warm LNG vapors up and minimum effectiveness foam thickness should be greater than the level of this frozen foam layer.

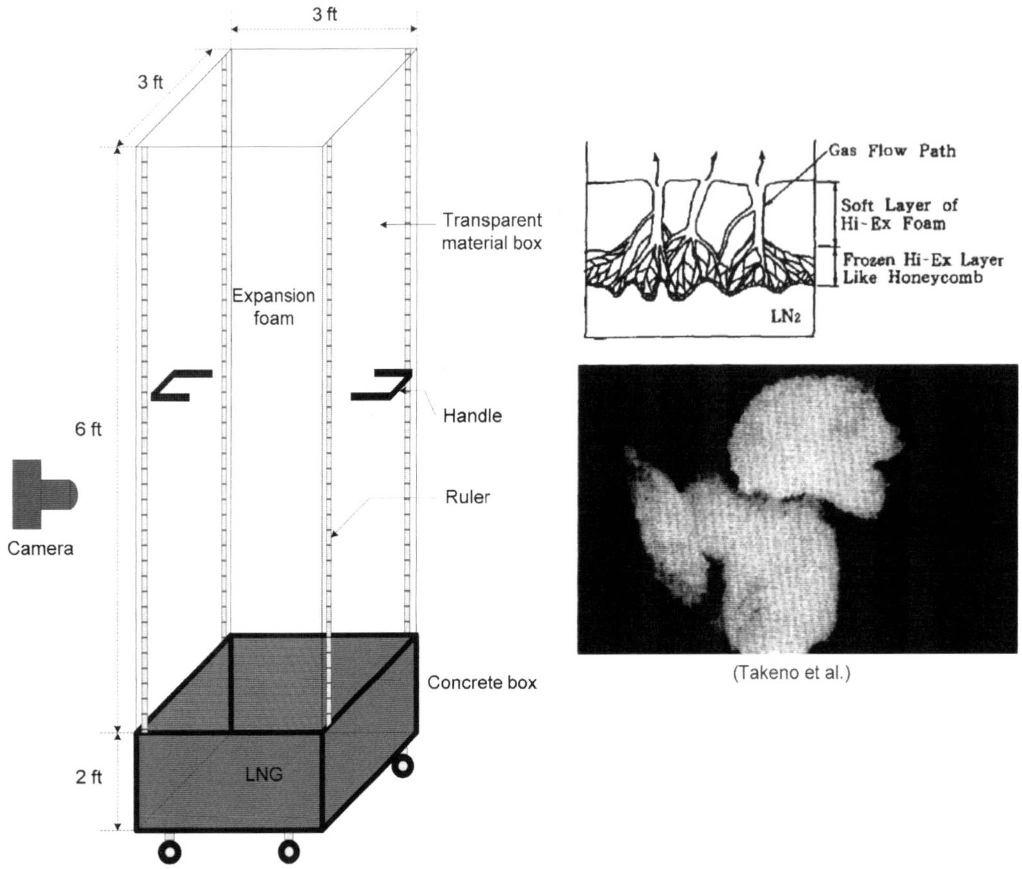

Figure 11 Transparent expansion foam box setup

4. Summary

From this research, we will achieve minimum effective foam thickness for vapor control as well as fire suppression by expansion foam for LNG in terms of heat transfer study and experimental data. The theoretical heat transfer study will provide the understanding and model of the heat transfer characteristics, and phenomena of expansion foam on LNG pool (or fire) and scale-up factor of foam thickness for application of real scale LNG facilities. Furthermore, the developed heat balance may be applied to the expansion foam applications of other flammable liquid facilities to estimate the minimum effective foam thickness. This experiment will provide temperature profile of LNG vapor inside and above the foam. The experiment can also produce field test data to validate the results of theoretical study. Consequently, this research may contribute to provide some application guidelines of expansion foam with the industry as well as academia and then increase the level of safety of flammable liquid facilities.

References

1. National Fire Protection Association (NFPA), *Standard for Low-, Medium-, and High-Expansion foam.* NFPA 11, 2002.
2. Takeno, K., et al., *Effects of high expansion foam dispersed onto leaked LNG on the atmospheric diffusion of vaporized gas.* Journal of Loss Prevention in the Process Industries, 1996. **9**(2): p. 125-133.
3. National Fire Protection Association (NFPA), *Standard for the Production, Storage, and Handling of Liquefied Natural Gas (LNG).* 2001. NFPA **59A**.

Biography of speaker

The Department of Chemical Engineering / Mary Kay O'Connor Process Safety Center
Texas A&M University, 3122 TAMU, College Station, TX 77843-3122
Phone: (979) 862-1310, Fax: (979) 458-1493
E-mail: g0y0859@chemail.tamu.edu

Yun, Geunwoong is a PhD student of the Chemical Engineering Department of Texas A&M University advised by Dr. Sam Mannan and a research assistant of the Mary Kay O'Connor Process Safety Center. He attended SungKyunKwan University and received his BS in chemical engineering, and also received a MS degree in mechanical engineering from the YonSei University in Rep. of Korea. Moreover, He obtained a MS degree in chemical engineering from the Texas A&M University in U.S.

Mr. Yun has worked for the Korea Gas Safety Corporation for 9 years as a safety engineer and research scientist for gas related facilities (including LNG terminals) and equipment. He was a national secretary for 5 years for ISO standardization in the field of gas cylinders and bursting discs.

Mr. Yun has the Certified Welding Inspector Certification issued by AWS (American Welding Society) and Design Corrosion Control Certification by NACE (The Corrosion Society of America). He received the Certification of LOPA Training as well as SIL (Safety Integrity Level) Verification by SIS-TECH. Furthermore, he has Gas Safety Expert Certificate and Occupational Safety Expert Certificate by the Human Resources Development Service of Korea.

89e Study on the Heat Transfer of Expansion Foam on LNG Pool

Geun Woong Yun and M. Sam Mannan. Chemical Engineering, Mary Kay O'Connor Process Safety Center, Texas A&M University, 812B Autumn circle, College station, TX 77840

Expansion foam is one of the important safety measures to control vapor dispersion and fire suppression. It has been recognized that the effectiveness of expansion foam may be achieved by reducing the LNG vapor density to make it positively buoyant by increasing its temperature adding heat from water in the foam to LNG vapors while penetrating through the foam.

The effectiveness of expansion foam for vapor dispersion control is very dependent on the thickness of foam to cover LNG pool. If the foam thickness is relatively small, it may not provide enough heat to LNG vapors up to reach positive buoyancy. In the LNG facilities, the possible incident scenarios might involve bigger releases due to its huge storage and processing capacity. Intuitively, in case of larger amounts of LNG release, it might require bigger foam depth because the larger LNG, the bigger needs of heat input from water in the foam by larger foam depth for the similar effect of vapor control. Therefore, it is very important to identify minimum effective foam thickness for vapor dispersion control. Consequently, heat transfer study on the LNG and expansion foam associated with field experiments is very necessary to obtain the required expansion foam depth for the safety of LNG or other flammable fuels.

Therefore, in this paper, heat transfer study between LNG and expansion foam is conducted and its results are compared with experimental results achieved by Mary Kay O'Connor Process Safety Center (MKOPSC) to validate the model. Finally, this generalized heat transfer study will allow us to estimate the effective foam depth for all kinds of LNG pool scenarios. Moreover, this paper will provide the basis for the estimation of minimum effective foam depth in case of foam application on fire, which is more complicated than the vapor control case.

LNG Safety and Security[1]

102a

Dmitry Volkov, Energy Analyst
Dr. Michelle M. Foss, Head and Chief Energy Economist
Ruzanna Makaryan, Senior Energy Analyst

Center for Energy Economics, Bureau of Economic Geology
Jackson School of Geosciences at the University of Texas at Austin
Houston, Texas

AIChE Spring Meeting, April 2009
9th Topical Conference on Natural Gas Utilization
Tampa, Florida, April 26-30, 2009

Introduction

LNG has been transported and used safely in the U.S. and worldwide for more than half a century. The U.S. has the largest number of LNG facilities in the world, scattered throughout the country and located near population centers where natural gas is needed. Our analysis of data on LNG safety and security indicates an excellent safety record, which we can contribute to several factors. First, the industry has evolved to ensure safe and secure operations. Technical and operational advances include everything from the engineering that underlies LNG facilities to operational procedures to technical competency of personnel. Second, the physical and chemical properties of LNG are such that risks and hazards are easily defined and incorporated into technology and operations. Third, a broad set of standards, codes and regulations applies to the LNG industry to further ensure safety. These have evolved through industry experience worldwide and affect LNG facilities and operations everywhere. Regulatory compliance provides transparency and accountability.

Beyond routine industrial hazards and safety considerations, LNG presents specific safety considerations. In the event of an accidental release of LNG, the safety zone around a facility protects neighboring communities from personal injury, property damage or fire. The one and only case of an accident that affected the public was in Cleveland, Ohio in 1944. Research stemming from the Cleveland incident has influenced safety standards used today. Indeed, during the past four decades, growth in LNG use worldwide has led to a number of technologies and practices which will be used in the U.S. and elsewhere in North America as the LNG industry expands.

Dynamics of LNG industry have significantly changed over the past few years. Explosive growth of interest to LNG as a new source of supply led to an avalanche of

[1] A lot of the material in this paper is taken from publications of the *Center for Energy Economics* and the complete *Guide to LNG in North America,* which provide extensive information to those interested in U.S. energy trends and security; LNG industry and market developments. The CEE web site, www.beg.utexas.edu/energyecon/lng provides links to industry, government and public information sources.

applications to develop regasification and liquefaction terminals, mostly in USA and Europe and in Asia, Africa and Australasia, accordingly. In many cases, especially on the regasification side, these plans met resistance of local communities, and companies turned to offshore versions of LNG terminals, which are considered to be more viable options in the densely populated areas. Raising demand for LNG tankers and freight rates resulted in new designs for bigger and more efficient LNG carriers. Finally, high price volatility of natural gas opened up numerous arbitrage opportunities, which resulted in partial discharge phenomenon. All these shifts in the LNG business paradigm required corresponding changes on the safety and security side. Monitoring of energy –related incident record, and LNG incident record in particular, allows giving full credit to the industry and regulatory authorities for a continuous excellent safety record.

Types of LNG Hazards[2]

The potential hazards of most concern to operators of LNG facilities and surrounding communities flow from the basic properties of natural gas. Primary containment, secondary containment, safeguard systems, and separation distance provide multiple layers of protection. These measures provide protection against hazards associated with LNG.

Explosion. An explosion happens when a substance rapidly changes its chemical state – *i.e.*, is ignited – or is uncontrollably released from a pressurized state. For an uncontrolled release to happen, there must be a structural failure – *i.e.*, something must puncture the container or the container must break from the inside. LNG tanks store the liquid at an extremely low temperature, so no pressure is required to maintain its liquid state. Sophisticated containment systems prevent ignition sources from coming in contact with the liquid. Since LNG is stored at atmospheric pressure – *i.e.*, not pressurized – a crack or puncture of the container will not create an immediate explosion.

Vapor Clouds. As LNG leaves a temperature-controlled container, it begins to warm up, returning the liquid to a gas. Initially, the gas is colder and heavier than the surrounding air. It creates a fog – a *vapor cloud* – above the released liquid. As the gas warms up, it mixes with the surrounding air and begins to disperse. The vapor cloud will only ignite if it encounters an ignition source while concentrated within its flammability range. Safety devices and operational procedures are intended to minimize the probability of a release and subsequent vapor cloud having an affect outside the facility boundary.

Freezing Liquid. If LNG is released, direct human contact with the cryogenic liquid will freeze the point of contact. Thus, containment systems surrounding an LNG storage tank are designed to contain up to 110 percent of the tank's contents. Containment systems also separate the tank from other equipment. Moreover, all facility personnel must wear gloves, face masks and other protective clothing as a protection from the

[2] Much of the material in this section is taken from the New York Energy Planning Board *Report on Issues Regarding the Existing New York Liquefied Natural Gas Moratorium*, November 1998.

freezing liquid when entering potentially hazardous areas. This potential hazard is restricted within the facility boundaries and does not affect neighboring communities.

Rollover. When LNG supplies of multiple densities are loaded into a tank one at a time, they do not mix at first. Instead, they layer themselves in unstable strata within the tank. After a period of time, these strata may spontaneously *rollover* to stabilize the liquid in the tank. As the lower LNG layer is heated by normal heat leak, it changes density until it finally becomes lighter than the upper layer. At that point, a liquid rollover would occur with a sudden vaporization of LNG that may be too large to be released through the normal tank pressure release valves. At some point, the excess pressure can result in cracks or other structural failures in the tank. To prevent stratification, operators unloading an LNG ship measure the density of the cargo and, if necessary, adjust their unloading procedures accordingly. LNG tanks have rollover protection systems, which include distributed temperature sensors and pump-around mixing systems.[3]

Rapid Phase Transition. When released on water, LNG floats – being less dense than water – and vaporizes. If large volumes of LNG are released on water, it may vaporize too quickly causing a rapid phase transition (RPT).[4] Water temperature and the presence of substances other than methane also affect the likelihood of an RPT. An RPT can only occur if there is mixing between the LNG and water. RPTs range from small *pops* to blasts large enough to potentially damage lightweight structures. Other liquids with widely differing temperatures and boiling points can create similar incidents when they come in contact with each other.

Earthquakes and Terrorism. In many aspects of daily life, *risk* of a certain event is very often perceived to be much different from reality. Sometimes potentially dangerous activities (such as driving a car or flying in a plane) can become so commonplace and accepted that the risk associated with those activities can be taken for granted. In other cases, the focus on worst-case events overshadows the real probability that such events will ever occur. In many such instances, worst-case scenarios are assumed without taking into consideration the numerous steps taken to prevent them. Risk is a combination of not only the consequence of an event, but also the probability of the event occurring. A high consequence event with a low probability of occurrence may be similar on a “risk basis” to a low consequence event with a high probability of occurrence.

Potential damage and injuries from an LNG incident would depend on initiating events, volume and location of LNG release, release rate, wind direction and speed, and other factors. However, the quantitative calculation of probabilities of such an event actually can only be done if sufficient data exists.

Unexpected risks are, of course, different from routine risks with regard to uncertainty about whether or when they could take place. There must be a general enforcement of

---

[3] Welker J. R. and Sliepcevich C.M., Radiation, Heat Flux, and Overpressure in LNG Tanks, Proceedings of the International Conference on LNG Importation and Terminal Safety, Boston (1972).

[4] Hashemi H.T., West H. H. and Sliepcevich C.M., LNG/Water Explosions: A Distributed Source, Proceedings of the 27th Annual Petroleum Mechanical Engineering Conference (1972).

security to protect all types of facilities, including LNG operations, and public places from acts of violence. With respect to unexpected risks such as terrorism, a system of safeguards is already in place.

LNG tanks, whether on ships, on land, or offshore, require exceptionally large amounts of force to cause damage. Because the amount of energy required to breach containment is so large, in almost all cases the major hazard presented by terrorists is a fire, not an explosion. If an aircraft crashed into an LNG facility, the impact would almost certainly cause a fire fueled initially by the aircraft fuel. It may also ignite the LNG, causing a larger fire at the facility. Emergency fire detection and protection at the LNG facility/ship would be used in such an event. Danger to the public from this type of event would be reduced or eliminated by the separation distance of the facility.

Rigorous siting reviews coupled with separation and distance requirements in U.S. safety codes[5] minimize risk to the public from land-based facilities. According to the opinion of Project Technical Liaison Associates, Inc. (PTL), "LNG land-based facilities are sited to very stringent design and construction codes and standards. These codes require that 'worst-case' accident scenarios be used in the siting and design of these facilities."[6]

When estimating the risk of LNG projects, the companies involved in LNG facilities consider the danger of strong ground movements and failures due to seismic activity, flooding and landslides in the area. The seismic design requirements are outlined in the NFPA 59-A 2001. Major earthquakes can cause severe damage if the facilities are not designed to withstand such events, so the companies conduct regional and site-specific studies to see if the areas are seismically active. These factors are then taken into account during planning and design stages. The design of LNG tanks can accommodate regional seismic activity in locations of potential risk. There are no known incidences of LNG storage tank failures due to seismic activity. In fact, in 1995, none of the LNG storage tanks in the Kobe, Japan area were damaged during a 6.8 earthquake on the Richter scale.

Japan is one of the world's largest users of LNG and has many LNG storage tanks. Table 1 below shows the different LNG facilities in the U.S. and Japan. Japan is also one of the more seismically active areas of the world. Damage to its LNG facilities from the most severe earthquakes has been limited to that of natural gas pipelines. CEE has a separate case study on Japan's long experience with LNG and safety record.[7]

---

[5] 49 CFR Part 193: http://cfr.law.cornell.edu/cfr/cfr.php?title=49&type=part&value=193

[6] Lewis, James P.; McClain, Sheila A. Project Technical Liaison Associates, Inc. (PTL): *LNG Security: Reality and Practical Approaches*, LNG: Economics & Technology Conference, January 2003.

[7] Contact the CEE for details and availability.

Table 1. LNG Facilities in the U.S. and Japan

Type of Facility	U.S.	Japan
Liquefaction Terminals	1	
Regasification Terminals	5	30
Peakshaving Facilities	57	
Satellite Storage Facilities (without liquefaction)	39	26
Others	12	
Total	114	56
Sources: EIA, Japan Gas Association		

Maritime Incidents. The history of the LNG industry has shown that maritime incidents with severe LNG releases are very rare. Over the industry's 50-year history of over 46,000 voyages, there has never been a spill from a ship into the water from either a collision or grounding. LNG ships are well-designed and well-maintained, which reduces the chances and severity of incidents. Their designs prevent breaching of cargo tanks and involvement of multiple tanks in accidents. Potential hazards could come from ignition of LNG pool fires or a vapor cloud.

Operational Incidents. Operational incidents, which are incidents resulting from human error, equipment failures, or both, can occur in any industry and any facilities. In the LNG facilities, incidents could happen during unloading, storage, vaporizing and pipeline transmission or other stages of production. Such errors could result in a spill or a fire. However, LNG facilities and ships have advanced monitoring and control systems that make an incident unlikely to occur compared to other releases. Consequences of the majority of potential incidents would be contained on site and managed before they could result in significant damage.

Safety and Security Issues along the LNG Value Chain

The U.S. and other countries maintain health, safety, and environment (HSE) policies and regulations that apply to all of these activities and sites as well as specialized policies, regulations, and industry standards targeted to specific needs and hazards. Worldwide, best practices for all of these activities have evolved and are becoming more firmly embedded in contractual and regulatory frameworks that establish the safety conditions of industry operations. The specific safety and security features embedded in the LNG value chain, as they pertain to the four elements of primary containment, secondary containment, safeguard systems and separation distances, are detailed below.

International standards and rules define containment with respect to types of structures and technologies in use. We use the term "containment" in this document to mean safe storage and isolation of LNG. Safe use of LNG, or any cryogenic substance, requires an understanding of how materials behave at cryogenic temperatures. For example, at extremely low temperatures, carbon steel loses its ductility and becomes brittle. The material selected for tanks, piping, and other equipment that comes in contact with LNG is critical. The use of high nickel content steels, aluminum, and stainless steels is costly but necessary to prevent embrittlement and material failures. High alloy steels composed of nine percent nickel and stainless steel typically are used for the inner tank of LNG storage tanks and for other LNG applications.

The major components of the LNG value chain include the following:

- *Natural gas production*, the process of finding and producing natural gas for delivery to a processing facility.
- *Liquefaction*, the conversion of natural gas into a liquid state so that it can be transported in ships.
- *Transportation*, the shipment of LNG in special purpose ships for delivery to markets.
- *Regasification*, conversion of the LNG back to the gaseous phase by passing the cryogenic liquid through *vaporizers*.
- *Distribution and delivery* of natural gas through the national natural *gas pipeline system* and distribution to end users.

There are over 250 LNG facilities worldwide, including 63 regasification terminals. The U.S. has the largest number of those with 114 active facilities. Natural gas is liquefied and stored at about 58 facilities in 25 states, including 96 connected to the U.S. natural gas pipeline grid. Massachusetts alone accounts for 14 major satellite facilities, or roughly 40 percent of all satellite facilities in the United States, and New Jersey has five satellite LNG facilities, the second highest in the U.S.

Siting of New LNG Terminals and Community Relations

Explosive growth of interest to LNG as a new source of supply led to an avalanche of applications to develop regasification and liquefaction terminals, mostly in USA and Europe and in Asia, Africa and Australasia, accordingly. In many cases, especially on the regasification side, these plans met resistance of local communities. Often success of the proposed projects hinged on the balance of benefits/cost, perceived by the communities.

The CEE-UT research team undertook a study effort to identify benefits and costs as well as associated issues pertaining to siting of 20 liquefied natural gas (LNG) import receiving and regasification terminals and associated infrastructure in five geographic regions of North America.[8] Ten issue dimensions were defined: safety/security, wetlands, fisheries, energy costs, roads, taxes, employment, air emissions, property value and other/intangibles. Results of the study showed that on average for all projects, the strongest tendencies to perceive costs were for the safety/security issues.

---

[8] Center for Energy Economics, Bureau of Economic Geology, The University of Texas at Austin, "Defining Community And Economic Benefits Associated With Large Energy Infrastructure Projects: LNG Case Study". CEE-UT, 2007-2009, Publication pending

One of the most counterintuitive finings was that communities near brown field sites (and more familiar with industrial activity) would demonstrate same or higher levels of concern with regards to safety and security. Another important finding was that safety and security issues are more difficult to address and can overwhelm other factors for certain projects and situations.

For onshore projects that have significant immediate site communities and for which internal waterways are used, safety and security is a considerable issue. For onshore projects that do not have significant immediate site communities or for which waterways are not considerably affected by the new LNG tanker traffic, concerns were reduced. For such communities and stakeholder groups, other issues are more important and have higher priority.

Safety and security concerns could upend the regulatory process or trigger substantial changes in project development as proponents seek effective mitigation strategies. Summary of the findings are shown in the Figure 1 below.

Figure 1. Stakeholder Characterization: Average for All Projects

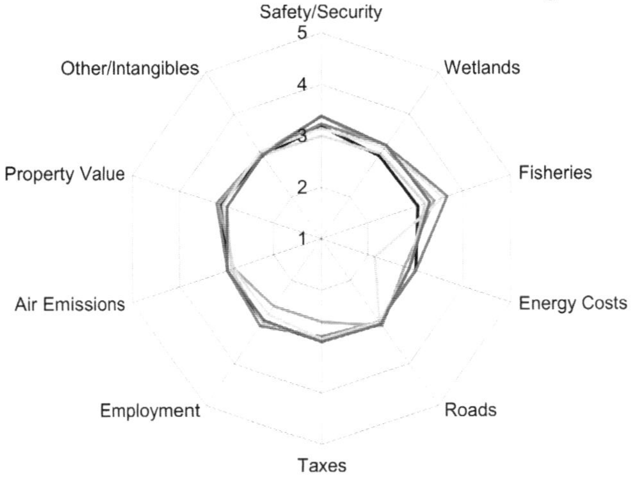

Legend

- Immediate Site Community
- Waterway Community
- Local Community (City)
- Greater Community (State)
- National Community (Federal)
- External Interest Group

Scale

1 – Perceived Benefit

3 – Indifferent

5 – Perceived Cost

At the outset of our study effort, we surmised that a potential cost-to-benefit conversion could be emergency response capacity building in both the immediate site community and for the waterway community. However, no evidence exists of strategies to foster perceptions of net benefits from this aspect of LNG import facility development. Our assumption was that improved emergency response capacity as a consequence of a new or expanded LNG facility could be meaningful especially to smaller communities. Reduced response times for injuries, illness and both natural and human induced hazards, disasters and other events ought to be valued. Clearly, the net benefit of expanded capacity does not outweigh the perceived cost of hosting the facility in the first place, and developers and other project proponents choose not to focus on what could be termed a "negative-positive". But the inability to bolster perceived net benefits via emergency response capacity also reflects the pervasive problem of misperception of risk, the tendency to focus on low probability-high consequence events rather than the everyday risks and incidents that take the overwhelming toll on human life and property.[9]

Finally, a clear mitigation strategy for perceived costs associated with safety and security is to move offshore with the associated higher costs and new risks and uncertainties. Many developers have refrained from onshore proposals altogether. However, offshore LNG facilities bear a unique set of challenges and considerations, as identified in our briefing paper.[10] Nor have offshore projects proved to be a panacea for developers; many have attracted attention on the same issue dimensions associated with supporting onshore facilities (such as pipeline landings and connections and wetlands) or become subject to new issue dimensions (such as fisheries and open loop revaporization).

Offshore LNG Terminals

Increased safety and security concerns after the tragic events of 9/11 brought to life the Maritime Transportation Security Act (MTSA) of 2002, which amended the Deepwater Port Act (DPA) of 1974. The MTSA provides guidance on security measures when dealing with offshore LNG receiving terminals. Offshore terminals were perceived to carry fewer security risks, and in the wave of new LNG terminal applications in the US in 2003-2008 were considered to be more viable options in the densely populated areas (which attracted most attention by the developers due to proximity to consumer markets) by both general public and industry. At the same time marine operations for offshore LNG facilities present new and different hazards and design specifications that must be dealt with and accommodated. This can increase the cost associated with LNG import operations.

Currently there are several offshore liquefied natural gas (LNG) receiving facilities with a handful more in the works. Additionally, several companies announced plans for

---

[9] In our discussions on the role of emergency response capacity, an analogy is the value placed on policing to reduce crime and the focus on crime rates in real estate investment decision making.
[10] CEE-UT, *Offshore LNG Receiving Terminals*, 2006.
http://www.beg.utexas.edu/energyecon/documents/CEE_offshore_LNG.pdf

offshore *LNG liquefaction facilities,* usually of a smaller scale than on-shore terminals. Offshore liquefaction facilities are considered in the remote areas with relatively marginal fields, yet the technology is still under development and actual developed projects are probably several years away.

Most proposed offshore LNG designs incorporate technologies that have been proven in other applications such as for onshore LNG receiving terminals, offshore oil and natural gas production platforms, and offshore crude oil receiving and storage terminals. By using proven technologies, technical and economic uncertainty is reduced and the resulting risks mitigated.

Water depth of proposed locations is usually the variable that determines the type of main structure to consider. When considering siting of LNG offshore terminals, a wide range of possible locations present themselves. LNG vessels typically have a draft of 38 feet and require at least an additional two to 5 feet of depth to provide sufficient clearance from the sea bottom for safe maneuvering. This means that the minimum water depth for siting LNG offshore terminals will be determined by the minimum depth of water required for the safe maneuvering of the LNG vessels, which is about 40 to 43 feet or approximately 14 meters. At present, there is no maximum depth of water that would limit the location of an offshore LNG terminal, but ultimate water depths for safe, economic operation are also determined by geometry of the sea floor, wave action, distance from shore, and other factors.

In order to mitigate security risks for offshore LNG terminals, the US Coast Guard has established a 500 m safety zone surrounding the facility to exclude ship traffic not related to the port operations, although this can be terminal and site specific. The requirement of this safety zone necessitates that the offshore LNG terminal be located away from shipping fairways, existing oil or gas platforms, other deepwater ports, and other areas of activity on the Outer Continental Shelf (OCS) to avoid interference with those activities. Other considerations of location, such as proximity to existing offshore and onshore pipeline distribution systems and support infrastructure, will also influence the cost effectiveness of a deepwater port.

The LNG terminal usually consists of several reinforced concrete gravity-based structures (GBSs). The GBSs support the control and maintenance buildings and utilities, regasification facilities, and LNG storage tanks to name the most important components. High-strength cement technology and steel reinforcing would be used to design the GBSs to safely withstand extreme stresses like the force of the Loop Current that is a permanent feature of the US Gulf of Mexico, severe wave loads caused by hurricanes or major storms, and other stress-inducing events including vessel impact.[11] If the GBS is sitting in about 60 to 80 ft of water, there will be about 70 to 90 ft of freeboard above the seawater level.

---

[11] US Coast Guard: Maritime Administration: Final Environmental Impact Statement for the Port Pelican Llc Deepwater Port License Application. August 2003.

One novel approach used for many other types of facilities is that of the construction of an offshore artificial island. This alternative provides the greatest “real estate” of the fixed structure alternatives. It provides space for LNG storage, multiple docking berths, air based vaporization and other space extensive services and processes. To date, only one project based on an offshore island has been announced in the US[12]. The concept allows for the use of onshore LNG tank designs, and other onshore technologies since space considerations are not as important as in the platform-based terminals.

To meet applicable safety standards, an emergency gas flare would be installed on a separate support structure adjacent to the end of the process areas. The flare would be oriented such that the prevailing winds would direct its plume away from the main facilities. If offshore, the flare would be accessed using an extended gangway. A flare header system would collect hydrocarbon flows from relief valves, tank blankets (air spaces around the tank with nitrogen and natural gas sensors), and miscellaneous sources and send them to a flare drum and then to the flare. The flare would be equipped with multiple pilots and electronic igniters.

As projects move further away offshore, water depths increase beyond those permissible for fixed structures and must consider floating facilities. In general, the same processes are considered for floating facilities: docking, offloading, storage, and regasification. Different processes can be part of the floating facility, for example, if storage and regasification are part of the facility, then they are commonly called a Floating Storage and Regasification Unit (FSRU). If storage is not incorporated into the floating facility, then a Floating Regasification Unit (FRU) is considered.

A FSRU LNG import terminal concept comprises of a purpose built moored[13] ship with several LNG ships shuttling between the export facility and the import site. The FSRU ship is typically between 350 to 400 meters long by up to 70 meters wide and normally does not have a complete propulsion system. There are applications in which rapid disconnection and relocation of the FSRU is a requirement. In these cases, a propulsion system will be part of the FSRU. Floating structures with storage capacity generally require an anchoring system and sufficient water depth (generally greater than 160 ft) to accommodate a flexible pipeline connection between the unit and the seafloor pipeline.

The FSRU consists of a double-hulled ship designed using normal shipbuilding blueprints and standards and can be constructed in a wide range of conventional ship yards worldwide. The regasification facilities are located on the main deck of the ship and are typically tailored to suit the specified gas send-out conditions. Since the FSRU is part ship, part storage tank, and part regasification unit, three separate design standards, guidance, and regulations must be satisfied. The vessel portion of the FSRU is subject to marine codes, the LNG storage tanks are subject to LNG storage and transfer rules, and the LNG regasification and send out processes are subject to

[12] Atlantic Sea Island Group’s Safe Harbor Energy Project
[13] FSRU’s are mostly permanently moored but may have to disconnect on occasion.

process standards and codes. Utilities and systems associated with FSRU operations include electric power generation and distribution, instrumentation and controls, and fire and safety systems.

The vessel is a turret moored floating receiving unit designed for loading LNG from a side-by-side moored LNG tanker in a relatively benign range of environmental conditions. The LNG carrier is moored alongside the FSRU with both vessels weathervaning around the FSRU's turret mooring.

In the event the FSRU is required to be located in harsher metocean conditions then the 'Stern to Bow' or 'Tandem' configuration can be used. While this tandem technology is still considered developmental, several leading industry equipment suppliers are actively advancing it. A key advantage of the FSRU concept is that it can be moored in a wide range of water depths. In shallow waters (approximately 65 to 100 feet), a jacket-based, soft yoke system can be used, in greater water depths a catenary based, turret mooring system can be employed. Both of these systems are weathervaning, allowing the FSRU ship's heading to rotate according to the direction and force of the wind.

An interesting case is that of the Excelerate Shipping Fleet which combines the capabilities of a conventional LNG tanker with that of an FSRU. Effectively, the Excelerate ships act as FSRUs when offloading natural gas directly into natural gas pipelines since it used the ship's facilities are used for storage and regasification. However, one important process is still lacking, the ship to ship transfer of LNG which in the case of the Excelerate ships is avoided since the LNG tanker and the LNG FSRU are in this case the same ship.

Another interesting example of a floating LNG facility is one that lacks local LNG storage (other than that of the LNG tanker) and allows for conventional LNG ships to dock[14]. In this case, conventional offloading arms are mounted on a purpose built floating structure (pontoon, keel and towers) which temporarily attaches to docking LNG tankers. Regasification is also performed on the floating structure with conventional regasification technology.

The exchange of heat with seawater in order to vaporize LNG is the most widely used technology in the world (although not so in the US). There are two main designs that use water as the primary source of heat: open rack vaporizers (ORVs) and shell and tube vaporizers (STV). In order to modularize the process and avoid freezing of the seawater, intermediate fluids (e.g. propane) can perform the direct heat exchange at low temperatures. The intermediate fluid would then exchange heat with the seawater. The most common technology used for LNG vaporization is Open Rack Vaporizers. The operating principle behind the ORVs is the heat exchange between the LNG located inside heat-conducting, zinc alloyed, finned panels and a falling film of seawater at atmospheric conditions flowing over it. The panels need to be cleaned periodically to remove any foreign matter that might have adhered to the panel surfaces.

---

[14] The recently announced project for the Bienville Offshore Energy Terminal in Alabama is an example: http://www.torplng.com

ORV and STV designs have been proven safe, have no moving parts and lack ignition sources. Similar heat exchangers are extensively used as heat sinks in power plants. In both cases, significant volumes of seawater are required to provide heat and thus large electrical loads are required for pumping the seawater[15]. In addition to the restrictions on the seawater discharge quality (intake-suspended materials and discharge-temperature[16]), there are possible environmental impacts from the use of biocide for the prevention of bio-fouling and the intake of sea fauna because of the large flow rates of water. The possible environmental impacts determine the dosage and schedule of biocide usage (usually chlorine based).

The possibility of the intake of sea fauna is addressed through specific design considerations such as positioning of the low velocity intake at depths where the impact is minimized. First, the inlet velocity is reduced by increasing the diameter of the intake. The lower velocities allow larger marine organisms the opportunity to swim from the intake and avoid impingement which could lead to injury. On the other hand, larger volumes (higher fluid speeds for a fixed inlet diameter) reduce the cooling of the seawater and thus reduce the impacts at discharge[17].

For the intake, the geometry, location, orientation and protection (via mesh screens) seek to minimize entrainment of smaller marine organisms such as eggs, larvae and young juveniles that cannot propel themselves free from the intake. Different approaches must be considered depending upon the location of the application. Great concern has been raised as to the impacts of open loop seawater vaporizers on fisheries in the US GOM due to the entrainment of the fish larvae. The US Clean Water Act (CWA)[18] requires that "the location, design, construction, and capacity of cooling water intake structures reflect the best technology available for minimizing adverse environmental impact"[19]. In order to evaluate the best technology available, adequate models and data need to be available and the uncertainty that govern them needs to be determined and taken into account. The fishery entrainment protocol developed by the USCG uses SEAMAP data to estimate the number of eggs and larvae entrained by an LNG facility and to provide an estimate of the future number of age-1 fish that are impacted. The protocol can be divided into two parts, a static part that looks at data to predict entrainment and one that, based on the first results, predicts fish lost to the coastal ecosystem.

The environmental impact statements developed by the USCG for proposed offshore LNG developments conclude that the general environmental impact would be minor and fall within acceptable limits. In addition, the estimated impact on fisheries would be minor. However, groups of stakeholders have raised concerns about the interpretation

---

[15] Thus there is a need for primer movers for the pumps which is limited on offshore facilities.

[16] The difference in temperature is between 5 and 15 C.

[17] There is no clear guideline for the environmental limit of cooled water discharge. There is a guideline put forward by the World Bank Group for power generation facilities that covers heated water discharge that provides a limit of 3 C at a distance of 100m from the discharge.

[18] Environmental Protection Agency (EPA) Section 316(b) of the Clean Water Act www.epa.gov

[19] The EPA is currently engaged in rulemaking for the discharge of cooling water used in condensers in power plants. In this case, the water that is discharged is hotter than ambient.

of the studies regarding the cumulative impacts if several offshore LNG terminals are built in the same region that would use seawater as the heating fluid.

While a long history and experience with offshore crude oil operations lends extensive support to offshore LNG developments, ranging from available technologies and designs to marine operations, safety, and environmental protection, the concept of offshore LNG is new and thus encompasses a number of uncertainties especially with regard to environmental impact, floating operations and floating LNG offloading.

New Trends in LNG Transportation (Shipping)

The critical link between liquefaction and regasification components of the LNG value chain is *shipping*. According to Maritime Business Strategies, there were 298 existing *LNG ships*, as of February 2009, with 79 more on order.[20] About 60 percent of the fleet is less than five years old. Older generation of LNG ships was designed to transport between 125,000 and 150,000 cubic meters (m^3) of LNG,[21] or about 2.8-3.1 billion standard cubic feet of natural gas. Currently, larger LNG ships with a capacity greater than 200,000 m^3 became a norm in the industry, and four ships of 263,000 and 270,000 cubic meters (m^3) capacity of LNG are in service already. The use of larger ships, which enable LNG value chain economics to improve and facilitate a larger supply base for the U.S. and other importing countries, is critical in determining how new baseload receiving terminals are designed as well as how existing facilities will be expanded. A typical ship measures some 900 feet in length, about 150 feet in width and has a 38-foot draft. LNG ships can be less polluting than other shipping vessels because they can burn natural gas, but may also substitute or supplement with fuel oil as an additional source for propulsion.

Engineering design for safety also applies to LNG ships. An onboard containment system stores the LNG, where it is kept at atmospheric pressure (to keep air from entering the tank) and at -256°F (-160°C). Existing LNG ship cargo containment systems reflect one of three designs. As of February 2009:

- Spherical (Moss) design accounts for 40 percent of the existing ships,
- Membrane design account for about 55 percent, and
- Self-supporting structural prismatic design account for about 5 percent.

Safety zones differ for ships in transit as opposed to ships in port. Port safety zones are established by the USCG and port captain, based on the specific risk factors at a given terminal. There are two purposes for safety zones for LNG ships – to minimize collision while the ship is underway, and at berth to protect surrounding property and personnel from hazards that could be associated with ignition. In the U.S., the use of safety zones around LNG ships began in 1971 at the Everett Terminal in Boston Harbor. Safety zones are established based on the specific circumstances, including navigational requirements, in a specific area.

In some ports, the USCG may require a tug escort and specified safety zones around LNG ships when a ship is underway to a U.S. receiving terminal. The USCG's intention

---

[20] Maritime Business Strategies, LLC: http://www.coltoncompany.com/.

[21] Typically, LNG ship size is designated by cubic meters of liquid capacity.

is to minimize disruption to area shipping and boating traffic while ensuring safe operations. Tugs assist in the safe docking of LNG ships. The USCG, which by Federal law has the responsibility for implementing safety regulations that apply to LNG marine operations in the U.S., is now part of the new U.S. Department of Homeland Security. For all vessels of special interest, including LNG ships, the USCG enforces strict measures towards terror threat protection. Measures taken to prevent terrorism on LNG facilities and ships worldwide include inspections and patrols, action plans for security breach, emergency communication systems and intelligence gathering. These same measures are also used at other critical facilities, operations and ships, such as passenger ships, oil tankers, containerships and other.

Currently, USCG responsibilities towards LNG shipments are directed by the Maritime Transportation Security Act of 2002, according to which it shares security burden with state and local organizations, as well as the LNG project operators. In April of 2008 the House of Representatives passed a H.R. 2830 - the Coast Guard Authorization Act of 2008 (the Senate version of the bill, S. 1892 has not been considered yet as of February 2009). The bill would appoint USCG as the only party responsible for security of an LNG vessel from its entering the service's security regime to the LNG terminal.

Past decade was marked by the meaningful increase of spot LNG trades. It was recently estimated that spot LNG reached 13 percent of total world LNG trade in 2006, compared to 2 percent at the end of 1990s.[22] Arbitrage opportunities, caused by the price differences in various markets at a certain point of time, created diversions of LNG ships under long-term contracts to other destinations, in which case the buyer might have ended up with a hefty penalty for not fulfilling the contract, yet most often the buyer and supplier would have split the profit resulted by LNG cargo diversion. They also produced a previously not-seen phenomenon in the LNG transportation – partial discharges, when LNG carrier would deliver and unload part of the cargo at one LNG receiving terminal, then proceed to another terminal and unload the rest of the cargo.

Also, the advent of LNG offshore terminals implies certain risks associated with tanks only partially filled with LNG. Carrying LNG in partially filled tanks could lead to sloshing - a violent motion of the fluid. Sloshing could lead to an increased high pressure of LNG on the tank walls, especially in an abnormally harsh wave environment.[23] Bureau Veritas and other classification societies have filling limitations for LNG cargo systems (see Table 2 below).

---

[22] TEEKAY LNG PARTNERS L.P., 20-F/A, December 31, 2007

[23] Mateusz Graczyk, Torgeir Moan, "A probabilistic assessment of design sloshing pressure time histories in LNG tanks". Ocean Engineering 35 (2008) 834–855

Table 2. Limitations for LNG cargo systems

System	Filling Limitation
Membrane	FL between 10%L-70/80%H not allowed
MOSS	Without limits due to spherical geometry
IHI-SPB	Possibility to arrange wash bulkheads
FL: filling level, L: length of the tank, H: height of the tank.	
Source: Adapted from L. Delorme, A. Souto Iglesias and S. Abril Perez, "Sloshing Loads Simulation In LNG Tankers With SPH". International Conference On Computational Methods In Marine Engineering, MARINE 2005	

Possible sloshing effect might require additional modifications to LNG cargo systems, especially taking into account increasing size of LNG carriers. Several engineering organizations, including Det Norske Veritas (DNV) and Norwegian Marine Technology Research Institute, have projects underway, researching this important safety issue.

Increased distances and LNG carrier capacities generated a need in propulsion systems of different designs, since steam turbine-based propulsions, fed by the naturally generated boil-off gas, are becoming less efficient. As noted in a recent study,[24] presently there are four major concept designs in the works: the low-speed diesel engine propulsion with reliquefaction, the medium-speed dual-fuel diesel electric (DFDE) propulsion, the low-speed dual-fuel diesel engine propulsion, and the gas turbine electric propulsion. These new systems might present significant hazard due to their need for fuel gas supply under pressure. High-pressure fuel gas leakages could result in fire or explosion. Classification societies issue safety requirements for the dual-fuel propulsion systems[25], which would typically include double-wall piping or ventilated pipe or ducts, automatic shut-off valves and master gas valves, gas detection and ventilation systems with independent redundancy and engine exhaust and scavenge system.

LNG Safety Record

LNG industry showed exceptional safety record over its history. In fact, recent study[26] showed that LNG industry proved to be the safest one among all energy-related sectors both in terms of social and economic costs in the past century (see Figure 2 below). The

---

[24] Daejun Chang, Taejin Rhee, Kiil Nam, Kwangpil Chang, Donghun Lee, Samheon Jeong, "A study on availability and safety of new propulsion systems for LNG carriers". Reliability Engineering and System Safety 93 (2008) 1877–1885

[25] IMO IGC code, international code for the construction and equipment of ships carrying liquefied gases in bulk; ABS, rules for building and classing steel vessel (Section 5-8-16); DNV, gas fuelled engine installations (Part 6, Chapter 13), ABS, guide for propulsion systems for LNG carriers. Source: Daejun Chang _, Taejin Rhee, Kiil Nam, Kwangpil Chang, Donghun Lee, Samheon Jeong, "A study on availability and safety of new propulsion systems for LNG carriers". Reliability Engineering and System Safety 93 (2008) 1877– 1885

[26] Benjamin K. Sovacool, "The costs of failure: A preliminary assessment of major energy accidents, 1907–2007". Energy Policy 36 (2008) 1802–1820

study analyzed major 279 incidents world-wide that have been responsible for $41 billion in property damage and 182,156 deaths.

Figure 2. Energy accident fatalities by source, 1907–2007

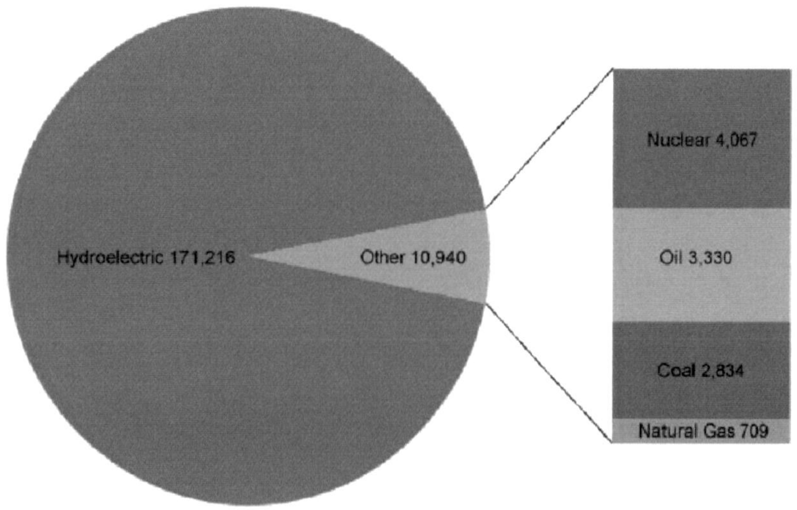

Source: Benjamin K. Sovacool, "The costs of failure: A preliminary assessment of major energy accidents, 1907–2007". Energy Policy 36 (2008) 1802–1820

With the exception of the 1944 Cleveland fire, all LNG-related injuries have occurred within an LNG facility. No death or serious incidents involving LNG have occurred in the United States since the Cove Point incident in 1979, although there were several incidents outside the US since then (see Table 3 below).

Table 3. Major Energy-related Incidents Worldwide, 1907-2007

	Facility	Date	Location	Description	Fatalities	Cost (in millions $)
1	Coal mine	December 6, 1907	Monongah, West Verginia USA	Underground explosion traps workers and destroys railroad	362	162
2	Coal mine	February 27, 1908	San Juan de Sabinas, Coahuila, Mexico	Mine shaft completely collapses	201	12
3	Coal mine	September 30, 1908	Palau Coal Mine, Coahuila, Mexico	Explosion and fire collapse multiple shafts	100	8

	Facility	Date	Location	Description	Fatalities	Cost (in millions $)
4	Coal mine	February 16, 1909	Stanley, England	Explosion and fire destroys entire mine	168	11
5	Coal mine	November 13, 1909	Cherry, Illinois, USA	Fire and explosion collapse multiple shafts	259	42
6	Coal mine	October 22, 1913	Dawson, New Mexico, USA	Fire induces explosion that buries workers	263	5
7	Coal mine	October 14, 1913	Cardiff, Wales	Mine shaft completely collapses	439	12
8	Coal mine	June 19, 1914	Hillcrest, Alberta, Canada	Fire and explosion collapse multiple shafts	189	7
9	Coal mine	June 5, 1919	Wilkes-Barre, Pennsylvania, USA	Underground explosion collapses facility	92	3
18	LNG Plant	October 20, 1944	Cleveland, Ohio, USA	Explosion at LNG Facility destroys 1 square mile of Cleveland	130	890
33	LNG facility	May 3, 1965	Canvey Island, UK	Explosion during LNG transfer operation	1	2
51	LNG facility	March 18, 1971	La Spezia LNG Import Terminal, Italy	The LNG vessel Esso Brega leaks 2000 tons of fuel	0	1
64	LNG facility	February 10, 1973	Staten Island, New York, USA	LNG pipeline leaks at industrial facility	40	15

	Facility	Date	Location	Description	Fatalities	Cost (in millions $)
				causing fire and explosion		
95	LNG facility	April 18, 1977	Arzew, Algeria	LNG releases from storage facility, causing fire and explosion	1	1
113	LNG facility	October 6, 1979	Covepoint, Maryland, USA	Fire and explosion at Covepoint LNG facility	1	9
183	LNG facility	May 10, 1988	Boston, Massachusetts, USA	LNG facility spills 30,000 gal	0	12
212	LNG facility	December 20, 1993	Bontang, Indonesia	LNG facility leaks fuel into underground sewer system	0	15
263	LNG facility	January 19, 2004	Skikda, Algeria	Explosion and fire occur at Skikda LNG facility	27	54

Source: Benjamin K. Sovacool, "The costs of failure: A preliminary assessment of major energy accidents, 1907–2007". Energy Policy 36 (2008) 1802–1820

As far as the safety of the LNG shipping, only few incidents on the record represent a remarkable achievement for the industry compared to other shipping sectors worldwide. Figure 3 below provides an insight into the LNG incidents by LNG carrier type.

Figure 3. LNG incidents before and after 1985 by ship type

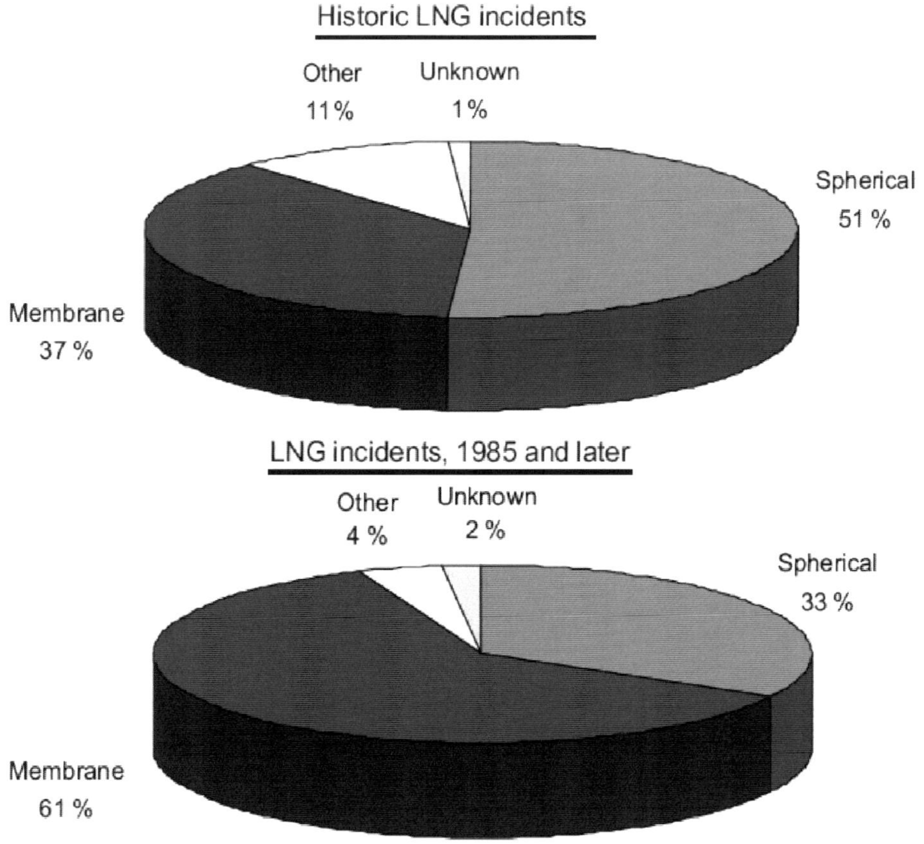

Source: Erik Vanem, Pedro Antao, Ivan Østvik, Francisco Del Castillo de Comas, "Analysing the risk of LNG carrier operations". Reliability Engineering and System Safety 93 (2008) 1328–1344

The recent study, commissioned by the European Commission, reassessed several major LNG associated hazards with a focus on safety and human life (risks to LNG crew as well as risks to crew and passengers onboard other vessels).[27] Collision risks were estimated to be twice as high as grounding, contact and leakage of LNG during loading/unloading operations, and three times as high as fire and explosion risks.

Conclusions

LNG has been handled safely for many years and the industry has maintained an enviable safety record. Engineering and design and increasing security measures are constantly improved to ensure the safety and security of LNG facilities and ships.

As of February 2009, the global LNG industry comprises 27 export (liquefaction) facilities, 64 receiving (regasification) terminals, and 298 ships, altogether handling more than 175 million metric tons of LNG every year. LNG has been safely delivered

[27] Erik Vanem, Pedro Antao, Ivan Østvik, Francisco Del Castillo de Comas, "Analysing the risk of LNG carrier operations". Reliability Engineering and System Safety 93 (2008) 1328–1344

via ocean-going transport for more than 50 years. During that time there have been more than 46,000 LNG ship voyages, covering more than 100 million miles, without any major incidents involving a major release of LNG either in port or on the high seas. LNG ships frequently transit high traffic density areas. For example, in 2000, one LNG cargo entered Tokyo Bay every 20 hours, on average, and one LNG cargo a week entered Boston harbor.[28]

The changing dynamics of LNG market brings to life new designs and technologies that are yet to obtain a proven safety record. Although classification societies and other industry organizations appear to be on track with these changes, attention of regulatory agencies could be required in some cases. Public perception on safety and security in many countries plays a highly important role in defining fate of LNG projects, and obtaining a positive one takes a lot of time and effort.

Numerous reviews and the extensive body of information and evidence that documents LNG industry safety records and practices support our conclusion that risks and hazards associated with LNG and LNG industrial facilities are manageable. They also show that LNG industry safety practices contribute toward reduced potential for catastrophic events such as might be associated with acts of terrorism. Overall, LNG safety is inherent in the properties of LNG, the technologies and operating practices that have evolved on the basis of understanding these properties, and regulatory requirements. Despite of many years of research, modeling and experimentation, many gaps exist in understanding risks associated with safety and security of LNG operations.

Publications of the *Center for Energy Economics* and the complete *Guide to LNG in North America* provide extensive information to those interested in U.S. energy trends and security; LNG industry and market developments. The CEE web site, www.beg.utexas.edu/energyecon/lng provides links to industry, government and public information sources. Companies with LNG operations maintain active public information offices, as do the federal agencies charged with regulatory and policy oversight.

References

1. New York Energy Planning Board, "Report on Issues Regarding the Existing New York Liquefied Natural Gas Moratorium", November 1998
2. Welker J. R. and Sliepcevich C.M., "Radiation, Heat Flux, and Overpressure in LNG Tanks", Proceedings of the International Conference on LNG Importation and Terminal Safety, Boston (1972).
3. Hashemi H.T., West H. H. and Sliepcevich C.M., "LNG/Water Explosions: A Distributed Source", Proceedings of the 27th Annual Petroleum Mechanical Engineering Conference (1972).

[28] Phil Bainbridge, VP BP Global LNG, *LNG in North America and the Global Context,* IELE/AIPN Meeting University of Houston, October 2002.

4. Lewis, James P.; McClain, Sheila A. Project Technical Liaison Associates, Inc. (PTL): "LNG Security: Reality and Practical Approaches", LNG: Economics & Technology Conference, January 2003
5. Sember, W.J. "ABS: Development of Guidelines for Classification of Offshore LNG Terminals", GASTECH 2002, Qatar, Oct. 2002.
6. Cates, Rusty, International Gas Consulting, Inc., "LNG - Hedging Your Bets," LNG: Economics & Technology Conference, January, 2003.
7. Benjamin K. Sovacool, "The costs of failure: A preliminary assessment of major energy accidents, 1907–2007". Energy Policy 36 (2008) 1802–1820
8. West, H.H. and Mannan, M.S. Texas A&M University, "LNG Safety Practice & Regulation: From 1944 East Ohio Tragedy to Today's Safety Record", AIChE meeting, April 2001.
9. Erik Vanem, Pedro Antao, Ivan Østvik, Francisco Del Castillo de Comas, "Analysing the risk of LNG carrier operations". Reliability Engineering and System Safety 93 (2008) 1328–1344
10. L. Delorme, A. Souto Iglesias and S. Abril Perez, "Sloshing Loads Simulation In LNG Tankers With SPH". International Conference On Computational Methods In Marine Engineering, MARINE 2005
11. Mateusz Graczyk, Torgeir Moan, "A probabilistic assessment of design sloshing pressure time histories in LNG tanks". Ocean Engineering 35 (2008) 834–855
12. Daejun Chang, Taejin Rhee, Kiil Nam, Kwangpil Chang, Donghun Lee, Samheon Jeong, "A study on availability and safety of new propulsion systems for LNG carriers". Reliability Engineering and System Safety 93 (2008) 1877– 1885

Biography of Speaker

Dmitry Volkov joined the CEE as a research assistant in October 2003. Later, upon completion of his MBA degree at the University of Houston, he joined the CEE as a full time research team member. At the CEE, Mr. Volkov is involved in various research projects, including LNG and Carbon Sequestration Value Chains, as well as studies of FSU energy sector development and International Energy Data Systems. He is an active member of the Association of International Petroleum Negotiators (AIPN), where he is involved with the LNG Sales Agreement committee and the International Association for Energy Economics. Mr. Volkov holds degrees from University of Houston (MBA in Accounting and Finance) and Moscow State University, Institute of Asian and African studies (B.S. Arabic Studies and History (Minor in International Relations). He speaks Arabic, English, French and Russian languages.
dmitry.volkov@beg.utexas.edu

Water Curtain Application for Forced Dispersion of LNG Vapor

102b

Morshed Rana and M. Sam Mannan
Mary Kay O'Connor Process Safety Center
Chemical Engineering Department
Texas A & M University, College Station, TX.

AIChE Spring Meeting, April 2009
9th Topical Conference on Natural Gas Utilization
Tampa, Florida, April 26-30, 2009

Abstract

Water curtain is currently a recognized technique to control and mitigate various hazards in the process industries. It has been used to absorb, dilute and disperse toxic or flammable vapor clouds, as well as to reduce heat radiation in case of vapor cloud fires. Water curtain has also been suggested as one of the economic and promising LNG vapor cloud control techniques. It is expected to create an effective barrier to reduce the "vapor cloud exclusion zone". Observations of the interaction of LNG vapor clouds with water spray curtains clearly demonstrate some enhanced dispersion. Detailed analysis of the complex mechanisms and the effect of important water spray features have not been studied experimentally in previous research. This paper presents a research methodology to study the physical phenomena involved and the effect of water curtain parameters during LNG vapor and water curtain interaction. The data from medium scale out-door experiments at the Brayton Fire School, Texas, are analyzed to understand the relative importance of induced mechanical mixing effects, dilution with air, and heat transfer between water droplets and the LNG. Analysis is presented in the paper to determine the effect of common types of water curtains in dispersing LNG vapor cloud effectively during an accidental release of LNG.

Introduction

Water spray curtains are usually simple, reliable, and inexpensive and can be either fixed or mobile. They are considered as economic and promising techniques to reduce flammable LNG vapor concentration. General physical processes involved in using a water curtain to control any vapor cloud include (i) mechanical effects of creating a barrier to the passage of a gas cloud as well as imparting momentum to it (ii) dilution of the gas cloud by air entrained by the water spray; (iii) thermal effects between the gas cloud, water droplets, and entrained air; and (iv) absorption of gas in water droplets with or without chemical reaction [1-2]. Actions of a water spray in mitigating a vapor cloud may consist of all or any combination of these mechanisms [3-6].

The ability of water curtains to show different effects initiated an interest in the utilization of water curtain as a potential LNG vapor cloud mitigation measure in the late 70's [7]. However, their effectiveness for the control of a specific hazard requires careful evaluation of credible release scenario, chemical characteristics, action mechanism and spray hydrodynamics (or characteristics). Several other variables can affect the efficiency of a water-spray curtain for the case of LNG release. Because of low molecular weight (methane) and extremely low temperature, LNG vapor cloud dispersion behaves differently from other dense gases [8]. Based on the physical phenomena mentioned earlier, water curtains are expected to enhance LNG vapor cloud dispersion mainly through three mechanisms: mechanical effects, dilution effects and thermal effects. As the solubility of natural gas in water is relatively low, the absorption effects can be ignored [9].

Though water curtain represents one promising technique to mitigate the hazard of accidental LNG release, the actual phenomena involved in LNG vapor and water drop interaction was not clearly established from previous research. The effectiveness, in terms of the degree of LNG vapor concentration reduction and temperature increase, of different types of water curtains of different types of nozzles (different in flow configurations and drop sizes), curtain heights and widths are yet remained undetermined. The optimum location of a water curtain from a potential release spot to show highest effectiveness also needs to be determined. That is why thus far a comprehensive design guideline of effective water curtain system for LNG vapor control could not be developed [7]. Indeed, it is also true that there is no simple set of optimum parameters, rather there will be a specific set of design parameters for an effective water curtain system for each particular type of scenario [10].

At present the Mary Kay O'Connor Process Safety Center (MKOPSC) is conducting research to understand the actual phenomenon of water curtain and LNG interaction from field tests to make the results directly applicable to the current LNG industry needs. This paper presents the out-door LNG spill experiments, conducted to determine the effectiveness of water curtains in dispersing LNG vapor clouds. LNG spill tests were conducted on the flat concrete ground and two types of water curtains were tested. This paper outlines a brief description of the tests and mainly focuses on test results. Results of the outdoor LNG spill tests have been discussed and analyzed in this paper to determine the effect of two types of water spray curtain in dispersing a small scale LNG vapor cloud through studying the phenomena of mechanical effects, dilution and temperature change.

Action mechanism and effect of spray characteristics

Water spray curtains are used for the dispersion of hazardous vapor clouds mainly using the feature of air entrainment by the spray. Entrained air comes into the cloud and reduces the gas cloud concentration downstream region of the water spray [11]. Thus a water spray can dilute a gas cloud by mixing it with the entrained air. Air entrainment takes place in the spray region as a result of momentum transfer from the water droplets to the ambient air. When the spray is activated, mixing and dilution result from the momentum transfer between the air, gas and water droplets. According to the design requirements, it is necessary to deliver air not only in sufficient quantity, but also at a sufficient velocity to ensure dilution [12]. Again, Air entrainment is a strong function of spray configuration, droplet size, droplet velocity, and spray arrangements. Large droplets entrain less air than smaller droplets but induce better mixing and dispersion. Although smaller droplets entrain more air, they have smaller terminal velocities which are almost constant and cause poor mixing and dilution due to less turbulence [13].

Mechanical effect of momentum and heat transfer effect are also considered when applying water curtain. It can impart momentum to the vapor cloud when interact because water is release from the nozzle at high pressure. This imparted momentum can disperse the vapor cloud upwards, downwards or sideways depending on the nozzle orientation. Heat transfer between water droplets, air and cold vapor can play an important role in dissipating cold clouds by enhancing their buoyancy.

The sprays produced by a nozzle can be categorized by their shapes as conical, fans, or fogs. The spray pattern that will be used is highly dependent on the planned application. Conical sprays can be further broken down into either full-cone or hollow-cone sprays. In the hollow cone pattern all of the spray is located at the surface of the cone produced by the nozzle and none of it is inside the cone. In the full cone pattern the liquid being sprayed fills the cone pattern produced by the nozzle. Conical sprays produce medium size droplets. Nozzles that spray water as a sheet of liquid produce a pattern known as a fan. The droplet sizes produced by fan nozzles are usually coarse. The fog nozzles atomize the liquid stream into a mass of very fine droplets [13].

Water spray curtains are mainly curtains of water droplets. Information of the size of the droplets produced by a spray nozzle is essential as it affects the action mechanisms. The droplet size of a certain type of curtain provides a certain surface area. When a water curtain control a vapor cloud by diluting it, then the volume of the droplets (i.e., its momentum) is important. A spray nozzle actually produces a range of droplet sizes. The simplest approach is to define a representative droplet diameter in terms of mean or median. The Sauter Mean Diameter (SMD) defines a droplet with mean surface area and volume for the whole spray. It is calculated from the sum of the droplet volumes in a given spray divided by the sum of their surface areas. Another commonly used representative diameter is the Volume Median Diameter. Half of the droplets in a spray have diameters greater than this value and half of the droplets smaller. Figure 1 shows effects of water droplet size in action mechanisms [13-14].

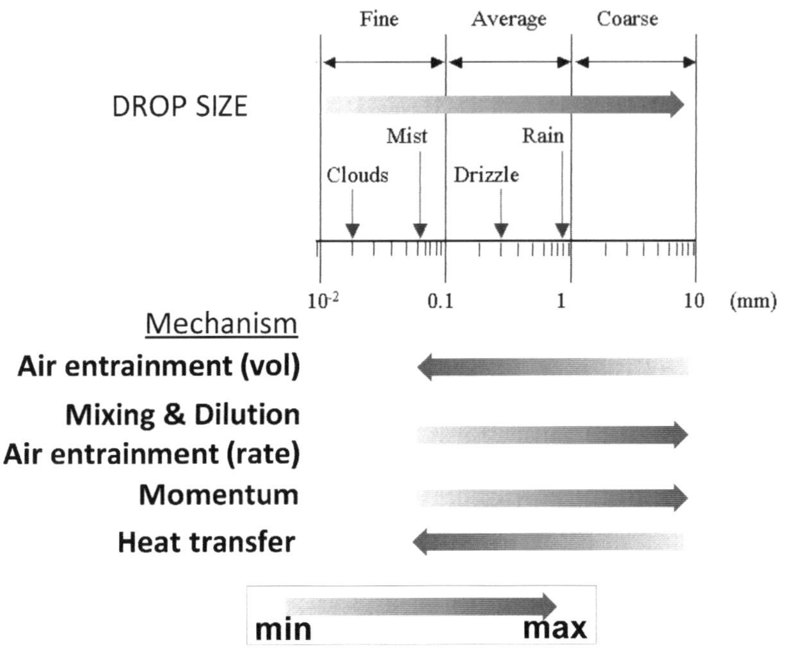

Figure 1 Effect of water droplet size on water curtain action mechanism

The spray flow direction is also an important factor to consider when dealing with a vapor cloud of specific nature. The primary purpose of a water curtain to control a flammable gas release is the attainment of a suitable dilution of the released gas, so that the cloud reach lower flammability limit rapidly before dispersing too far. When as dealing with a toxic cloud water curtains are used to ensure a partial abatement of the released compound by chemico-physical absorption, in addition to the dilution phenomenon. Usually downward water curtains are used for heavier than air flammable and toxic gases to contain the gas cloud in an area. Downward curtains create a barrier to the gas cloud by the air vortex created by the spray and force the gas downwards [15]. In the LNG industry, upward-pointing water curtains should be used to facilitate LNG dispersion. Though LNG is a heavier than air vapor cloud, it becomes lighter when gets diluted and warmer. For this water curtains with an upward flow should show better effectiveness to control LNG than those with a downward flow. The momentum from an upward spray can enhance air entrainment through the lower surface of the LNG cloud and influences the mixing between the LNG cloud and entrained air. Thus a better dilution could be achieved. The upward momentum imparted to the LNG cloud also pushes it from the bottom and facilitates the cloud's dispersion upward [5]. However care should be given to prevent water entering the LNG pool, as it will increase vaporization rate.

Experiment

Outdoor small-scale experiments are designed to spill LNG onto confined land and use different types of water spray curtains individually or combined to control the LNG vapor cloud. Experiments are designed to be carried out at the LNG test facility at the Brayton

Fire School (BFS) at Texas A&M University. LNG is released continuously from a pipe to simulate a continuous LNG spill into a dike. LNG cloud dispersion behavior is observed and measured. After achieving a steady state plume, vertically upward pointing water spray curtain, placed perpendicularly to the prevailing wind direction and downwind of the spill source, is activated to disperse the cloud. Continuous observation and measurements at different points before and after the water curtain area are interpreted to study the droplet-vapor interaction phenomena and to evaluate the effectiveness of the water curtain in controlling and dispersing the LNG vapor cloud.

The purpose of the experiments is to study both natural dispersion of LNG and forced dispersion by water curtain by observation and measurement of LNG vapor cloud concentration and temperature change. To identify the physical mechanism and to evaluate the effectiveness of the water spray curtain in term of phenomena; water pressure, water flow, LNG flow, air velocity and meteorological condition are also measure. Overall the experiment consists of the LNG source, LNG spill, water spray nozzle and water curtain, placements of the measurement points and meteorological measurements. The vapor concentration and temperature are measured with multiple infrared (IR) hydrocarbon point gas detectors and K-type thermocouples. The gas detectors and the thermocouples are installed on tripod poles at different elevations. Poles are placed downwind distances from the LNG spill area to measure concentration and temperature at different downwind distances and elevations. The liquid (LNG) temperature is measured with N-type thermocouples. For the observation of vapor cloud and water movement multiple video cameras like regular digital, hydrocarbon and IR flame detector video cameras are used. Measurements of meteorological data such as wind speed, wind direction, temperature, and humidity are performed with weather station and anemometers. The following figure shows the sensors, detectors and water curtain placements for the experiments.

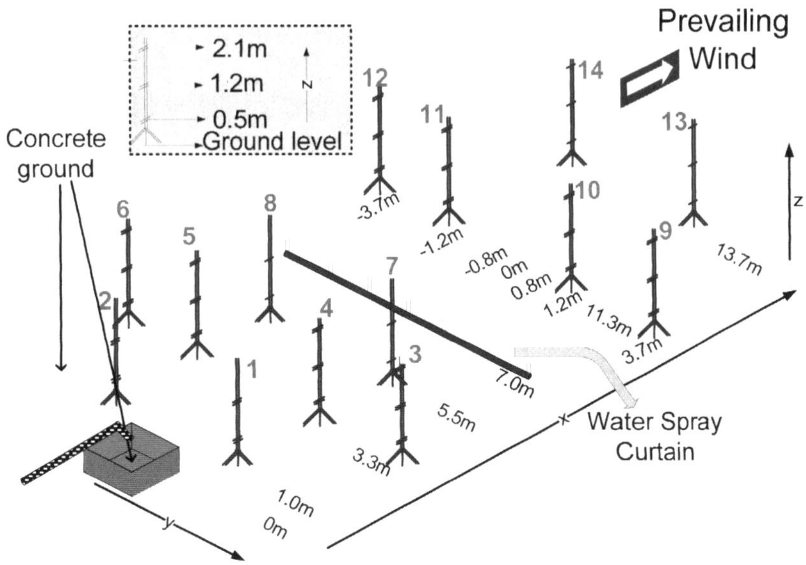

Figure 2 Experimental set up [16]

Two types of portable upward directed spray curtains were employed in both tests. One was constructed with several full cone, spiral, 60° nozzles installed on a carbon pipe. The other one was portable water curtain device which produces 180° flat fan shaped spray pattern. Figure 3 shows pictures of water curtain used.

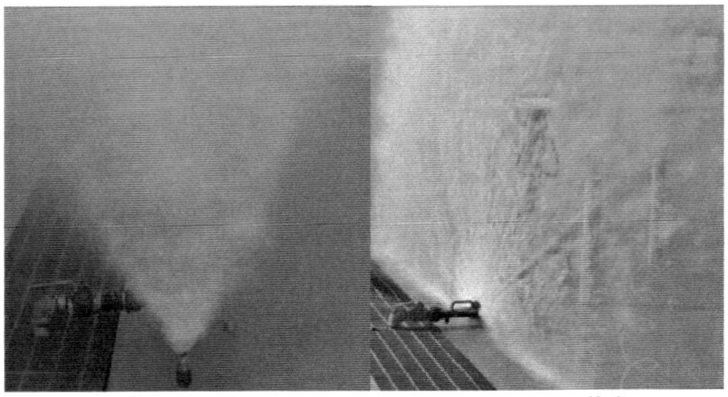

(a) (b)

Figure 3 Photographs of water sprays used: (a) full cone spray, (b) flat fan spray [17]

Results and Discussion

Spill experiments were conducted in 2007. Water curtains were directed vertically upward and positioned perpendicularly to the prevailing wind direction, at a fixed downwind distance from the spill location. During the tests, the wind speed averaged 2 meters per second (4.5 miles per hour) and blew mainly from the south-west to south-east region. The average ambient temperature during the tests was 295.4K and humidity was 25.5% [16].

Analysis of the experimental data is performed to determine the concentration profile in downwind direction at a certain height. Ground level (0.5m) concentration profile is preferred for the analysis to study the forced dispersion of LNG vapor cloud in reducing CH_4 concentration. Downwind concentration at higher elevation (2.1m) in this analysis gives an idea on the momentum effect of the water curtains to disperse LNG vapor upward. Previous presented data in the paper show concentrations at certain height and downwind locations at different times. This analysis selects point values of concentrations at different downwind distances by averaging concentration data of 1 minute before and after the water curtain, and after discontinuation of LNG flow from the pipe. Figure 4 through Figure 7 show the downwind concentration profiles at two different elevations for two tests.

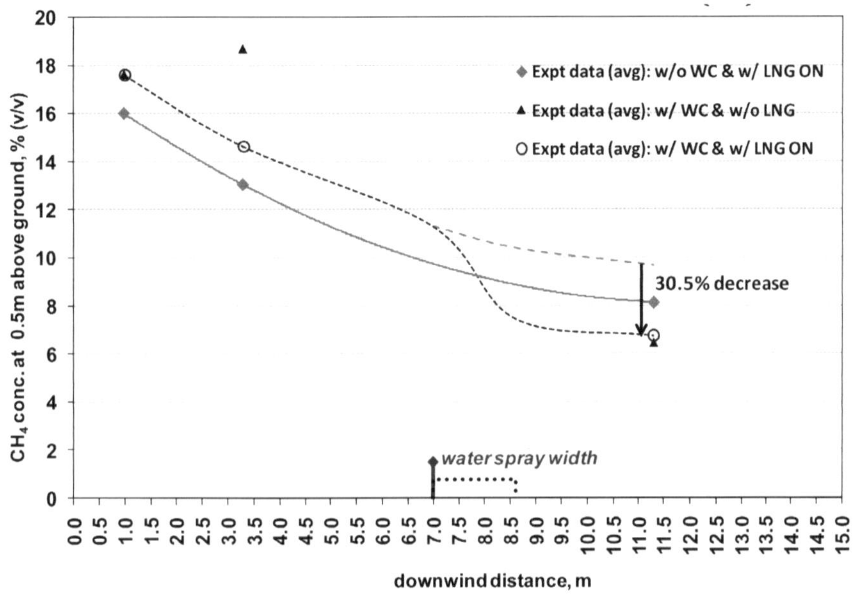

Figure 4 Downwind concentration at 0.5m elevation: full cone curtain

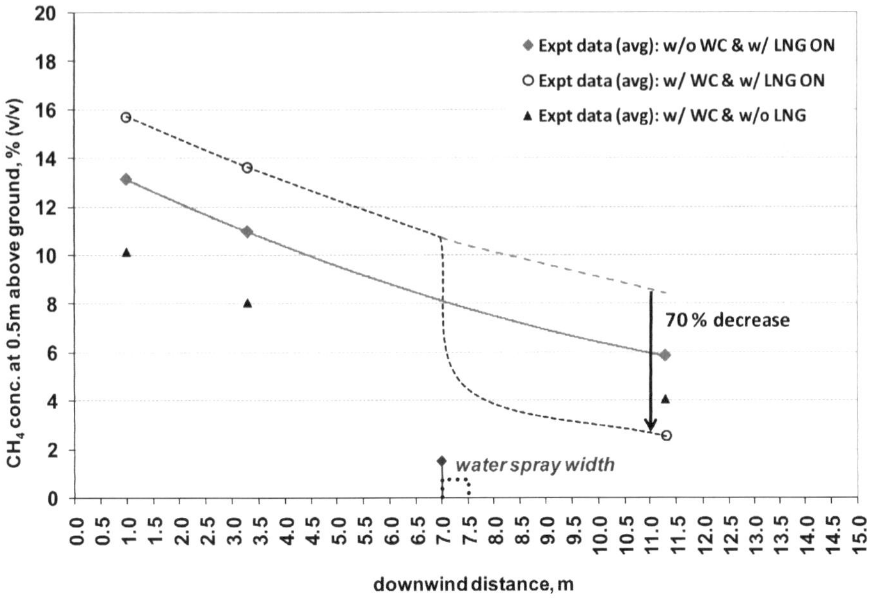

Figure 5 Downwind concentration at 0.5m elevation: flat fan curtain

Ground level concentration data before the water curtain application (natural dispersion) clearly show similar trends for downwind concentration profile for both the tests. Two degree polynomial trend lines (R^2=1) represent the average data for the natural dispersion case. To study the data after the water curtain application (both natural and forced dispersion) these trends are utilized to determine the concentration that would be expected after the curtain region (at 11.3m downwind) if the water curtains were not active. The expected concentration profiles are shown in the figures with light blue dotted lines. The dark blue dotted lines are divided into natural and forced dispersion trends. The average downwind lengths, defined as water curtain width, for the full cone and flat fan water curtains were observed during the experiments as 1.6 m and 0.5m respectively. Forced dispersions should only be observed when the vapor passes through the water curtain width. As it is already mentioned earlier that LNG vapor concentration eventually decreases as it travels downwind distances, the water curtains are expected to enhance this reduction of concentration by forced dispersion. From the expected and actual data at this position the additional decrease of LNG concentration by the force of water curtain is determined. For the full cone curtain the additional decrease in concentration is 30.5% where as for the flat fan curtain it is 70%. In Figure 4results for the water curtain application after discontinuing liquid flow shows similar results like water curtain application with continuous liquid flow. However, in Figure 5 results for similar situation failed to show any decrease than the natural dispersion. The causes of the effectiveness of the curtains in decreasing CH_4 concentration can be determined from the following analysis of higher elevation concentration. Again, careful observation of the above figures can also state that by placing the curtains closer to the spill location as well as increasing the spray width could give much higher reduction in concentrations.

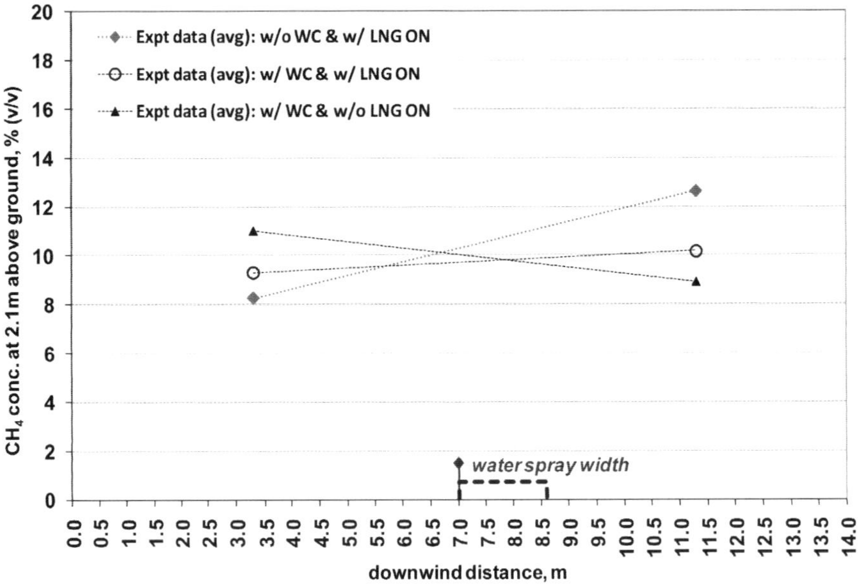

Figure 6 Downwind concentration at 2.1m elevation: full cone curtain

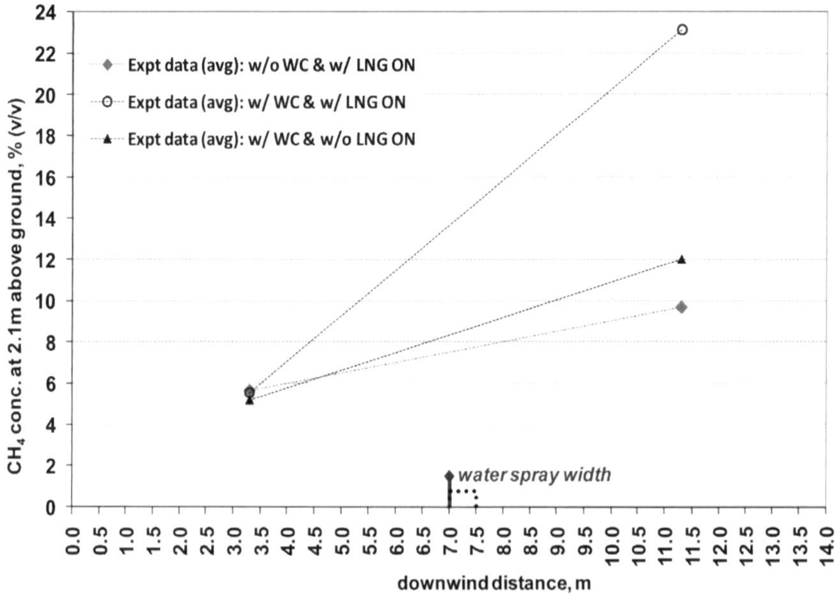

Figure 7 Downwind concentration at 2.1m elevation: flat fan curtain

Both the figures show that concentration of CH_4 increases at higher elevation as it travels forward naturally and that is obvious. For the flat fan curtain action the concentration increased to much more higher value when both the water and LNG flow was continued than natural dispersion situation (Figure 7). Concentration increase during the discontinued LNG flow is lower even though it is higher than natural dispersion situation. However during the full cone curtain action (Figure 6) the concentration is actually reduced for both LNG on and off situations. These phenomena indicates than full cone actually reduced CH_4 methane concentration by mixing it with air and the flat fan imparted so much momentum to the cloud that the ground level concentration becomes very low and higher level concentration increased to a very high value. It is clear that both of the curtains reduced the ground level concentration. The full cone curtain (a) mixes vapor efficiently; (b) provides inefficient momentum to the cloud to lift it upward; and the flat fan curtain (a) imparts very high momentum; (b) provides very inefficient mixing with air.

Conclusion

The preliminary results indicate that the water sprays reduced the LNG concentration as well as pushed the vapor upward because of the mechanisms of (1) dilution with air; (2) momentum; and (3) heat provided to the gas cloud by the water curtain. However, the tests could not show a significant change in LNG vapor concentration by the water curtains. Data and results show that the upward directed water curtains are able to push the LNG vapor upward due to momentum effect and the strength depends on water pressure, droplet size and flow patterns. Flat fan spray created a high, solid like barrier in the path of the cloud and pushed vapor upward. So the main mechanism of the fan type in reducing the methane concentration is momentum effect, i.e. by pushing the vapor upward. On the other hand the full cone spray could not provide enough height and strong barrier for the LNG cloud due to less water pressure, and smaller droplet. So some vapor managed to escape over the curtain and through the gaps between the nozzles.

The full cone fan spray could entrain more air than the flat fan curtain due to smaller drop size (larger surface area), spiral flow pattern and wider spray area. It was visualized that the full cone curtain created more turbulence close to the spray region than the flat fan spray. So the cone type reduced the methane concentration mainly by dilution, i.e. mixing methane with entrained air. The flat fan spray could not show significant turbulence, i.e. air entrainment, because of thinner spray width and larger droplet sizes (smaller surface area).

Average temperature data indicates that water curtain provide heat to the mixed gas cloud and the full cone curtain provided more heat to the cloud than the flat fan. However it was observed that methane concentration became somewhat little higher further downwind than right after the spray region, for all the tests. This indicated that LNG vapor was not warm enough to become positively (/neutrally) buoyant. So instead of dispersing upward by the curtain action, LNG vapor came back to the ground. Ineffective heat transfer between water, LNG and air could not dilute the LNG vapor significantly. It could be implied from the results that the heat from both curtains was not adequate. This might be because of the water drop sizes and droplet terminal velocities. Both the droplet size and droplet velocity were not adequate enough to create large surface area and higher contact time to heat the vapor adequately.

Though the curtains reduced the methane concentration, it never reached below ½ LFL for any of the test. This might be because of the curtains position was far away from the spill and could not provide enough coverage, portion of the LNG vapor cloud passed the curtain and disperse downwind without any interaction with the spray. Shorter distance between the curtains and spill location could provide better coverage and LNG-water spray interaction to reduce the methane concentration to significantly lower values.
Water sprays of different flow patterns and droplet sizes control the LNG cloud in terms of different physical processes. It may be possible to develop a curtain system with higher effectiveness by combining both the full cone and the flat fan. More definitive results could be obtained with more rigorous theoretical analysis and modeling by considering the droplet sizes. However, to establish a definitive guideline on water curtain application and design for LNG cloud dispersion, some more experiments of these curtains positioned at different locations, for longer time and at different wind conditions and theoretical studies are also needed.

References

1. G. Dusserre, A. Dandrieux and O. Thomas, "The DVS Model: a New Concept for Heavy Gas Dispersion by Water Curtain," Environmental Modeling and Software, 18 (2003), 253-259.

2. K. Hald, J. M. Buchlin, A. Dandrieux and G. Dusserre, "Heavy Gas Dispersion by Water Spray Curtains: A Research Methodology, Journal of Loss Prevention in the Process Industries, 18 (2005), 506-511.
3. P. A. C. Moore and W. D. Rees, "Forced Dispersion of Gases by Water and Steam," North Western Branch Papers, IChemE, (1981), 4.1-4.14.
4. J. McQuaid and R. D. Fitzpatrick, "The Uses and Limitations of Water-Spray Barriers," North Western Branch Papers, IChemE, (1981), 5, 1.1-1.13.
5. J. McQuaid and R. D. Fitzpatrick, "Air Entrainment by Water Sprays: Strategies for Application to the Dispersion of Gas Plumes," Journal of Occupational Accidents, 5 (1983), 121-133.
6. K. Moodie, "The Use of Water Spray Barriers to Disperse Spill in the Event of Heavy Gases – The Performance Characteristics of Full-Scale Water Spray Barriers when Dispersing Spills of Heavy Gases," Plant Operation Progress, 4, (1985), 234-241
7. S. E. Atallah, Guzman and J. N. Shah, "Water Spray Barriers for LNG Vapor Mitigation," Risk and Industrial Safety Consultants, Inc. Report No 4015.1, (November 1988), Prepared for Gas Research Institute.
8. M. J. Ivings, S. F. Jagger, C. J. Lea and D. M. Webber, "Evaluating Vapor Dispersion Models for Safety Analysis of LNG Facilities," Technical Report for the Fire Protection Research Foundation, Health and Safety Laboratory, (April 2007).
9. A. C. Rojey, Jaffret and N. Marshall, "Natural Gas: Production, Processing, Transport," Editions TECHNIP, (1997), 205-207.
10. N. C. Harris, "The Design of Effective Water Sprays – What We Need To Know," North Western Branch Papers, IChemE, (1981), 7.1-7.5.
11. M. vaan Doorn, "The Control and Dispersion of Hazardous Gas Clouds with Water Sprays," PhD Thesis, PhD Thesis, University of Delft, Holland (September 1981).
12. V. M. Fthenakis, "Modeling of Water Spraying of Toxic Gas Releases," Ph.D. Thesis, Fluid Dynamics and Atmospheric Science Department, New York University: NY, (March 1991).
13. Guidelines for Post release Mitigation Technology in the Chemical Process Industry, CCPS, AIChE, (1997), 58-90.
14. S. E. Gant, "CFD Modelling of Water Spray Barriers," Health and Safety Laboratory, (2006), Report No. HSL/2006/7.
15. E. Palazzi, F. Curro, R. Pastorino and B. Fabiano, "Liquid Spray Curtains Design to Contain and Mitigate Toxic and Flammable Jets and Releases," Proceedings of 11th International Loss Prevention Symposium, Loss Prevention and Safety Promotion in the Process Industries, (2004), 3127-3136.
16. M. A. Rana, B. Cormier, J. Suardin, Y. Zhang and M.S. Mannan, "Experimental Study of Effective Water Spray Curtain Application in Dispersing Liquefied Natural Gas Vapor Clouds," Process Safety Progress, 27 (4), (2008).
17. M. A. Rana, B. Cormier, J. Suardin, Y. Zhang and M.S. Mannan, "LNG Vapor Cloud Dispersion with Water Spray Curtain," Proceedings of International Symposium and World Conference on Safety of Oil and Gas Industry, (October 28-29, 2008), Mary Kay O'Connor Process Safety Center, College Station, TX.

Biography of the Speaker

Morshed A. Rana
Research Assistant & PhD Candidate
Mary Kay O'Connor Process Safety Center
Department of Chemical Engineering, Texas A&M University, 3122 TAMU,
College Station, TX 77843-3122
Phone: (979) 862-1310, Fax: (979) 458-1493
E-mail: morshed.rana@chemail.tamu.edu

Rana, Morshed A. is a research assistant of the Mary Kay O'Connor Process Safety Center and PhD candidate of the Chemical Engineering Department of Texas A&M University advised by Dr. Sam Mannan. He attended Bangladesh University of Engineering and Technology and received his BS in chemical engineering. He obtained MS degree in chemical engineering from the University of South Alabama, Mobile, AL in 2004.

Mr. Rana joined Mary Kay O'Connor Process Safety Center in August 2005 and began conducting research under the guidance of Dr. Sam Mannan. His research topic is to study LNG vapor dispersion behavior using water spray curtain. Currently he is leading a team of three PhD students to conduct outdoor spill tests with LNG. He is also involved in other research in process safety area. He participated several industrial projects which include PHA and HAZOP study, P&ID and PFDs developments, QRA, FHA, consequence modeling, CFD as well as process simulation.

LNG Vapor Dispersion Modeling with CFX

102d

Ruifeng Qi, Graduate Research Assistant
Dedy Ng, PhD, Assistant Research Scientist
William J. Rogers, PhD, Laboratory Director
M. Sam Mannan, PhD, PE, CSP, Director
Mary Kay O'Connor Process Safety Center

Artie McFerrin Department of Chemical Engineering Department
Texas A&M University System
College Station, Texas 77843-3122, USA

AIChE Spring Meeting, April 2009
9th Topical Conference on Natural Gas Utilization
Tampa, Florida, April 26-30, 2009

Abstract

Regulation 49 CFR 193 and Standard NFPA 49A require using validated consequence models to determine vapor cloud dispersion exclusion zones for accidental LNG releases. One approach that is being developed in industry for exclusion zone determination and LNG vapor dispersion modeling is computational fluid dynamics (CFD) codes. The main advantages of CFD codes are that they have the capability to define the source term, and allow for the representation of complex geometry and its effects on vapor dispersion.

This paper presents the use of CFD codes (ANSYS CFX) in modeling LNG vapor dispersion in the atmosphere. Some key parameters which are essential inputs to CFD simulation setup were discussed. Sensitivity analysis was conducted to illustrate the impact on the simulation results of mesh length scale and turbulence intensity in the source term. Moreover, experimental data from a medium-scale LNG spill test performed in Brayton Fire Training Field were used for validating CFX prediction results. By comparing test data with simulation results, it was shown that CFX model was able to well describe the physical behavior of LNG vapor dispersion and showed a good correlation with test data in predicting the downwind distance to half the lower flammable limit (LFL).

1. Introduction

As LNG importing terminal and facility construction increases due to the growth of demand for importing LNG overseas, concerns about the potential hazards that LNG spill could pose have been raised. One of the major hazards is formation of flammable vapor cloud. The vapor cloud drifts downwind near the ground for a certain time until it completely warms up and dissipates in the atmosphere. To ensure the safety of the public in adjacent populated area, validated consequence models are recommended to determine potential hazardous area in case of an accidental LNG spill.

Most integral models that are currently used cannot take into account terrain and congestion density of obstacles in potential LNG spill scenarios, whose effects on vapor cloud and turbulence cannot be neglected in vapor dispersion modeling. The need for advanced models based on Computational Fluid Dynamics (CFD) has been proposed and accepted.

BP Global SPU Gas and Mary Kay O'Connor Process Safety Center (MKOPSC) have jointly established a R&D program on LNG spill emergency response and hazard control. A series of medium scale field tests have been performed at Brayton Fire Training Field (BFTF), which is affiliated with Texas A&M University System. As part of this program, experimental data collected from these tests are utilized to study the physical behavior of vapor dispersion following an LNG spill on water or on land and to validate consequence model prediction results.

The objective of this paper is to present a CFD (CFX) simulation for a BFTF test performed by MKOPSC. Essential inputs associated with domain and boundary conditions to set up a simulation are discussed. Sensitivity analysis is conducted to illustrate the impact on the simulation results of mesh length scale and turbulence intensity in the source term. Simulation results are validated by experimental data from a BFTF test. Through comparing test data with simulation results, it has been shown that CFX is able to describe the physical behavior of LNG vapor dispersion in the atmosphere and its prediction results of distance to half Lower flammable limit (LFL) shows a good fit into test data.

2. Simulation Setup with CFX

The following table lists all the essential input variables for LNG vapor dispersion simulation using CFX.

Table 1. Summary of input variables for LNG vapor dispersion simulation

Components	Inputs (Parameters)
Geometry creation	3-D geometry (terrain and obstacles)
Simulation type	Transient / steady state
Domain	Fluids: air and CH_4
	Turbulence model: k-epsilon
	Heat transfer option: total energy
	Buoyant flow
Opening boundary condition (Atmosphere)	Fluid: air
	Wind direction and velocity profile
	Temperature profile
	Turbulence profile
Inlet boundary condition (LNG pool)	Fluid:CH_4
	Vapor evaporation velocity
	Vapor temperature
	Turbulence

Wall boundary condition (Ground and geometry)	Influence on flow: no slip
	Surface roughness height
	Surface temperature

2.1 Domain and mesh

Domains are the enclosure space within which the transport equations are solved. CFX is a general purpose CFD software, therefore a list of physical models needs to be identified to characterize the physical behavior of LNG vapor dispersion. When LNG is spilled on land or water, it forms a pool which generates a fog-like LNG vapor cloud. The dispersion of vapor cloud in the atmosphere goes through three stages—negative buoyancy, neutral buoyancy and positive buoyancy, depending on the density of vapor cloud. Thus, buoyancy model is needed to capture the density change caused by the variation of vapor temperature. Buoyancy effects can be simulated with full buoyancy model in which buoyancy reference density can be set as 1.225 g/m^3. Similarly, a turbulence model needs to be identified to predict the effects of turbulence in fluid flow. CFX contains a large variety of turbulence models. Some models are based on Reynolds Averaged Navier-Stokes (RANS) equations such as k-epsilon model, while others are mainly composed of Large Eddy Simulation (LES) and Detached Eddy Simulation (DES) turbulence models. In this study, k-epsilon model has been chosen because it is widely used in industry and proved numerically robust but with less computational cost than LES or DES. A heat transfer model is also selected to represent the heat transfer throughout fluids within the domain. This model should take into account both thermal energy and kinetic energy, which could be addressed in CFX by total energy model [1]. Once the reference pressure is set, all the pressure inputs in setting up the simulation will be specified with respect to reference value. In this case, the reference pressure is recommended to be 1 atm, which includes the effects of hydrostatic pressure.

The domain is discretized into small volume elements before simulation. This process is called mesh generation. The size of mesh elements, or mesh length scale more or less affects the simulation results. A mesh independence solution is often recommended to eliminate this influence and yield higher accuracy, while it is time and computation consuming. An approach to deal with this issue is running the simulation with two or three gradually decreasing mesh size and comparing the results to estimate the accuracy.

2.2 Boundary conditions

2.2.1 Atmospheric condition

Atmospheric surface layer is the region of interest where LNG vapor dispersion occurs. Anay Luketa-Hanlin gives a comprehensive review of the description of atmospheric characteristics [2]. For practical application in setting up the atmospheric boundary, the Monin-Obukhov similarity theory provides a suitable method for describing wind velocity, temperature and turbulence profile in the surface layer if only limited measurements on different height above the ground are available. This theory was proposed by Monin and Obukhov in 1954 based on the assumptions that the air flow is horizontally homogeneous and the mean variables are only depend on height z [3]. The wind velocity and potential temperature gradient functions along the height z are given as follows [4]:

$$\frac{\partial U}{\partial z} = \frac{U_*}{\kappa \cdot z} \cdot \phi_m\left(\frac{z}{L}\right) \quad (1)$$

$$\frac{\partial \theta}{\partial z} = \frac{\theta_*}{\kappa \cdot z} \cdot \phi_h\left(\frac{z}{L}\right) \quad (2)$$

where parameters L, U_* and θ_* are Monin-Obukhov length, friction velocity and scaling potential temperature, respectively. κ is the Von Karman constant which has a value of 0.41.

Integrating equations above from z to z_0 gives

$$U(z) = \frac{U_*}{\kappa}\left[\ln\left(\frac{z}{z_0}\right) - \psi_m\left(\frac{z}{L}\right)\right] \quad (3)$$

$$\theta(z) = \theta_0 + \frac{\theta_*}{\kappa}\left[\ln\left(\frac{z}{z_0}\right) - \psi_h\left(\frac{z}{L}\right)\right] \quad (4)$$

where

$$\psi_m\left(\frac{z}{L}\right) = \int_{z_0}^{z}\left(\frac{1}{z} + \frac{\phi_m\left(\frac{z}{L}\right)}{z}\right)dz \quad (5)$$

$$\psi_h\left(\frac{z}{L}\right) = \int_{z_0}^{z}\left(\frac{1}{z} + \frac{\phi_h\left(\frac{z}{L}\right)}{z}\right)dz \quad (6)$$

where z_0 is the surface roughness length, and θ_0 is the potential temperature at z_0.

The potential temperature is related to ambient temperature and pressure by

$$\theta = T\left(\frac{P_0}{P}\right)^{\mu} \quad (7)$$

Where

T is the actual temperature,
P is the actual pressure,
P_0 is the standard reference pressure, and
μ is a constant taken at 0.285.

Usually in the atmospheric surface layer, potential and actual temperatures in absolute unites do not differ by more than 10% [3].

The Cartesian components of wind velocity in three directions are then expressed by introducing wind direction angle α.

$$U_x(z) = U(z) * cos(\alpha) \tag{8}$$

$$U_y(z) = U(z) * sin(\alpha) \tag{9}$$

$$U_z(z) = 0 \tag{10}$$

Air flow in the atmospheric surface layer is type of external flow, which means the air flows over objects rather than through enclosure. In such case, the CFX cannot automatically calculate the turbulence characteristics, such as turbulence kinetic energy and eddy dissipation rate. This limitation can be fixed by setting values of these turbulence parameters for atmospheric boundary. One approach to represent the turbulence kinetic energy k and energy dissipation rate ϵ is relating them with Monin-Obukhov length, which is shown below [5]:

$$k(z) = 5.48U_*^2 \left[\frac{\phi_h\left(\frac{z}{L}\right)}{\phi_m\left(\frac{z}{L}\right)}\right]^{1/2} \tag{11}$$

$$\epsilon(z) = \frac{U_*^3}{\kappa \cdot z}\phi_h\left(\frac{z}{L}\right) \tag{12}$$

The functions $\phi_h\left(\frac{z}{L}\right), \phi_m\left(\frac{z}{L}\right), \psi_m\left(\frac{z}{L}\right), \psi_m\left(\frac{z}{L}\right)$ can be expressed as a function of Monin-Obukhov length:

When L>0

$$\phi_m\left(\frac{z}{L}\right) = \phi_h\left(\frac{z}{L}\right) = 1 + 5\frac{z}{L} \tag{13}$$

$$\psi_m\left(\frac{z}{L}\right) = \psi_h\left(\frac{z}{L}\right) = -5\frac{z}{L} \tag{14}$$

When L<0

$$\phi_m\left(\frac{z}{L}\right) = \frac{1}{x} \tag{15}$$

$$\phi_h\left(\frac{z}{L}\right) = \frac{1}{x^2} \tag{16}$$

$$\phi_\epsilon\left(\frac{z}{L}\right) = 1 - \frac{z}{L} \tag{17}$$

$$\psi_m\left(\frac{z}{L}\right) = 2log_e\left[\frac{1+x}{2}\right] + log_e\left[\frac{1+x^2}{2}\right] - 2tan^{-1}(x) + \frac{\pi}{2} \tag{18}$$

$$\psi_h\left(\frac{z}{L}\right) = 2log_e\left[\frac{1+x^2}{2}\right] \tag{19}$$

with

$$x = \left(1 - \frac{16z}{L}\right)^{1/4} \tag{20}$$

2.2.2 LNG pool

The LNG pool can be specified as an inlet boundary, where LNG vapor flows into the domain. The mean vapor velocity across the pool entering the domain can be determined by

$$v_g = \frac{\dot{m}_{liq}}{\rho_g \cdot D} \tag{21}$$

Where

$\dot{m}_{liq}$ is the LNG mass flow rate,
ρ_g is LNG vapor density at boiling point, and
D is LNG pool diameter.

The turbulence above the pool induced by vapor evaporation should be specified due to the requirement of using k- ε model, which could be modeled through turbulence kinetic energy and dissipation rate [2].

$$k = \frac{3}{2}\left(v_g T_i\right)^2 \tag{22}$$

$$\varepsilon = C_\mu^{3/4} \frac{k^{3/2}}{0.07D} \tag{23}$$

where D is LNG pool diameter, v_g is the mean vapor velocity, C_μ is the turbulence constant (0.09), and Ti is the turbulence intensity parameter which is always between 1% and 10%.

The vapor temperature in this boundary is always recommended to be set at 110K, which is LNG boiling point under the normal condition.

2.2.3 Ground

The ground is always set to be no-slip boundary, which means the velocity on the ground surface is zero. Either heat flux or surface temperature should be specified at the ground boundary. The heat flux from the ground to the atmosphere could be positive or negative depending on the temperature difference between ground and atmosphere. Terrain and obstacles are always represented as a feature of the ground.

3. Results and discussions

A pit (24ft*24ft*8ft) prefilled with approximately 8ft of water at the BFTF was used to perform the LNG spill test, which is shown in Figure 1. The LNG was spilled on water through a 3 inch diameter pipe at a flow rate around 40 gallon/min. Gas detectors are placed downwind to measure vapor concentration at different positions. Geometric construction and meshing details on CFX are shown in Figure.2.

Figure 1. Test representation

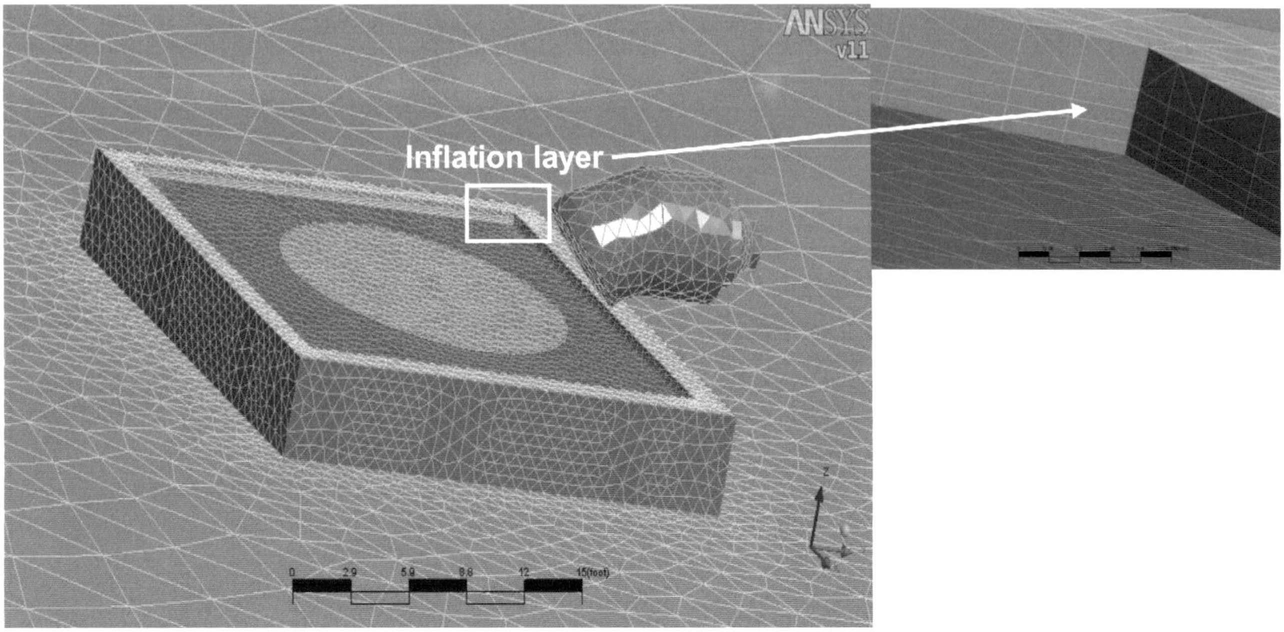

Figure 2. Geometry construction and meshing details

The characteristics of the BFTF test are presented in Table.2.

Table 2. Summary of collected data from the test

LNG flow rate, m^3/min	0.15
LNG pool diameter, m	2.13*
Average wind speed @ 3 m, m/s	1.8
Average wind speed @ 10 m, m/s	2.4
Average wind direction @ 3 m, degree	77**
Average wind direction @ 10 m, degree	95.1**
Temperature @ 3 m, K	299.65
Temperature @ 10 m, k	299.15
Absolute air pressure, mb	998.6
Relative humidity, %	64.5
Stability class	C
Monin-Obukhov length, m	20.45
Roughness height, m	0.001

*estimated from on-site photos

**0 degree is true north

Based on the data provided in Table 2, CFX simulations were performed with the boundary conditions mentioned above. Each simulation was solved in a steady state with convergence criterion set as RMS residual less than $1e^{-4}$. All simulations were run on an AMD Turion X2 Dual-Core Processor with 2GB memory. Table 3 shows the comparison between experimental data and CFX simulations result on downwind distance to half the LFL.

Table 3. Experimental data and CFX simulation result on ½ LFL

	½ LFL downwind @ 4ft, ft
Experimental data	21.85±6.5
CFX simulation result	24

As described earlier, the selection of mesh size (mesh length scale) can affect the simulation accuracy. In order to select an adequate mesh size in CFX simulation, a series of runs was performed with the same inputs setup but with different mesh length scales and meshing approach. Table 4 shows the meshing information used for these three runs.

Table 4. Mesh length scales and meshing approach for Run1-3

	Run 1	Run 2	Run 3
Maximum body spacing, ft	23	11	11
Inflation	No	No	Yes
Particular face spacing	No	No	Yes
Total number of nodes	2674	10422	20350

Total number of tetrahedral	9483	48659	97312
Total number of pyramids	0	0	52
Total number of prisms	0	0	2282
Total number of elements	9483	48659	99646
Total running time, s	197.6	502	968

Figure 3 shows the contours of vapor fraction for these runs at 4 ft elevation. As seen in Figure 3, the methane volume concentration can be monitored in order to observe the influence of mesh size. Generally the 5 and 2.5 % volume concentration levels are the range of interest for this study since the k-epsilon model is based on a time averaged formulation, while the true LFL may lie between 2.5 and 5 % in the simulated concentration levels. When these vapor contours do not change gradually by varying the mesh length (below 1%), an adequate accuracy is obtained.

Run1 Run 2 Run 3

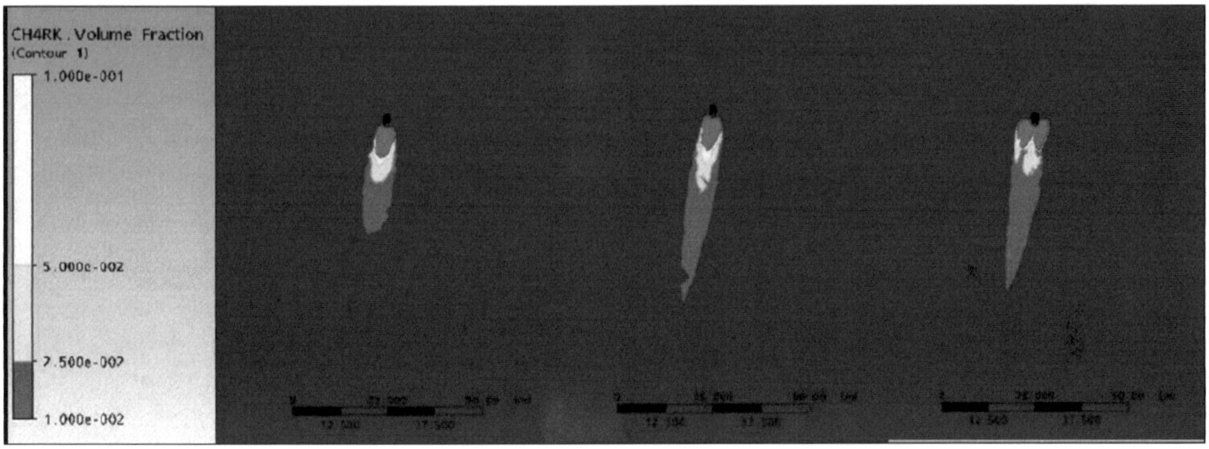

Figure. 3 Vapor fraction contours at 4ft elevation

Figure 4 shows the effect of turbulence intensity (at 1% and 10%) in the source term. Changing the turbulence intensity will influence the LFL distance and the shape of subsequent vapor cloud.

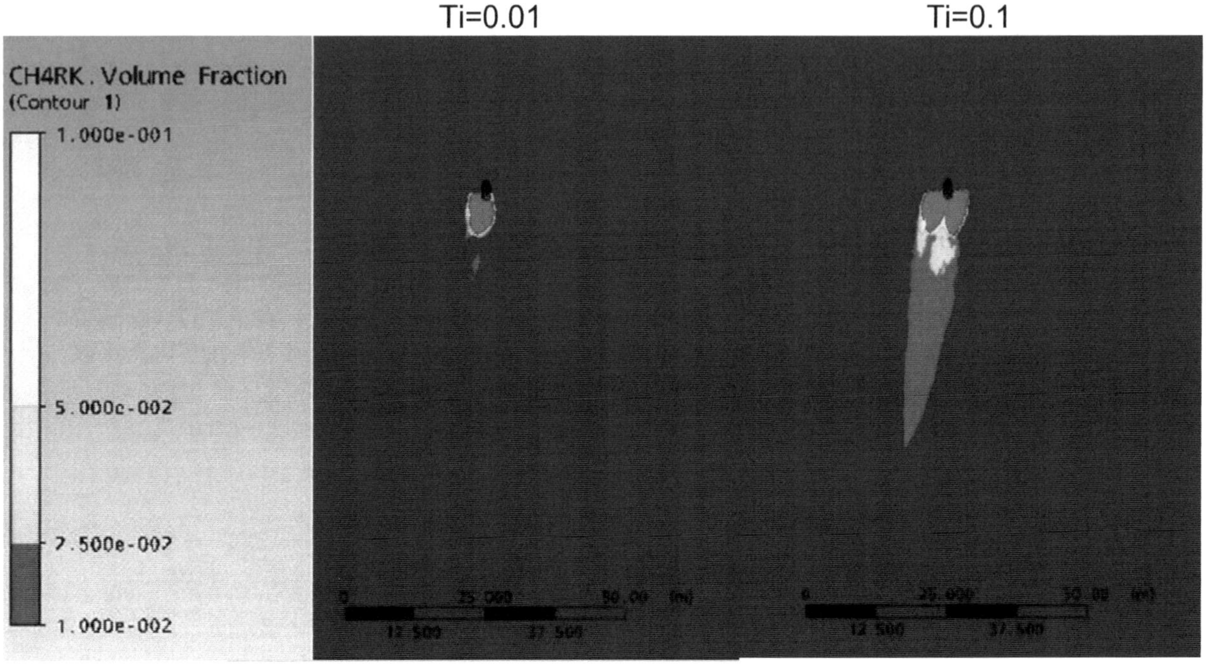

Figure 4. Vapor fraction contours at 4ft elevation for different turbulence intensity

4. Conclusions

In this work, we investigated the LNG vapor dispersion modeling using CFX combined with a medium-scale LNG spill test onto water in Brayton Fire Training Field. This paper identified key parameters needed to setup CFD simulation. It was found that some of these parameters such as mesh size and turbulent intensity in the source term will affect the simulation result significantly. We recommended that the simulated results must be compared at different mesh size (mesh length scale) in order to obtain higher accuracy. Since turbulent mixing of vapor cloud in the source area will be either through buoyancy, shear, or other mechanisms, it is recommended to take into account turbulence intensity in the source term for simulating subsequent vapor dispersion. Future work will be carried out to perform the same type of analysis for LNG release on ground.

5. Acknowledgement

The authors would like to thank BP Global Gas SPU for their financial support and guidance in these tests. The authors would like to thank the support of Brayton Fire Training Field of Texas A&M University System for providing support in the execution of these tests.

References

1. ANSYS CFX-11 Manual, 2006.
2. A. Luketa-Hanlin, R.P. Koopman and D.L. Ermak, “On the application of computational fluid dynamics codes for liquefied natural gas dispersion”, Journal of Hazardous Materials, 140(2007), 504-517.
3. S.P. Arya, “Introduction to micometeorology”, Academic Press, 2001.
4. T.C. Brown, R.T. Cederwall, S.T. Chan, D.L. Ermak, R.P. Koopman, K.C. Lamson, J.W. McClure and L.K. Morris, “Falcon Series Data Report 1987 LNG Vapor Barrier Verification Field Trials”, 1990.
5. H.A. Panofsky and J.A. Dutton, “Atmospheric turbulence”, New York John Wiley & Sons,1984

Biography of Speaker

Ruifeng Qi is a PhD student at the Artie McFerrin Department of Chemical Engineering, Texas A&M University. He has been involved in LNG emergency response and consequence modeling since January 2008 which is sponsored by BP Global Gas SPU. His research involves LNG tests which are usually carried out at the Brayton fire Training Field at Texas A&M University. He has been focusing on modeling LNG vapor dispersion with Computational Fluid Dynamics. ruifeng.qi@chemail.tamu.edu

Have You Considered High-Pressure LNG Releases?

102e

Sanjeev Saraf, Ph.D.
Subodh Medhekar, Ph.D., P.E., CRE
En-Hua Yang, Ph.D.

Exponent Inc.
10850 Richmond Ave., No,175
Houston TX 77042
saraf@exponent.com

AIChE Spring Meeting, April 2009
9th Topical Conference on Natural Gas Utilization
Tampa, Florida, April 26-30, 2009

Introduction

LNG import terminals in the U.S. follow NFPA-59A and 49CFR193 standards for facility siting. As a part of the siting studies, LNG regasification facilities report thermal radiation exclusion zones and flammable vapor exclusion zones:

- Flammable vapor hazard zones are based on a design spill from a single accidental source, usually a guillotine rupture.
- Thermal radiation hazards are based on tank fire.

These events used for determining exclusion zones, pipe rupture and tank fire, have a low probability. Surprisingly, the current standards have no requirements to model consequences from high-pressure LNG releases which are more likely. Furthermore, the current codes do not reference Vapor Cloud Explosions (VCE) which directly affects plant layout and siting.

There is high-pressure processing equipment in LNG import and export terminals (e.g. vaporizers in LNG import terminals) that can result in aerosolized release of LNG and can potentially lead to an explosion. Therefore, in addition to the consequence modeling performed for code compliance, facilities must model high pressure releases and VCEs for:

- Optimizing plant layout
- Deciding risk mitigation measures, and
- Terminal/building siting

Failure Rates and Scenarios Modeled

The LNG regasification terminals in the U.S. are mandated by 49 CFR 193 and NFPA-59A to determine thermal radiation and flammable vapor exclusion distances based on certain release scenarios.

The design spill used for determining flammable vapor exclusion distances is the flow from any single accidental leakage source for 10 minutes for other process and transfer areas. This definition is open for interpretation as to what is an acceptable single accidental leakage source.

Usually a full line rupture is used as the basis for determining flammable vapor cloud distance to ½ LFL. Thermal radiation exclusion distances are generally based on an LNG container (tank) impoundment fire. Typically DEGADIS is used to determine distance to ½ LFL and LNGFIRE is used to determine distance to threshold radiation values.

Now let us look at failure rates of these two scenarios – full line rupture and tank failure.

Below are the failure rates for tank failures and pipelines based on failure data for LNG facilities reported in the 1989 CCPS publications.[1]

Equipment	Failure rate	Comment
Tank	1 in 200 years	2 tank failures occurred over 1,809,000 operating hours
Cryogenic pipeline	1 in 50 years	2 pipeline failures occurred over 132,877 ft-yrs. The failure rate reported here is based on 0.25-mile pipeline.

One can argue that with advances in fabrication techniques and increased maintenance efficiency failure rates will further reduce. Therefore, it is safe to conclude that a catastrophic failure of a tank or a full line rupture have a low probability and exclusion distances based on such scenarios will lead to conservative exclusion zones.

While tank failure and full line rupture scenarios may be an appropriate basis to determine exclusion distances, a more likely scenario is a leak through pipelines and equipment. The existing "design spill" approach and most safety studies do not address high pressure releases. Most of the work in LNG consequence modeling appears to be focused on LNG dispersion and pool fire. Surprisingly, none of the codes/standards reference Vapor Cloud Explosion (VCE) which can potentially result from high-pressure releases. High pressure releases can occur at a few locations within a regasification plant – unloading arm of the LNG tanker, send-out pumps and LNG vaporizer. Typical pressures are indicated in the following flow diagram of a standard regasification plant.

[1] Center for Chemical Process Safety (CCPS), "Guidelines for Process Equipment Reliability Data," 1989.

Therefore, the consequence modeling performed for siting, plant layout and risk mitigation measures for a LNG facility is distinct from code compliance and has to account for high pressure release and VCEs.

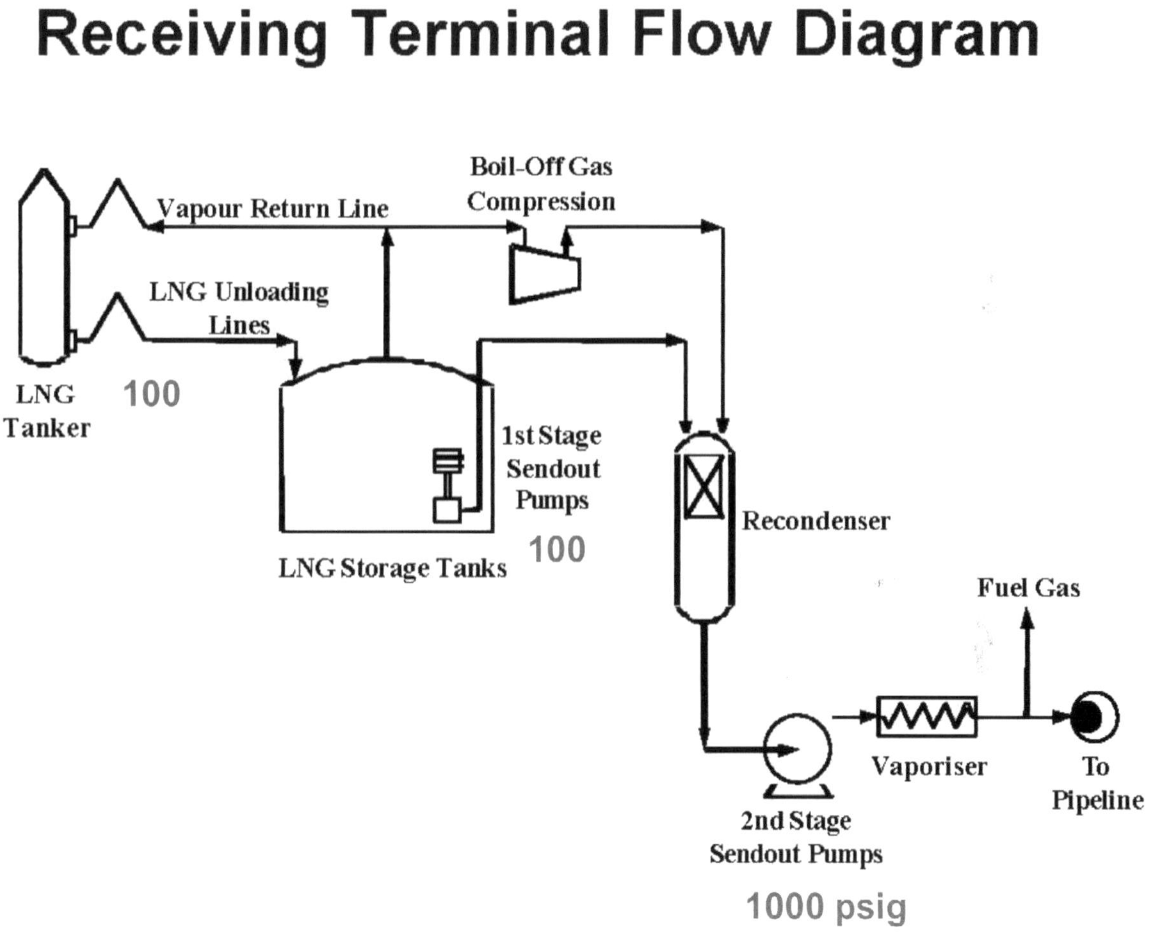

Figure 1. Typical LNG Regasification Terminal (Source: Tarlowski et al LNG Terminals – Recent Developments)

Consequences of High Pressure Releases

A high pressure LNG release can result either in a 2-phase flashing jet or a high-pressure jet depending on the release conditions. A 2-phase flashing jet results in droplet or aerosol formation which can easily ignite since minimum ignition energy is inversely proportional to the mean droplet diameter. Similarly, the flame speed increases within an aerosol jet with increasing density of smaller particles and in certain situations this can lead to an explosion. This phenomenon is observed in mist explosions.

In this paper, we estimate flame dimensions due to an immediate ignition following a high-pressure release. We model these jet fires to determine flame size for natural gas releases at 100-psig and 1000-psig.

For modeling jet fire consequences, semi-empirical mathematical models are employed because these models predict flame properties of interest to safety studies with practical accuracy. The jet fire is represented by a cone as shown in Figure 2. The flammable portion of the jet is assumed to be a frustum and dimensions of the frustum are calculated using empirical correlations.

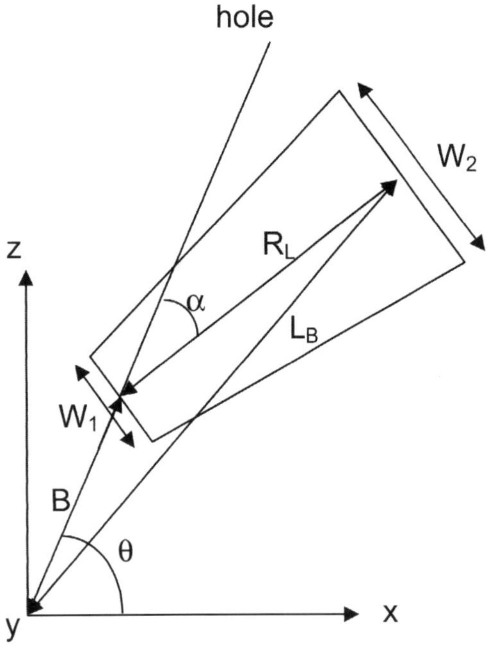

Figure 2. Characteristic dimension of a jet fire

LB - Flame Length
RL – Frustum lengths
W1 and W2 – Frustum base and tip widths,
α – Angle between the frustum and release axes
B – Lift-off distance of the frustum from the release plane

We used three different models to calculate the jet dimension. The Chamberlain model[2] was originally developed for modeling jet fires resulting from vertical/inclined (i.e. angle of inclination > 45° from the horizontal) vapor phase releases. Cook et al[3] modified the Chamberlain model in order to account for changes in effluent buoyant behavior following liquid or two-phase release. Johnson et al[4] extended the model to specifically simulate jet fire resulting from horizontal/near horizontal vapor phase releases. The Chamberlain, Cook, and Johnson models account for the influence of wind speed, entrainment rate and crosswind effects on flame characteristics. Table 1 summarizes the applicability of the three jet fire models.

Table 1: Applicability of the jet fire models

Model	Vapor release	Two-phase release	Liquid release	Horizontal/near horizontal release	Vertical/inclined release
Chamberlain	☑				☑
Cook et al.	☑	☑	☑		☑
Johnson et al.	☑			☑	

We incorporated Chamberlain, Cook, and Johnson models in a spreadsheet to compute jet fire dimensions.

To validate the three models, field data reported by Chamberlain[2], Bennett et al.[5], and Selby and Burgan[6] were used. Figures 3(a), 3(b), and 3(c) show the variations of predicted frustum length, flame tip width, and angle between release and frustum axes of a vertical vapor release with field data reported by Chamberlain[2] (error bar represents the range for observed field data).

As seen from Figure 3, the model for pure vapor jet flames resulting from vertical releases show good agreement with field data.

---

[2] Chamberlain, G.A., "Developments in Design Methods for Predicting Thermal Radiation from Flares," Chem. Eng. Res. Des., Vol.65, pp.299-309, 1987.

[3] Cook, J., Bahrami, Z., and Whitehouse, R.J., "A Comprehensive Program for Calculation of Flame Radiation Levels," J. Loss Prev. Process Ind., Vol.3, pp.150-155, 1990.

[4] Johnson, A.D., Brightwell, H.M., and Carsley, A.J., "A Model for Predicting the Thermal Radiation Hazard from Large Scale Horizontally Released Natural Gas Jet Fires," Trans. IChemE., Vol.72, Part B, pp.157-166, 1994.

[5] Bennett, J.F., Cowley, L.T., Davenport, J.N., and Rowson, J.J., "Large Scale Natural Gas and LPG Jet Fires Final Report to the CEC," TNER 91.022, 1991.

[6] Selby C.A., and Burgan, B.A., "Blast and Fire Engineering for Topside Structures-Phase 2, Final Summary Report," The Steel Construction Institute, Silwood Park, pp. I-A7 to I-A18, 1998.

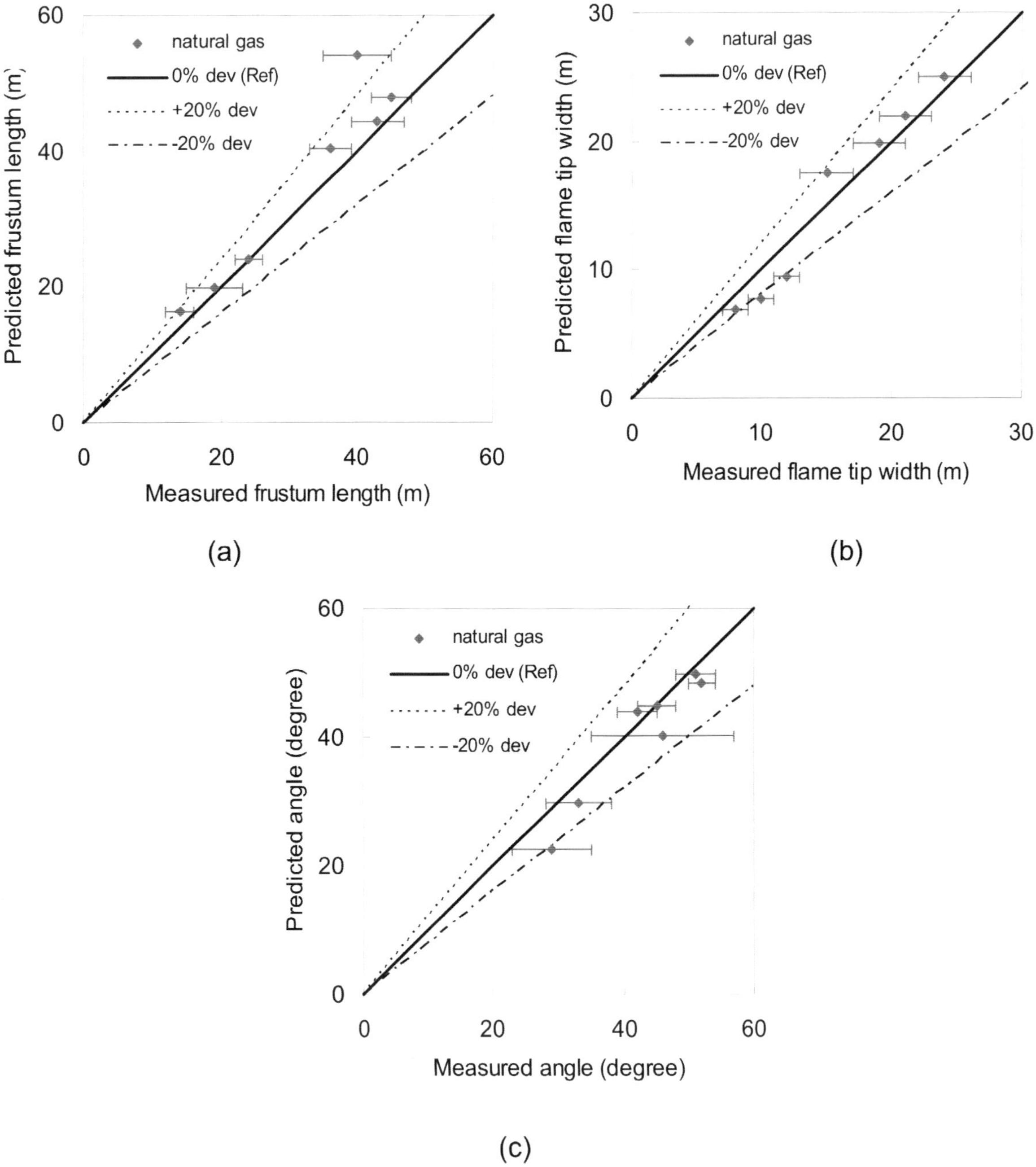

Figure 3. Predicted (a) frustum length, (b) flame tip width, and (c) angle between release and frustum axes of a vertical vapor release versus field data reported by Chamberlain[2]

Figure 4 shows predicted flame length for a horizontal vapor, two-phase, or liquid releases with field data reported by Bennett et al[5] and Selby and Burgan.[6] Based on available field data, the models generally predict flame length to within ±20% of the measured values.

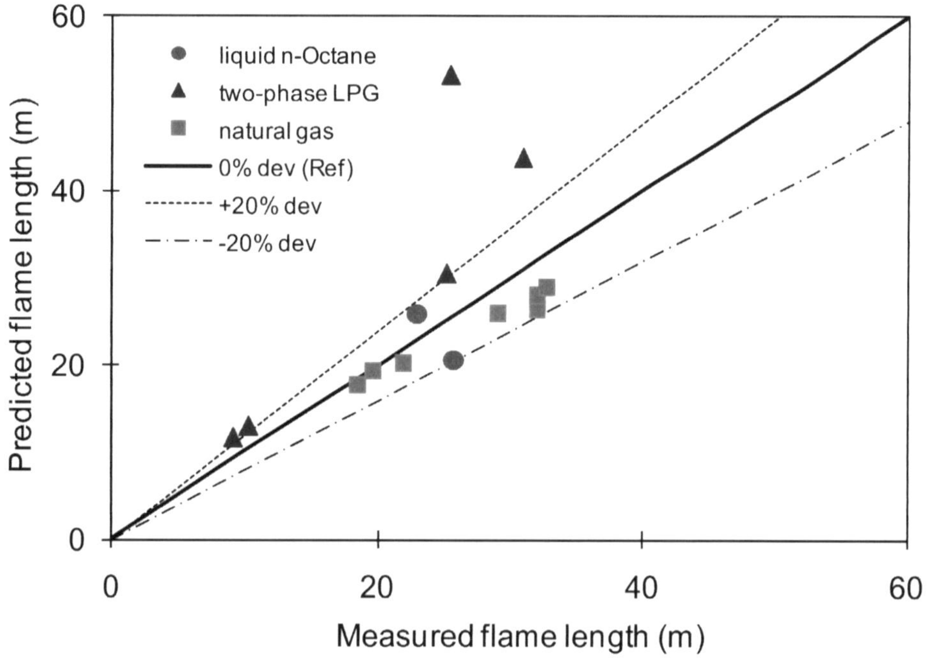

Figure 4. Predicted flame length of a horizontal release versus field data reported by Bennett et al.[5] and Selby and Burgan[6]

We calculated jet fire dimensions for 2-inch horizontal high-pressure LNG releases at 100-psig and 1000-psig. Table 2 summarizes the calculated results and Figure 5 illustrates the shape and dimension of the resulting jet flames. As can be seen, 1000-psig LNG release results in a much bigger jet flame with longer flame length and wider frustum width. The flame length can reach approximately 48 m (157 ft) which is longer than commonly used spacing requirements.

Table 2. Calculated jet flame from 2-in. high-pressure (100-psig and 1000-psig) LNG releases

LNG release pressure (psig)	Flame length, L_B (ft)	Frustum lift off, B (ft)	Frustum length, R_L (ft)	Frustum base width, W_2 (ft)	Frustum tip width, W_1 (ft)
100	58	18	40	3	9
1000	157	58	102	15	31

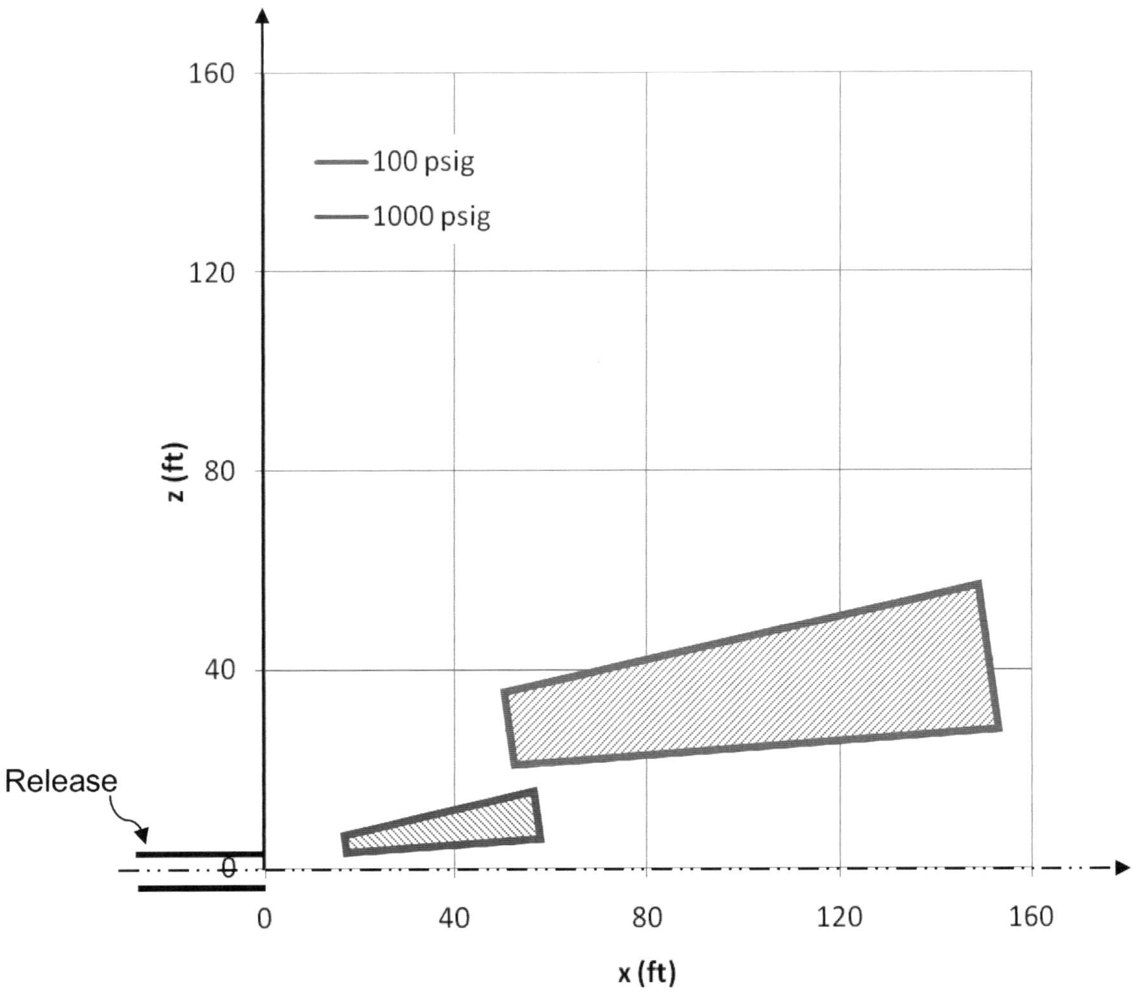

Figure 5. Jet flame from 2-inch high-pressure LNG releases

Thus, a 2-inch natural gas release from 100-psig and 1000-psig can result in a jet fire covering 50-150 ft.

Implications for LNG Facilities

Immediate Ignition

Fire and Gas detectors

Methane being extremely flammable, the fire and gas detection (FGD) system in LNG facilities should be capable of early detection of flammable gas leading to shutdown/isolation and depressurization.

The existing codes and regulations do not provide guidelines on the type of FGD and their locations. Usually catalytic combustion type gas sensors are used for detecting LNG vapors; however, for these combustion gas sensors to work the vapors within flammable range have to reach the gas sensor and by then the release may have already spread within the facility. Therefore, LNG facilities should consider Ultrasonic gas detectors – these detect hydrocarbon gas leaks by sensing the distinct high frequency acoustic ultrasound emitted by all high-pressure gas leaks. While ultrasonic sensors provide early detection of leaks, they are unable to distinguish the ultrasonic activity generated from other spurious noise that is generated by mechanical equipment in close proximity.

To systematically devise a fire and gas detection system for a LNG facility, we recommend considering the following performance characteristics for the detection system:

- Area Covered
- Response time
- Reliability
- Cost

Late Ignition

Siting

The U.S. LNG codes and standards address facility siting by recommending separation distances between equipment. The spacing is intended to reduce escalation from potential fire and eliminate potential ignition sources for vapors releases. However, none of the codes address the semi-congestion levels in and around the plant which can potentially lead to a VCE. The gas released can migrate and following a late ignition can lead to a fire and explosion. The most important parameter affecting explosion is the size and degree of semi-confinement. Therefore, at the design stage it is recommended to optimize the plant layout for minimizing congestion and meeting operational/maintenance requirements. Consequence models that account for semi-confinement can also be used for deciding possible layout for equipment and pipelines. If there are occupied buildings within the regasification terminal we recommend performing siting studies in accordance with API 752.

The API Recommended Practice 752 provides a protocol to assess and manage the hazards associated with locating the buildings within process plants. We recommend that a preliminary evaluation (based on a simple check list) be performed to determine if a particular process/LNG building needs further analysis. Then a credible-worst case analysis be performed (which should also include the typical high pressure release scenarios as described in this paper) to predict the hazard footprints. If these hazard footprints reach/impact the building in question, a more detailed “risk assessment” needs to be performed. This additional risk assessment should include an assessment of the “likelihoods” of various scenarios along with their “consequences” or impacts to the building. Finally, a corporate specific “risk acceptance” criteria should be used to determine if the current building siting is appropriate or whether additional mitigation measures are required.

AUTHOR INDEX

AUTHOR INDEX

American Institute of Chemical Engineers
3 Park Avenue
New York, New York 10016-5991

proceedings.com